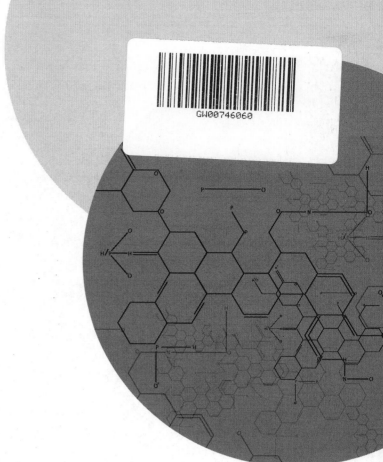

Collins

OCR Gateway GCSE (9–1)
Chemistry

Teacher Pack

Pam Large
Lyn Nicholls
Series editor: Ed Walsh

GW00746060

William Collins' dream of knowledge for all began with the publication of his first book in 1819. A self-educated mill worker, he not only enriched millions of lives, but also founded a flourishing publishing house. Today, staying true to this spirit, Collins books are packed with inspiration, innovation and practical expertise. They place you at the centre of a world of possibility and give you exactly what you need to explore it.

Collins. Freedom to teach

HarperCollins Publishers
1 London Bridge Street
London SE1 9GF

**Browse the complete Collins catalogue at
www.collins.co.uk**

First edition 2016

10 9 8 7 6 5 4 3 2 1

© HarperCollins*Publishers* 2016

ISBN 978-0-00-815103-4

Collins® is a registered trademark of HarperCollins Publishers Limited

www.collins.co.uk

A catalogue record for this book is available from the British Library

Commissioned by Lucy Rowland, Lizzie Catford and Joanna Ramsay
Edited by Hamish Baxter
Project managed by Elektra Media Ltd
Development and copy edited by Ray Loughlin
Proofread by Dr Alison Craig
Illustrated by Jerry Fowler and Geoff Jones
Typeset by Jouve India Private Limited
Cover design by We are Laura
Printed in Great Britain by Martins the Printers
Cover images © Shutterstock/watchara, Shutterstock/Everett Historical

All rights reserved. No part of this book may be reproduced, stored in a retrieval system, or transmitted in any form or by any means, electronic, mechanical, photocopying, recording or otherwise, without the prior permission in writing of the Publisher. This book is sold subject to the conditions that it shall not, by way of trade or otherwise, be lent, re-sold, hired out or otherwise circulated without the Publisher's prior consent in any form of binding or cover other than that in which it is published and without a similar condition including this condition being imposed on the subsequent purchaser.
HarperCollins does not warrant that www.collins.co.uk or any other website mentioned in this title will be provided uninterrupted, that any website will be error free, that defects will be corrected, or that the website or the server that makes it available are free of viruses or bugs. For full terms and conditions please refer to the site terms provided on the website.

Contents

6 Global challenges

Introduction to the course

Welcome to the Collins GCSE Science course. These materials have been developed by a team of experts drawing on many years' experience in teaching, curriculum development and educational publishing. They have been produced to support you, the professional in the classroom, teach more effectively and your students to make good progress.

The aim is to both develop students' interest in science and enable them to attain grades that reflect a mastery in the subject. The course is not intended to be prescriptive or restrictive, shoe horning teachers into a very particular way of teaching, but to be used in a range of ways. On the other hand, it doesn't offer a bewildering array of options which require a lot of filtering and selection of activities.

We are aware that specifications sometimes seem to change at an alarming rate. We have tried to reflect both the key features of the new specifications and the classic elements of high quality teaching.

Key changes for the courses introduced in 2016

Non-modular

The new specifications have a non-unitised structure. Content can be covered in any order deemed appropriate. This doesn't mean that the order doesn't matter. There are a number of principles that can be used when selecting a 'running order', including:

- deciding which topics are more accessible for students at an earlier stage in their GCSE studies
- identifying which topics need to be covered to give access to which other concepts
- using topics that have possibly greater appeal, due to the role of practical work or engaging contexts, to sustain interest over the whole course.

However, what is also important is to build in assessment, tracking and intervention. You need to know if students are developing an understanding that at least corresponds to their ability and to be able to respond.

The Collins course offers a particular route. The exams are designed to assess one part of the content in Paper 1 and the other part in Paper 2. One approach is to teach the 'Paper 1 content' first and then the 'Paper 2 content'. This means that a past Paper 1 could be used as a 'half way assessment point'.

However, the materials don't have to be used in this order. It may be decided that the needs of students are not best served by concentrating all the material on one topic in one place, but rather by revisiting it later on.

Maths

Science has always had a strong relationship with maths and this continues. What has been strengthened in this set of courses is the specificity of the location of the skills and the role of the skills in exams. It has been made very clear in the specifications which skills relate to which topics and the exams will disadvantage candidates who can't demonstrate mastery of the relevant skills.

In the course we have taken a 'two pronged approach'. Mathematical skills are used in context as key concepts are covered. They are reflected in questions that are set and students are shown how they are drawn upon at particular points. However, we know that for some students this won't be sufficient. Therefore, we have developed other spreads that will focus on a particular set of mathematical skills and explore how they develop. These always place the skills into a scientific context, so it is clear why they are relevant, but the text is led by the developing mathematical skill.

Practical skills

Practical work has a strong role in supporting students' developing understanding of science, and many people involved in the British system are passionate about ensuring this role is maintained. In recent years much of the debate is about how it should be assessed, and successive specifications have used different approaches.

In this series there is no direct assessment of practical skills. Instead the approach is being taken that enquiry will be put in a better place if students are assessed on their mastery of skills and processes in the final exams. A minimum of 15% of marks will be allocated to questions relating to the stipulated practical investigations. However, these won't necessarily be AO1 questions; as well as understanding and recalling the procedures, candidates will also be expected to apply the skills to other contexts and to interpret and evaluate evidence.

As well as covering the effective running of the investigations themselves, the Collins course also offers guidance on placing these in a wider context, so that students can see how their skills and understanding can be used and assessed.

Synoptic questions

Another of the key changes is the introduction of synoptic questions, drawing upon more than one set of ideas. The justification behind this is that scientists often need to draw upon ideas and processes from different areas of science. Again, this is reflected in the assessment approach developed.

Extended written responses

Unlike the previous set of specifications, QWC (Quality of Written Communication) is no longer being assessed in science. However, there is a requirement for candidates to be able to display an ability to develop and sustain a longer response. This is reflected in the questions in the end of topic tests as well as in the Collins Connect digital assessment materials.

Assessment

Grades

The grading system is being changed to a numbered system, in which 1 is the lowest grade awarded and 9 the highest. There is no one-to-one correlation between grades in the incoming and outgoing systems, but the following features are being deployed:

- The new grade 4 is equivalent to the old grade C.
- The new grade 7 is equivalent to the old grade A.
- The threshold for performance indicators will be the new grade 5.

Endorsed publications are no longer permitted to carry specific reference to how certain grades may be attained. Nevertheless, although the calibration has altered, what has not changed is what constitutes a better quality of response. Each of the spreads in the Student Book is designed to present students with an increasing challenge and provide access to higher grades. However, we have also been mindful when developing the resources that a wide range of students will be using them, and so there is a focus upon accessibility as well.

Attribution grids

There has long been an important role attached to the use of end of topic tests. When students have been taught something, we (and they) want to know how well they've understood it and so we've included tests. However, we have developed these further, both in terms of structure and use.

In order to be useful, we think that as far as possible tests should be authentic to the form and function of assessment in GCSE. In other words, if students do well in the tests, this should correspond to doing well in the final exams. This is important, both to improve the validity of any data generated and also in terms of communicating to students what it feels like to answer the kind of questions that they will face in the exam. These aren't mock exams, but they should have a similar balance of components.

Each of the end of topic tests has been written to an attribution grid, so that the basis is the same. The tests are marked out of 40 and have ten marks at each of four different levels of demand. Broadly speaking, Foundation Tier students should do the first two bands and Higher Tier students the final three bands. If they do so, each group will be answering questions which have a 2:2:1 ratio of assessment objectives 1, 2 and 3, which mirror the make-up of the final exams. This is important. AO1 questions focus on knowledge and understanding and are sometimes over-represented in internal tests. As many marks go on AO2 (application). Another 20% go on AO3 (interpretation and evaluation); students need to be aware of this. The constant proportions mean that it is possible to analyse student performance against AO and modify subsequent teaching accordingly.

Similarly, the tests also reflect a mix of response types (objective test questions, short written responses and longer written responses), and the presence of questions relating to mathematical skills and to the stipulated practicals.

Intervention

The composition of the tests is designed to inform the process of intervention. They will obviously provide a global score which can be used to draw conclusions about attainment in that topic. However, intervention is then somewhat problematical; if the group has moved on to another topic, the only option might be to set up additional sessions elsewhere in the school day. These may be successful but it is useful to be able to move analysis beyond the perspective of looking for which areas of content cause problems. Some do, but this isn't the only source of underperformance.

With the end of chapter tests (and the Collins Connect tests), the attribution grids mean that student performance can be analysed in other ways (see above), including against:

- assessment objectives
- style of response
- command of mathematical skills
- mastery of enquiry skills.

As these are all generic components of science education, whichever topic is being addressed next, there will be opportunities to focus upon improving performance.

It will then be possible in the next module test to see if the analysis of marks indicates that this has been addressed.

How to use these resources

Flexibility

The first thing to say is that the resources are designed to be used flexibly. There is no assumption that all students will do all of the activities suggested in the order indicated or that no other resources or learning activities will be used. The materials have an open and transparent structure, which clearly indicates key components and supports the teacher to select activities that are right for their students.

Features

Student Book

Topic introduction: every chapter starts with the page on the left showing students ideas they have met previously that they will be making use of, and a page on the right indicates why the ideas in the forthcoming chapter are both interesting and important. The purpose of this is to engage students, indicate that they already know something about it but persuade them that it is worth engaging with.

Main content spread: each spread includes material that is ramped in challenge. There is no set grade equivalence for the start and end but the spread always starts with more accessible material. Each of the sections includes questions to embed understanding of the ideas, learning objectives, key words and a 'did you know?' feature to engage students further.

Key concept spread: in every topic we have identified a key concept and revisited this in a slightly different way. These concepts have a kind of 'gatekeeper' function: if students grasp these it will give them access to several other areas of content and so they are worth focusing upon. The highlighting of the spread means they can easily be referred back to, either from further on in the same chapter or subsequent ones.

Practical spread: the specification indicates particular investigations that students should have undertaken. Questions relating to these will feature in terminal examinations; however, these will be not only AO1 (knowledge and understanding) questions but also AO2 (application) and AO3 (interpretation and evaluation) questions. These spreads have been written to support students to take ideas explored in these experiments and place them in a wider context, thus reflecting the range of skills that examiners will be assessing.

Maths skill spread: mathematical skills have long been a key part of GCSE Science courses and exams. However, the proportion and challenge of these questions is being increased. We have developed spreads to reflect this. Maths is still embedded in the main content, with skills being drawn upon as necessary. However, some students will need more than this, and so these additional features have been written. Each takes a particular set of maths skills and shows how they are used in a scientific context, taken from that topic. As the spread becomes more challenging, it shows how the mathematics can be used to answer more complex questions.

Check your progress spread: at the end of each chapter the main ideas are set out as progressive outcomes. In essence they show what it looks like to be better in the use of a particular idea. They can be used in various ways but students self-assessing is one of the more usual ones.

Worked examples: these provide a commentary on typical student responses to GCSE-style questions, showing what has been done well and how further improvement might be achieved.

End of topic questions: these are written to an attribution grid as described above. The questions provide a broad coverage of ideas within the topic, items with different levels of demand (both in terms of cognitive and conceptual complexity) and needing different types of response. This means that they provide varying degrees of challenge and in various ways; they should indicate both what students can do and also what the next steps in their learning need to be.

Teacher Pack

The support for individual lessons includes features such as learning objectives, outcomes, working scientifically focus, maths skills focus, key words and resources needed. The outline for the lesson structure is based around a five-stage learning cycle; this proposes that in every lesson there should be these components:

- Engage: in every lesson students need to be engaged, so that they become involved in the lesson and see it as being of interest and relevance.
- Challenge and develop: students will already have a certain level of understanding of a topic. The idea of this phase is to challenge this and move their thinking on. This is what will drive progress; students should see that their current mode of thinking can be developed so that they can make sense of more complex ideas or a wider range of contexts. This might be a demonstration, question to discuss or video clip.
- Explain: this is where the teacher uses some means of offering an explanation. It could be formal explanation, a demonstration or commentary on a diagram; it might use targeted questioning. It should offer a way of making sense of a problem or phenomenon. The Student Book often plays a crucial role in this phase.
- Consolidate and apply: this is a crucial phase in which students embed the ideas. It should consist of opportunities to 'take on board' new ideas and own them. Written questions and discussion may support this. The worksheets often have a significant role here.
- Extend: students may be ready to extend their learning to something broader or more challenging.

In the ideas for learning activities there is a choice and the options relate to planning provision for students working at different levels.

Teacher Pack Resources

Each lesson is supported by worksheets, which provide activities for students to apply their understanding. They have a ramped structure, offering different levels of challenge.

Practical worksheets support students with planning, carrying out and writing up practical and investigation work.

Technicians' notes provide equipment lists and set-up instructions for practicals.

Collins Connect

Collins Connect makes Collins GCSE Science available at home and at school online and can be used as a front-of-class teaching tool, as well as a way to set homework and tests.

OCR GCSE (9–1) Science on Collins Connect contains:

- Bookview – an interactive digital version of the Student Books. It is ideal for whiteboard use and gives total fluidity between digital and print, with a page-for-page match.
- Quick starters – activities that pose an interesting question to hook students in at the start of a lesson. You can click to show a hint and then click to show the answer.

- Videos and animations – to support you to make real world links and model abstract or challenging concepts.
- Slideshow presentations – available in PowerPoint so you can edit and adapt them to suit the needs of your class.
- This Teacher Pack – so you can access the lesson plans and supporting resources online.
- Homework activities for every lesson – you can assign automarked homework quizzes to your class. Students can log in from home to complete the activities and get immediate feedback. For every lesson there are also 'creative' homework sheets with more open-ended activities.

You can also assess and track students' progress in GCSE Science using the full suite of digital resources on Collins Connect. Collins Connect provides regular and timely assessments to help you analyse student performance.

- The digital tests come in two versions: 'fully automarked' and 'automarked plus teacher response'. The fully automarked tests enable you to get a very quick gauge of students' knowledge and understanding. The 'automarked plus teacher response' tests include longer written response questions, giving students practice of the full range of question styles they will encounter in the final exams. All tests come with a mark scheme.
- End of chapter tests help you to review students' progress on a topic-by-topic basis.
- End of teaching block tests (at the end of every two chapters) provide additional common review checkpoints to help you review progress through the linear course and check students' retention of earlier learning.
- End of year and end of course tests help students to prepare and practice for the final exams.

Differentiation

Differentiation is a key feature of the student books, the lesson plans and the worksheets.

- The student books support differentiation with their ramped structure in each spread. The early section offers an accessible 'entry point' and later text is more challenging. This is reflected in the questions, both in the spreads and at the end of each topic. The progress checking spreads also support an understanding of progressing to higher levels of understanding.
- The teacher packs support differentiation by offering learning activities at different levels of challenge for different phases of the lesson, thus making it easier for teachers to tailor provision to a particular class.
- The worksheets also have a ramped structure.

Four stage plan

One of the ways of running the GCSE course is to use a four-stage plan. The idea behind this is that:

- some topics need to be taught before others and so should be placed earlier in the course
- with some topics there is no preferred running order and departments might want to avoid all groups in the same year doing exactly the same topic at the same time
- having what might be termed 'common assessment points' enables periodic assessment and reporting of progress to take place at points at which all students have covered the same ground – the same assessment can be used and students switched between groups if necessary.
- To provide additional common assessment points, Collins OCR Chemistry can be divided into four teaching blocks:

  o Teaching block 1: chapters 1 and 2
  o Teaching block 2: chapter 3
  o Teaching block 3: chapters 4 and 5
  o Teaching block 4: chapter 6

The teaching blocks have been created to provide assessment at regular intervals although chapters are not divided evenly. End of teaching block tests are provided on Collins Connect and can be used as common assessment points.

The order of the teaching blocks can be altered, if you prefer to teach content from Paper 1 and 2 over the duration of the course. The four-stage plan can be represented diagrammatically as follows:

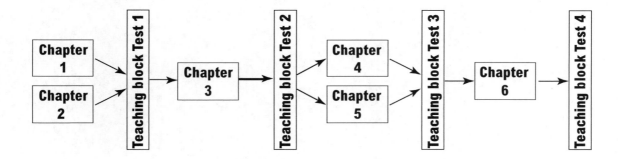

Particles: Introduction

When and how to use these pages

This unit builds on the idea that the atom is the building block of matter – it underpins a lot of work in chemistry.

Overview of the unit

The students will learn about the structure of atoms and use a range of different models to describe both the atoms themselves and the physical and chemical properties of the solids, liquids and gases they form. They will also explore how atomic models developed as new data became available. In the process they will learn how testing predictions can support or refute new scientific ideas.

The unit offers a number of opportunities for students to use mathematics. They visualise and represent two-dimensional (2D) and 3D forms, use prefixes and powers of ten to describe the sizes of atoms and their components, and convert decimals to standard form and vice versa.

Obstacles to learning

Atoms and their sizes are abstract concepts and are hard to visualise. The use of models and analogies may help some students to recognise that atoms are very small, that most of their volume is made up of empty space and that forces hold them together to make 3D structures. The use of physical models and diagrams will also help students to classify substances as solids, liquids or gases and their particles as atoms, ions or molecules.

Practicals in this unit

In this unit students will do the following practical work:

- Observe a demonstration of iodine sublimation.

- Observe a demonstration of 'Rutherford and Bohr's nuclear model'.

- Use microscopes to measure the sizes of sodium chloride crystals and sand crystals.

 © HarperCollins*Publishers* Limited 2016

Chapter 1: Atomic structure and the periodic table

	Lesson title	Overarching objectives
1	Three states of matter	Use data to predict the state of a substance and use ideas about particles to explain state changes
2	Changing ideas about atoms	Describe how ideas about atoms changed as new evidence became available.
3	Modelling the atom	Model atoms as positive nuclei surrounded by negative electrons.
4	Key concept: Sizes of particles and orders of magnitude	Use orders of magnitude to compare sizes.
5	Relating charges and masses	Describe the similarities and differences between protons, neutrons and electrons.
6	Subatomic particles	Use atomic numbers and mass numbers to determine the numbers of subatomic particles in atoms.
7	Maths skills: Standard form and making estimates	Convert decimals to standard form and vice versa.

Lesson 1: Three states of matter

Lesson overview

OCR Specification reference

OCR C1.1

Learning objectives

- Use data to predict the states of substances.
- Explain the changes of state.
- Use state symbols in chemical equations.
- Explain the limitations of the particle model.

Learning outcomes

- Use melting point and boiling point data to predict whether a substance is solid, liquid or gas at a given temperature. [O1]
- Use ideas about the particle theory of matter to explain melting, freezing, boiling, condensing and evaporation. [O2]
- Use state symbols (s), (l), (g) and (aq) in balanced chemical equations. [O3]

Skills development

- WS 1.2 Use a variety of models, such as representational, spatial, descriptive, computational and mathematical, to solve problems, make predictions and develop scientific explanations and understanding of familiar and unfamiliar facts.
- WS 3.5 Interpret observations and other data, including identifying patterns and trends, making inferences and drawing conclusions.
- WS 4.1 Use scientific vocabulary, terminology and definitions.

Maths focus Visualise and represent two-dimensional (2D) and 3D forms, including 2D representations of 3D objects

Resources needed Practical sheet 1.1; Worksheets 1.1.1 and 1.1.2; Technician's notes 1.1

Key vocabulary changes of state, condensing, limitations, particle

Teaching and learning

Engage

- Demonstrate the sublimation of iodine using Practical sheet 1.1. Note that this needs to be carried out in a fume cupboard. Ask students to write down why this is unusual and keep their ideas until the end of the lesson. [O1]

Challenge and develop

Low demand

- Students can work in pairs. Write the terms 'solid', 'liquid', 'gas', 'freezing', 'melting', 'boiling' and 'condensing' on the board. Allow students 2 minutes to **establish** the links between the terms. They can **share** their ideas with the class and then complete question 1 in Worksheet 1.1.1. [O1]

- Students can read the information given in the table in Section 1.1 in the Student Book and **identify** the unknown substances. If students find this difficult, especially handling negative numbers, a temperature line is useful to plot the temperatures and provide a visual representation. Students can answer question 2 in Worksheet 1.1.1. [O1]

Explain

Standard demand

- Students can use Figure 1.3 in the Student Book. Discuss what the diagrams represent and establish that in solids the particles are touching and vibrating, in liquids some are touching and they are tumbling over each other, and in gases they are moving in all directions and are far apart. Students need to understand that the kinetic energy of the particles increases from solid to liquid to gas.

  Explain that the individual particles in solids, liquids and gases do not have the same bulk properties as materials. They can read about the changes involved in freezing, melting, boiling and condensation in the Student Book and then work in pairs to **explain** to each other what they have just read. [O2]

- Students can answer question 1 on Worksheet 1.1.2. You could use poly(styrene) spheres to model the changes of state if appropriate. [O2]

High demand

- Allow students to work in pairs to **discuss** the limitations of the simple model used to show solids, liquids and gases. They can check their ideas with Section 1.1 in the Student Book. Establish that this simple model shows no forces between the spheres, represents all particles as spheres and does not show movement.

- Remind students that chemical reactions frequently involve different states of matter. Explain that this information can be shown in balanced chemical equations by using the notation (s), (l), (g) and (aq). You may need to explain the difference between (l) and (aq). Write the following equation on the board:

  $Zn(s) + CuSO_4(aq) \rightarrow Cu(s) + ZnSO_4(aq)$.

  Use the state symbols to discuss what students might expect to observe in this reaction. They can answer question 2 on Worksheet 1.1.2. [O3]

Consolidate and apply

- Remind students about the demonstration at the beginning of the lesson when iodine was heated. They may need to revisit question 1 on Worksheet 1.1.1. Students can **explain** why this is unusual behaviour. [O1]

- They can **discuss** in pairs what changes are happening to the iodine particles when they sublime [O2] and write **an** equation for the change of state. [O3]

- Students can **describe** what happens to the molecules in ice as they are heated through to steam. [O2]

Extend

Ask students able to progress further to **list** the usefulness of the model used to describe the three states of matter. Section 1.1 in the Student Book will give them some clues. [O2]

Plenary suggestions

Ask students to write down three ideas they have **learned** during this lesson. They can **share** their ideas in groups to **compile** a master list of facts and **rank** them in order of importance.

Divide the class into three groups, named 'solid', 'liquid' and 'gas'. Ask questions to which the answers are solid, liquid or gas. Students can 'claim' their questions.

Answers to Worksheet 1.1.1

1. a) 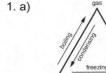 b)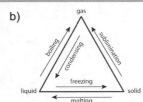

2. a) Liquid; b) Liquid; c), Zirconium, yttrium;

d) Boron, neon, yttrium, e) Boron, niobium, zirconium, yttrium

Answers to Worksheet 1.1.2

1. a) *Melting*: decrease, increases, increases; *Freezing*: increase, decreases, decreases; *Boiling*: decrease, increases, increases; *Condensing*: increase, decreases, decreases; b) They have stronger attractive forces between particles; 2. a) Solid particles form/precipitate; b) Solid copper carbonate is heated; carbon dioxide gas is given off; and solid copper oxide is formed; c) $Cu(s) + 2AgNO_3(aq) \rightarrow Cu(NO_3)_2(aq) + 2Ag(s)$

Lesson 2: Changing ideas about atoms

Lesson overview

OCR Specification reference

OCR C1.2

Learning objectives

- Learn how models of the atom changed as scientists gathered more data.
- Consider the data Rutherford and Marsden collected.
- Link their data to our model of the atom.

Learning outcomes

- Describe how and why the atomic model has changed over time. [O1]
- Describe the difference between the plum pudding model of the atom and the nuclear model of the atom. [O2]
- Describe why the new evidence from the scattering experiment led to a change in the atomic model. [O3]

Skills development

- WS 1.1 Explain why new data led to changes in a model.
- WS 1.2 Match features of a model to data from experiments.
- WS 1.3 Explain why the data needed to answer scientific questions may be incomplete.

Resources needed Equipment as listed in the Technician's notes; Worksheets 1.2.1, 1.2.2 and 1.2.3; Technician's notes 1.2

Digital resources Presentation 1.2 'Name that model'

Key vocabulary electron shell, Ernest Rutherford, Geiger and Marsden experiment, J. J. Thompson, James Chadwick, John Dalton, Niels Bohr

Teaching and learning

Engage

- Display the three atomic models described in the Technician's notes. Explain that we started with Dalton's atomic model, but the model had to be changed whenever scientists found evidence that didn't fit the model they were using. [O1]

Challenge and develop

- Give out Worksheet 1.2.1. Students work in pairs to cut-out the cards. Then read Section 1.2 in the Student Book to **find out** when each scientist worked and **arrange** the cards in chronological order. [O1]

Low demand

- Pairs stick the cards in chronological order onto an A3 sheet, and then **draw** each scientist's atomic model below their card. [O1]

Standard and high demand

- Pairs use the cards and information from Section 1.2 in the Student Book to **create** a timeline showing each scientist's contribution to the development of the atomic model. [O1]

Explain

Low and standard demand

- Display Thomson's plum pudding model and the Rutherford–Bohr nuclear atom. Challenge students to find two things that are the same about them and two things that are different. [O2]

- Demonstrate Geiger and Marsden's scattering experiment by throwing model alpha particles at the Rutherford–Bohr atomic model. Most will pass straight through as Rutherford expected but any that hit the positive nucleus will bounce back. Rutherford found this amazing. His calculations showed that a bounce back would only be possible if almost the entire mass of the atom was squashed into a tiny nucleus. There must be a vast empty space between this nucleus and the electrons. [O3]

- Show a video to reinforce the significance of the scattering experiment [Google search string:] New Syllabus 2.1 Aim: Rutherford's gold foil experiment. [O3]

Standard demand

- Complete Worksheet 1.2.2. [O3]

High demand

- Display Thomson's plum pudding model and the Rutherford–Bohr nuclear atom. Challenge students to **describe** why the new evidence from Geiger and Marsden's scattering experiment led to a change in the atomic model. Individuals could **explore** the question briefly, and then **discuss** their ideas in pairs and fours before they **summarise** them in writing. [O3]

Consolidate and apply

- Show a video to reinforce the steps in the development of ideas about the atom [Google search string:] The 2400-year search for the atom. [O1, O2, O3]

Low demand

- Complete Worksheet 1.2.3. [O1, O2]

Standard and high demand

- Students work together in pairs to **decide** which scientist made the biggest contribution to our understanding of atoms and why. Each pair then joins up with another pair to **explain** and **compare** ideas and they elect a spokesperson to **feedback** to the rest of the class. [O2, O3]

Extend

Ask students able to progress further to **find out** more about Geiger and Marsden's scattering experiment and **suggest** why it could not have been done 100 years earlier. [O1]

Plenary suggestion

Run Presentation 1.2 'Name that model' as a whole class quiz.

Answers to Worksheet 1.2.1

The correct order is 1. Democritus, 400 BC; 2. Dalton, 1803; 3. Thomson, 1897; 4. Geiger and Marsden, 1909; 5. Rutherford, 1911; 6. Bohr, 1913

Answers to Worksheet 1.2.2

1. The gold foil has only a few layers of atoms and most of an atom's volume is empty space; 2. The tiny positive nuclei in the gold atoms were repelling the positively charged alpha particles and causing them to change course; 3. The positive charge would be spread too thinly to repel the alpha particles

Answers to Worksheet 1.2.3

1. Dalton, Thompson, Rutherford, Bohr; 2. Thompson; 3. Dalton; 4. Bohr; 5. Thompson; 6. Rutherford and Bohr

Lesson 3: Modelling the atom

Lesson overview

OCR Specification reference

OCR C1.2

Learning objectives

- Explore the structure of atoms.
- Consider the sizes of atoms.
- Explore the way atomic radius changes with position in the periodic table.

Learning outcomes

- Describe atoms as positive nuclei surrounded by shells of negative electrons. [O1]
- Recognise that the nucleus is very small but contains most of an atom's mass. [O2]
- Describe trends in atomic radius down a group and across a period. [O3]

Skills development

- WS 1.2 Recognise, draw and interpret models of atoms.
- WS 3.5 Describe trends in graphical data.
- WS 4.4 Use prefixes and powers of ten to compare sizes of atoms.

Maths focus Use prefixes and powers of 10; plot two variables from data

Resources needed Model atom as described in the Technician's notes; periodic tables; graph paper; Worksheet 1.3; Technician's notes 1.3

Digital resources Presentation 1.3 'Helium'; Graph plotter 1.3

Key vocabulary charge, electron, nucleus, electron shell

Teaching and learning

Engage

- **Use** Presentation 1.3 'Helium' to introduce helium as a gas made of individual atoms.
- **Ideas hothouse.** Ask students to work in pairs and **list** things they know or can **deduce** about helium atoms. Then ask the pairs to join together into fours and then sixes and eights to **construct** an agreed list of points. Ask one person from each group to **report** back to the class and collate their ideas. [O1]
- Explain that although we can draw atoms as solid spheres or circles, real atoms are mostly empty space.

Challenge and develop

- Read the 'Atoms' and 'More on atoms' sections in the Student Book. [O1, O2]
- Introduce the model atom described in the Technician's notes and highlight the small positive nucleus, the 'shell' of negative electrons represented by the hoop and the empty space between the two where there is nothing at all. [O1, O2]
- Students work in pairs to **list** ways in which the model helium atom is the same as a real atom and ways in which it is different. Then they join with another pair to **explain** and **compare** ideas and **agree** on three main points. [O1, O2]

- Survey students' ideas and bring out the idea that both the real and the model atoms contain a central positive nucleus surrounded by a shell of negative electrons. However, real atoms are spherical and if a real helium atom was magnified to give it a diameter of 1 m, its nucleus would only be about 0.1 mm in diameter – too small for us to see it on the model. Also, the model has air between the nucleus and the electrons but in real atoms that space is completely empty. [O1, O2]

- Use a video clip to review the sizes of atoms. [Google search string:] Just How Small Is an Atom? – Jonathan Bergmann. [O2]

Explain

Low demand

- Students complete Worksheet 1.3. [O1, O2]

Standard and high demand

- Students draw a labelled diagram of a helium atom and **list** differences between their drawing and a real atom. [O1, O2]

Consolidate and apply

- Draw student's attention to the table of atomic radii in the Student Book. [O3]

- Low demand: Students use Graph Plotter 1.3 to answer questions 4 and 5 in the Student Book. [O3]

- Standard and high demand: Students use graph paper to answer questions 4 and 5 in the Student Book. [O3]

Extend

Ask students able to progress further to use the patterns they investigated to **predict** which element has the largest atoms and which has the smallest. Then use the Internet to check their predictions. [O3]

Plenary suggestion

What do I know? Ask students to write down one thing about atoms they are sure of, one thing they are unsure of and one thing they need to know more about.

Answers to Worksheet 1.3

1. The following parts should be labelled: positive nucleus; negative electrons; empty space; 2. Any two differences: real helium atoms are spherical rather than flat; the nucleus is much smaller in a real atom; there is nothing between the nucleus and the electron shell in a real helium atom.

 © HarperCollins*Publishers* Limited 2016

Lesson 4: Key concept: Sizes of particles and orders of magnitude

Lesson overview

OCR Specification reference

OCR C1.2

Learning objectives

- Identify the scale of measurements of length.
- Explain the conversion of small lengths to metres.
- Explain the relative sizes of electrons, nuclei and atoms.

Learning outcomes

- Know the relationship between metres, millimetres, micrometres and nanometres using decimal form and standard form. [O1]
- Be able to convert measurements used to describe atoms and subatomic particles to metres. [O2]
- Apply knowledge of atoms and sub-atomic particles to their relative sizes and carry out calculations based on the size of small particles. [O3]

Skills development

- WS 3.1 Present observations and other data using appropriate methods.
- WS Use SI units (e.g. kg, g, mg, m, mm, kJ and J) and IUPAC chemical nomenclature unless inappropriate.
- Use prefixes and powers of 10 for orders of magnitude (e.g. tera, giga, mega, kilo, centi, milli, micro and nano).

Maths focus Recognise and use expressions in decimal form; recognise and use expressions in standard form; make order of magnitude calculations.

Resources needed Practical sheet 1.4; Worksheets 1.4.1 and 1.4.2; Technician's notes 1.4

Key vocabulary magnitude, diameter, radius, nanometre

Teaching and learning

Engage

- Provide each student with a 2 cm^2 piece of graph paper and ask them to fold it in half and in half again repeatedly until they no longer can. Ask them to write down the length and/or width of the folded paper after each fold, using suitable units. Discuss the units they have used. You could ask students to **estimate** how many atoms wide their final folded paper measures. They can keep their answers till the end of the lesson. [O1]
- Present students with a range of laboratory/school objects with a range of sizes (e.g. a football pitch to chalk dust). They can work in pairs to **decide** the most suitable units to use to measure the objects. [O1]

Challenge and develop

Low demand

- Students can use Practical sheet 1.4 to **measure** the size of sodium chloride and sand crystals. Emphasise that they only need a few crystals of each substance and these need to be well separated on the slide. Check their measurements. The evaluation section will introduce micrometres. [O1]

Standard demand

- Students can read 'Orders of magnitude' in Section 1.4 of the Student Book to find out about the relationship between metres, millimetres, micrometers and nanometres. Explain the use of standard form, if necessary.

Standard form is a number in the form of $A \times 10^x$, where A is between 1 and 10. Students can answer question 1 on Worksheet 1.4.1, and questions 1 and 2 in Section 1.4 of the Student Book. [O1]

Explain

- Explain that carbon nanotubes are about 1 nm wide, but that these are made from many carbon atoms bonded together. Smaller units are needed to measure the sizes of atoms. Introduce the picometre (pm). Students can read 'Atoms and ions' in Section 1.4 of the Student Book. They can complete the table on Worksheet 1.4.2. Emphasise that each unit in the table is 1000 times larger than the previous. [O2]

High demand

- Students can answer question 2 on Worksheet 1.4.1. This requires them to use appropriate units for measurements of sub-atomic particles. [O3]

- Students can answer questions 3 and 4 in Section 1.4 of the Student Book.

- Ask students to read the 'Radius at the atomic level' in Section 1.4 of the Student Book. Discuss the reasons for differences between the diameters of atoms and their ions. They can answer questions 5 and 6 in Section 1.4 of the Student Book.

Consolidate and apply

- Tell students that one Buckminsterfullerene molecule has a radius of 1 nm. They can **convert** this into metres using decimal numbers and standard form.

- Refer back to the 'Engage' activity in which students folded graph paper and **estimated** how many atoms would fit across it. Remind them that atoms are measured in picometres and that carbon atoms have a diameter of 10 pm. They can now make a more accurate estimate of the number of carbon atoms that could fit across the folded paper. [O2, O3]

- Tell students that the diameter of a gold atom is 288.4 pm. They can **convert** this measurement in pm to m, mm and nm using decimal numbers and standard form. [O2, O3]

Extend

Tell students able to progress further that an electron is 100 000 times smaller than the nucleus of a carbon atom (question 2 in Worksheet 1.4.1). They can **calculate** the size of an electron in attometres. [O3]

Plenary suggestions

Students can work in groups of five or six. Each writes a question involving units for measuring lengths. They can **test** their questions on each other. This can be made more challenging by asking them to write questions involving unit conversions.

Ask students to work in groups and **decide** on one thing they are sure of, one thing they are uncertain of and one thing they need to know more about. Groups can double up and **prepare** an agreed list before **sharing** it with the class.

Answers to Worksheet 1.4.1

1. a)(i) Micrometres; (ii) Metres; (iii) Micrometres or millimetres; (iv) Millimetres; b)(i) 1×10^3; (ii) 1×10^{-3}; (iii) 1×10^9; (iv) 1×10^{-9}; (v) 1×10^3; c) Units included are multiples of 1000 only

Missing answers for Question 2

Answers to Worksheet 1.4.2

1. *(Left to right)*: mm, 0.001, 1×10^{-3}; μm, 0.000001, 1×10^{-6}; nm, 0.000000001, 1×10^{-9}; pm, 0.000000000001, 1×10^{-12}; fm, 0.000000000000001, 1×10^{-15}; am, 0.000000000000000001, 1×10^{-18}

Lesson 5: Relating charges and masses

Lesson overview

OCR Specification reference

OCR C1.2

Learning objectives

- Compare protons, neutrons and electrons.
- Find out why atoms are neutral.
- Relate the number of charged particles in atoms to their position in the periodic table.

Learning outcomes

- Use a periodic table to find the number of protons or electrons in an atom. [O1]
- Explain why atoms are neutral. [O2]
- Recall the relative masses and charges of protons, neutrons and electrons. [O3]

Skills development

- WS 1.2 Use models in explanations.
- WS 1.2 Recognise, draw and interpret diagrams.
- WS 3.5 Recognise patterns in data.

Maths focus Subtracting negative numbers

Resources needed Periodic tables; equipment as listed in the Technician's notes; Worksheet 1.5; Technician's notes 1.5

Digital resources Presentation 1.5 'One of these is not like the others'

Key vocabulary atomic number, electron, neutral, neutron, proton, symbol

Teaching and learning

Engage

- Display the Rutherford–Bohr model of a helium atom used in Lesson 1.3. Point out the two positive charges in the nucleus and the two electrons in a shell around the outside. Challenge students to **suggest** what makes atoms of hydrogen different from helium atoms. [O1]

Challenge and develop

- Use red balls to model the two protons that make up the positive nucleus in helium. Then use two green balls to model the rest of helium's nucleus – two neutrons – and two blue balls to model its two electrons. Stress that these models are not very realistic. Real protons and neutrons have 20 000 times more mass than electrons. [O2]

- Use a video clip to review the structure of helium [Google search string:] BBC Bitesize – GCSE chemistry – Atomic structure. [O1, O2, O3]

- Emphasise that different elements have different numbers of subatomic particles. The number of protons is crucial. Atoms with the same number of protons always belong to the same element. Hydrogen has only one proton so it only needs one electron to balance the charge on its nucleus. [O1, O2]

Explain

Low demand

- Read 'Structure of atoms' and 'Masses and charges' in Section 1.5 in the Student Book and answer questions 1–4. [O1, O2, O3]

Standard and high demand

- Complete questions 1–4 in Worksheet 1.5. [O1, O2, O3]

Consolidate and apply

- Use the Rutherford–Bohr model to introduce the idea that the nucleus never changes during a chemical reaction. However, metallic elements lose electrons when their atoms react to form compounds – making their atoms become charged. [O2]

- Use model protons and electrons to show how losing one electron leaves a lithium atom with an overall positive charge. This new charged particle is called an ion. [O2]

- Students complete question 5 in Worksheet 1.5. [O2]

Extend

Ask students able to progress further to **describe** any patterns they spotted in the table in question 5 from Worksheet 1.5.

Plenary suggestion

Use Presentation 1.5, 'One of these is not like the others' to display three particles and ask groups to say why one of them is not like the others. Encourage students to find as many possible answers as they can.

Answers to Worksheet 1.5

1. Protons in the centre; electrons in the top-left section; neutrons in the middle-right; 2. They have different numbers of protons (neutrons and electrons); 3. Electrons; 4. The atomic number shows the number of protons; which is the same as the number of electrons in a neutral atom; 5.

	Atomic number	Protons	Electrons	Charge
Li atom	3	3	3	0
Li ion	3	3	2	+1
Be atom	4	4	4	0
Be ion	4	4	2	+2
Na atom	11	11	11	0
Na ion	11	11	10	+1
Mg atom	12	12	12	0
Mg ion	12	12	10	+2
Al atom	13	13	13	0
Al ion	13	13	10	+3

© HarperCollins*Publishers* Limited 2016

Lesson 6: Subatomic particles

Lesson overview

OCR Specification reference

OCR C1.2

Learning objectives

- Find out what the periodic table tells us about each element's atoms.
- Learn what isotopes are.
- Use symbols to represent isotopes.

Learning outcomes

- Use atomic numbers and mass numbers to find the number of subatomic particles in an atom. [O1]
- Explain what isotopes are and how they differ. [O2]
- Use symbols to represent isotopes. [O3]

Skills development

- WS 1.2 Use models in explanations.
- WS 1.2 Recognise, draw and interpret diagrams.
- WS 4.1 Use scientific definitions.

Resources needed Periodic tables; Worksheets 1.6.1 and 1.6.2

Digital resources Presentation 1.6 'Isotopes'

Key vocabulary atomic mass, isotope, neutrons, protons

Teaching and learning

Engage

- Display slide 1 of Presentation 1.6 'Isotopes' to prompt pairs to **discuss** what they know about atoms and isotopes. Slides 2–4 can be used to clarify student's ideas. [O2]

Challenge and develop

- Display slide 5 of Presentation 1.6 'Isotopes' and demonstrate how to calculate numbers of neutrons from mass and atomic numbers. [O1]
- Use a video clip to review how to use atomic numbers and mass numbers to find the number of subatomic particles in an atom [Google search string:] BBC Bitesize – GCSE chemistry – How mass and atomic numbers explain atomic structure. (Avoid the isotopes video in the same series because its description of isotopes could cause misconceptions.) [O1]

Explain

Low and standard demand

- Students use Worksheet 1.6.1 to **calculate** the number of neutrons in atoms. [O1]

High demand

- Students use the 'Neutrons' section of Worksheet 1.6.2 to **compare** the neutron to proton ratio in a range of atoms. [O1]

Consolidate and apply

- Display slide 5 of Presentation 1.6 'Isotopes' and introduce the term 'isotopes' to describe atoms of the same element with different numbers of neutrons. [O2]

Low demand

- Read 'Isotopes' from Section 1.6 in the Student Book; write the formula of each isotope of hydrogen, and draw a diagram of its atom. [O2, O3]

Standard demand

- Read 'Isotopes' from Section 1.6 in the Student Book and answer questions 3–5. [O2, O3]

High demand

- Students use the 'Isotopes' section of Worksheet 1.6.2 to practice using symbols to represent isotopes. [O3]

Extend

Ask students able to progress further to **consider** that some elements have only radioactive isotopes. **Find out** which elements these are.

Plenary suggestion

Ask students to write down three things they know about atoms, and then move into groups to **compile** a master list with the most important at the top. The most important one can then be **shared** with the rest of the class.

Answers to Worksheet 1.6.1

Atom	Symbol	Mass number	Atomic number	Neutrons
Hydrogen	H	1	1	0
Helium	He	4	2	2
Lithium	Li	7	3	4
Beryllium	Be	9	4	5
Boron	B	11	5	6
Carbon	C	12	6	6
Nitrogen	N	14	7	7
Oxygen	O	16	8	8
Fluorine	F	19	9	10
Neon	Ne	20	10	10
Sodium	Na	23	11	12
Magnesium	Mg	24	12	12
Aluminium	Al	27	13	14
Silicon	Si	28	14	14
Phosphorus	P	31	15	16
Sulfur	S	32	16	16
Chlorine	Cl	35	17	18
Argon	Ar	40	18	22
Potassium	K	39	19	20
Calcium	Ca	40	20	20

Answers to Worksheet 1.6.2

Neutrons: 1. Li, Be, B, F, Na, Al, P, Cl, Ar, K; 2. Only H
Isotopes: 1. $^{32}_{16}S$ $^{35}_{17}Cl$ $^{40}_{18}Ar$ $^{39}_{19}K$ $^{40}_{20}Ca$;

2. Carbon-12 has six protons, six electrons, six neutrons; Carbon-13 has six protons, six electrons, seven neutrons; Carbon-14 has six protons, six electrons, eight neutrons

 © HarperCollins*Publishers* Limited 2016

Lesson 7: Maths skills: Standard form and making estimates

Lesson overview

OCR Specification reference

OCR C1.2

Learning objectives

- Consider the sizes of particles.
- Use numbers in standard form to compare sizes.
- Use numbers in standard form in calculations.

Learning outcomes

- Recognise numbers written in standard form. [O1]
- Convert decimals to standard form and vice versa. [O2]
- Use standard form to make estimates. [O3]

Skills development

- WS 4.2 Recognise the importance of scientific quantities.
- WS 4.4 Use powers of 10 for orders of magnitude.

Maths focus Recognise and use numbers in decimal and standard form

Resources needed Equipment as listed in the Technician's notes; Worksheet 1.7; Technician's notes 1.7

Digital resources Presentation 1.7 'Powers of 10'

Key vocabulary standard form, decimal point

Teaching and learning

Engage

- Ask pairs to sort cards cut from Worksheet 1.7 into sets of three. Check that students **recognise** that the long numbers are decimals and the powers of 10 are the same numbers written in standard form. [O1, O2, O3]

Challenge and develop

- Use a video clip to show the powers of 10 between the sizes of the largest and smallest objects in the universe. [Google search string:] Powers of Ten™ (1977). [O1]

- Use Presentation 1.7 'Powers of 10' to show how numbers move up the hierarchy to the left of the decimal point each time they are multiplied by 10, and down the hierarchy to the right of the decimal point when they are divided by 10. [O1, O2]

Low demand

- Pairs **arrange** the sets of cards used at the start of the lesson so that each decimal number covers the number in standard form. Then they take turns to **convert** each decimal into standard form. They could then turn the decimal card over and **convert** the numbers they have written in standard form back to decimals. [O2]

Standard and high demand

- Ask students to read Section 1.7 of the Student Book and answer questions 1–4. [O2]

Explain

- Students **explain** to each other how to convert numbers from decimals to standard form and vice versa. [O2]

Consolidate and apply

- Remind students how to multiply and divide numbers written in standard form. [O3]

Low and standard demand

- Students complete question 5 from the Student Book. [O3]

High demand

- Students complete question 6 from the Student Book. [O3]

Extend

Ask students able to progress further to **research** Avogadro's number [Google search string:] The History of Avogadro's Number with Bill Bryson. [O1]

Plenary suggestion

Ask groups to **create** an outline for a short version of the 'Powers of 10' video for use in primary schools. It should only cover sizes from 10^2 m to 10^{-2} m. What images would they choose for each scene?

Answers to Worksheet 1.7

The correct sets are: AHN; BJL; CGO; DIK; EFM

When and how to use these pages: Check your progress, Worked example and End of chapter test

Check your progress

Check your progress is a summary of what students should know and be able to do when they have completed the chapter. Check your progress is organised in three columns to show how ideas and skills progress in sophistication. Students aiming for top grades need to have mastered all the skills and ideas articulated in the final column (shaded pink in the Student book).

Check your progress can be used for individual or class revision using any combination of the suggestions below:

- Ask students to construct a mind map linking the points in Check your progress
- Work through Check your progress as a class and note the points that need further discussion
- Ask the students to tick the boxes on the Check your progress worksheet (Teacher Pack CD). Any points they have not been confident to tick they should revisit in the Student Book.
- Ask students to do further research on the different points listed in Check your progress
- Students work in pairs and ask each other what points they think they can do and why they think they can do those, and not others

Worked example

The worked example talks students through a series of exam-style questions. Sample student answers are provided, which are annotated to show how they could be improved.

- Give students the Worked example worksheet (Teacher Pack CD). The annotation boxes on this are blank. Ask students to discuss and write their own improvements before reviewing the annotated Worked example in the Student Book. This can be done as an individual, group or class activity.

End of chapter test

The End of chapter test gives students the opportunity to practice answering the different types of questions that they will encounter in their final exams. You can use the Marking grid provided in this Teacher Pack or on the CD Rom to analyse results. This shows the Assessment Objective for each question, so you can review trends and see individual student and class performance in answering questions for the different Assessment Objectives and to highlight areas for improvement.

- Questions could be used as a test once you have completed the chapter
- Questions could be worked through as part of a revision lesson
- Ask Students to mark each other's work and then talk through the mark scheme provided
- As a class, make a list of questions that most students did not get right. Work through these as a class.

Chapter 1: Particles

Marking Grid for Single Chemistry End of Chapter 1 Test Version 3816

Student Name	Tier / Category	Question
	Getting Started [Foundation Tier]	Q. 1 (AO1) 1 mark
		Q. 2 (AO1) 2 marks
		Q. 3 (AO1) 2 marks
		Q. 4 (AO1) 1 mark
		Q. 5 (AO2) 1 mark
		Q. 6 (AO1) 1 mark
		Q. 7 (AO1) 2 marks
	Going Further	Q. 8 (AO2) 1 mark
		Q. 9 (AO1) 1 mark
		Q. 10 (AO2) 2 marks
		Q. 11 (AO3) 4 marks
		Q. 12 (AO2) 2 marks
	More Challenging [Higher Tier]	Q. 13 (AO1) 1 mark
		Q. 14 (AO2) 1 mark
		Q. 15 (AO1) 2 marks
		Q. 16 (AO2) 2 marks
		Q. 17 (AO2) 4 marks
	Most demanding	Q. 18 (AO2) 2 marks
		Q. 19 (AO3) 4 marks
		Q. 20 (AO2) 4 marks
		Percentage

Check your progress

You should be able to:

☐ use data to predict the states of substances	→ ☐ explain the changes of state	→ ☐ use state symbols in chemical equations
☐ describe a change of state as a physical change	→ ☐ explain the difference between a physical change and a chemical change	→ ☐ explain how chemical changes involve rearrangement of bonds between particles
☐ explain in terms of the particle model the differences between the states of matter	→ ☐ explain in terms of the particle model the distinction between physical changes and chemical changes	→ ☐ explain the limitations of the particle model when particles are represented by inelastic spheres
☐ describe the work of Dalton and Thomson	→ ☐ explain how the work of Rutherford changed ideas about the atom	→ ☐ explain how the theoretical ideas of Bohr changed the idea of electron structure
☐ explain that early models of the atom did not have shells with electrons	→ ☐ explain that early models of atoms developed as new evidence became available	→ ☐ explain why the scattering experiment led to a change in the atomic model
☐ draw a diagram of a small nucleus containing protons and neutrons with orbiting electrons at a distance	→ ☐ describe that the nuclear radius is much smaller than that of the atom and that most of the mass in the nucleus	→ ☐ explain why the atom is mostly empty space
☐ recall the relative sizes of everyday objects and compare these to the relative sizes of atoms	→ ☐ recall the typical size (order of magnitude) of atoms and small molecules	→ ☐ recall that typical atomic radii and bond length are in the order of 10^{-10} m
☐ recall the relative charges and approximate relative masses of protons, neutrons and electrons	→ ☐ calculate the atomic masses of elements from the numbers of protons and neutrons	→ ☐ calculate numbers of protons, neutrons and electrons in atoms given the atomic number and mass number
☐ describe the difference between isotopes of an element	→ ☐ calculate numbers of protons, neutrons and electrons in atoms given the atomic number and mass number of isotopes	→ ☐ complete data tables showing atomic numbers, mass numbers and numbers of subatomic particles from symbols
☐ describe the difference between an atom and an ion	→ ☐ calculate numbers of protons, neutrons and electrons in ions given the atomic number and mass number of isotopes	→ ☐ represent numbers of subatomic particles of isotopes using standard symbols

© HarperCollins*Publishers* Limited 2016

Worked example

1 a Draw diagrams to show how substances change from solids to liquids.

The particles move faster when they are heated and break away from the solid structure to move more freely.

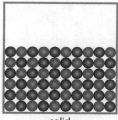 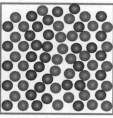

solid   liquid

> This answer shows both diagrams and has an added explanation.

b Describe what kind of process is happening during this change and explain whether this is a physical change or a chemical change.

When a solid turns into a liquid it is called 'melting'. This is a physical change because the particles don't join together.

> The process is correct and an explanation is given of why it is not a chemical reaction. The student should add that no new substance is made and that the substance will go back to the same solid if cooled.

2 Describe the structure of an atom that has atomic number 3 and mass number 7.

This atom has 3 protons and 3 electrons around them and 1 neutron.

> The numbers of protons and electrons are correct. The number of protons is equal to the number of electrons and is the atomic number of an element.
>
> The number of neutrons is incorrect and should be 7. The mass number is not equal to the sum of the numbers of protons, electrons and neutrons – it is equal to the sum of the numbers of just the protons and neutrons.

3 a Describe what is meant by 'isotopes' and draw an isotope of hydrogen that has 2 neutrons.

Isotopes are atoms that have the same number of protons but different numbers of neutrons. This means that their atomic number is the same but their mass numbers are different.

> The description is correct and the diagram is correct.

b Describe the charges and masses of the subatomic particles you have drawn and explain how they relate to the atomic number and mass number.

The red dot is a proton that has a positive charge and a mass of 1. The blue dot is an electron that has no mass and a negative charge. The green dots are neutrons with no charge and a mass of 1. The atomic number is 1 and the mass number is 4.

> The three descriptions of the charge and mass of the proton, neutrons and electron are correct. The atomic number is correct. However, even though the description of the mass of the electron is correctly given as 0, this has been included in the mass number (4). There are 2 neutrons and 1 proton so the mass number is 3.

20   © HarperCollins*Publishers* Limited 2016

Elements, compounds and mixtures: Introduction

When and how to use these pages

This unit builds on ideas about particles introduced in Chapter 1.

Overview of the unit

The students will learn to distinguish pure substances from mixtures and explore ways of separating different mixtures. They will also explore the different types of bonding – ionic, covalent and metallic – in pure substances. They will find out how to represent both ionic and covalent bonding using dot-and-cross diagrams. They will use suitable diagrams to show metallic bonding and find out about delocalised electrons. They will learn how to calculate the charge on the ions in an ionic compound. Students will then explore the types of structures produced by the different types of bonding. This includes giant ionic structures, small molecules, polymers, giant covalent structures, and metals and alloys.

Students next investigate the properties of each type of substance, concentrating on explanations involving the bonding and structure. This includes knowledge of intermolecular forces. They study examples of giant covalent substances, namely diamond and graphite. Students find out about their specific properties and explain them in terms of bonding and structure. Finally, they explore the new developments of graphene and fullerenes, and of nanoparticles. This concentrates on their structure, properties and possible uses.

Students will carry out experiments, describe practical procedures, select techniques for a particular purpose and explain their choices. They will also make and record observations, describe patterns in data and use data to make predictions.

This unit offers a number of opportunities for the students to use different models to represent the structure of substances. This includes ball-and-stick models, space-filling models and two-dimensional (2D) and 3D representations. An appreciation of the limitations of these models is developed. They are encouraged to explore how the properties of the individual particles in a substance are different from the bulk properties of that substance. Throughout the chapter, students use ideas about bonding and structure to explain and predict the properties of substances.

The unit offers a number of opportunities for students to use mathematics. They visualise and represent two-dimensional (2D) and 3D forms, use ratios to write simple formulae, practice using an appropriate number of significant figures.

Obstacles to learning

Students may need extra guidance with the following terms and concepts:

- Drawing 3D structures, especially ionic lattices, is difficult for many students. These can be traced if required.

- Many students misuse the word 'molecule', incorrectly applying it to ionic compounds. They need to be aware that only covalent substances can form molecules and that not all covalent substances form molecules. Similarly a formula for a substance, such as $NaCl$ or H_2O, gives different information for ionic or covalent substances.

- A common error in drawing dot-and-cross diagrams to represent covalent bonds is either not showing a pair of shared electrons and/or showing one shared electron only, or ending up with more or less than the required number of electrons in the outer shell. Doing a final electron count helps avoiding this error. Students will only be required to use atoms with two or eight electrons in their outer shells.

- Students need a clear understanding of how the type of bonding relates to the type of substance. The multiple examples in this chapter should help to establish this link.

Practicals in this unit

In this unit students will do the following practical work:

- Devise procedures to obtain crystals of copper sulfate from a mixture of copper sulfate solution and sand

- Separate copper powder and sodium chloride using filtration and evaporation

- Make chromatograms of coloured inks and food colourings

- Investigate how paper chromatography can be used in forensic science

- Test a range of oxides in water with universal indicator solution

- Model the electron arrangements in atoms

- Model the changes in electron arrangements when atoms react to form ions

- Investigate the properties of ionic compounds.

- Stretching polybags.

- Investigating a tin-lead alloy

- Model the arrangement of elements in the periodic table.

Students will observe:

- How adding sodium chloride to water changes its melting point and boiling Burning magnesium.

- Burning magnesium

- Crystals of sodium chloride under the microscope

- Growing silver crystals under the microscope.

- Sublimation of iodine.

- The electrical conductivity of molten ionic compounds.

- Silicon dioxide under the microscope.

- Products containing nanoparticles

	Lesson title	Overarching objectives
1	Pure substances	Suggest ways of separating mixtures.
2	Relative formula mass	Use atomic masses to calculate relative formula masses
3	Mixtures	Understand how different techniques are used to separate mixtures.
4	Formulations	Recognise that formulations designed for specific uses are mixtures.
5	Chromatography	Use chromatographic methods to distinguish pure substances from impure substances, and calculate R_f values.
6	Practical: Investigate how paper chromatography can be used in forensic science to identify an ink mixture used in a forgery	Investigate how paper chromatography can be used in forensic science to identify an ink mixture used in a forgery.
7	Maths skills: Use an appropriate number of significant figures	Using an appropriate number of significant figures.
8	Comparing metals and non-metals	Distinguish metals from non-metals using their physical and chemical properties.
9	Electron structure	Describe the arrangement of electrons in shells or energy levels.
10	Metals and non-metals	Explain the differences between metals from non-metals.
11	Chemical bonds	Describe the three main types of bonding and how electrons are used in each.
12	Ionic bonding	Draw dot-and-cross diagrams to represent ionic bonds and calculate the charge on the ions.
13	Ionic compounds	Know the structure of ionic compounds and be able to work out their empirical formulae.
14	Properties of ionic compounds	Relate the properties of ionic compounds to the forces between ions.

15	Properties of small molecules	Relate the properties of small molecules to the strengths of covalent bonds and the intermolecular forces between them..
16	Covalent bonding	Draw dot-and-cross diagrams for small molecules and be able to work out their molecular formulae from models and diagrams.
17	Giant covalent structures	Explain why substances with giant covalent structures are solids at room temperature.
18	Polymer structures	Describe the structure of polymers and explain their melting points in terms of the size of the polymer molecule and the intermolecular forces.
19	Metallic bonding	Describe that metals have a giant structure consisting of metal ions and delocalised electrons.
20	Properties of metals and alloys	Explain the properties of metals in terms of their structure and bonding and why we use alloys.
21	Key concept: The outer electrons	Model bonding in terms of sharing or transferring electrons.
22	The periodic table	Describe the link between electronic structure and an element's position in the periodic table.
23	Developing the periodic table	Describe how Mendeleev developed the periodic table.
24	Diamond	Explain the properties of diamond in terms of its covalent bonding and tetrahedral structure.
25	Graphite	Describe how graphite differs from diamond and explain why graphite is soft and slippery and conducts electricity.
26	Graphene and fullerenes	Describe the structure of graphene and fullerenes and know some properties and uses.
27	Nanoparticles, their properties and uses	Understand the size of nanoparticles and evaluate the benefits of using them against their possible harmful effects.
28	Maths skills: Using ratios in mixture, empirical formulae and balanced equations	Use ratios to describe the composition of mixtures and the empirical formulae of compounds.

 © HarperCollins*Publishers* Limited 2016

Lesson 1: Key concept: Pure substances

Lesson overview

OCR Specification reference

OCR C2.1

Learning objectives

- Describe, explain and exemplify processes of separation.
- Suggest separation and purification techniques for mixtures.
- Distinguish pure and impure substances using melting point and boiling point data

Learning outcomes

- Describe and explain why filtration, crystallisation, simple distillation, fractional distillation and chromatography are used to separate mixtures. [O1]
- Devise a practical procedure to separate a given mixture using filtration, crystallisation, simple distillation, fractional distillation or chromatography. [O2]
- Know how the melting points and boiling points of pure and impure substances differ and use such data to identify pure and impure substances. [O3]

Skills development

- WS 2.2 Plan experiments or devise procedures to make observations, produce or characterise a substance, test hypotheses, check data or explore phenomena.
- WS 2.3 Carry out experiments appropriately having due regard for the correct manipulation of apparatus, the accuracy of measurements and health and safety considerations.
- WS 4.1 Use scientific vocabulary, terminology and definitions.

Resources needed A range of pure and impure substances, such as deionised/distilled water, copper sulfate solution, milk, soft drink, soap, detergent, copper, cosmetics (products with the word 'pure' in the packaging would be useful); Practical sheets 2.1.1 and 2.1.2; Worksheets 2.1.1 and 2.1.2; Technician's notes 2.1.1 and 2.1.2

Key vocabulary pure, impure, melting point, boiling point

Teaching and learning

Engage

- Show students a range of pure and impure substances as suggested above in 'Resources needed'. Ask students to work in pairs and decide which are pure substances and which are impure substances. They can use Section 2.1 in the Student Book to help if needed. Discuss their ideas and deduce that the everyday understanding of 'pure' is not always the same as the scientific meaning. Explain that a pure substance, in chemistry, consists of one element or compound only. [O1]
- Ask students what we call impure substances and **deduce** that they are mixtures. Revisit the properties of a mixture if necessary. [O1]

Challenge and develop

Low demand

- Ask students to list the different methods they have met that are used to separate the components of a mixture. These should include filtration, simple distillation (to separate two miscible liquids), fractional distillation (to separate more than two miscible liquids, such as crude oil), crystallisation and chromatography. You may need to explain crystallisation and chromatography. They can answer question 1 in Worksheet 2.1.1. Encourage them to use the terms 'soluble', 'insoluble', 'solution', 'solvent' and 'solute' [O1]

Standard demand

- Introduce students to Practical sheet 2.1.1. They are to plan and carry out separation procedures to obtain pure crystals of copper sulfate. The scenario is a copper leach mine at which sulfuric acid is pumped into old

25

copper mines to react with the copper ores. The resulting mixture of copper sulfate, water and sulfuric acid is pumped to the surface. Students can work in pairs to **plan** their procedures. These should include filtration and crystallisation. They must write a risk assessment. Check their plans and allow students to carry out their experiments. The solutions will need to be left until next lesson to crystallise. Do not allow students to touch the copper sulfate solution or the crystals. [O2]

- Students can answer question 1 in Worksheet 2.1.2. This provides additional practice in identifying suitable separation techniques. [O2]

Explain

High demand

- Ask students to recall any melting points and boiling points they know about. They should recall those of water and the boiling point of ethanol. Explain that pure substances have definite melting and boiling points and these can be used to identify a substance. These temperatures can also be used to check purity. You could use Practical sheet 2.1.2 to demonstrate the changes to the melting point and boiling point of water when salt is added. [O3]

- Students can answer questions 5 and 6 in Section 2.1 of the Student Book and answer question 2 in Worksheet 2.1.2. [O3]

Consolidate and apply

- Ask students to **explain** why salt is spread on icy roads in the winter. [O3]

- Ask them to **explain** why a pharmaceutical company might measure the melting point of a batch of medicine. [O3]

Extend

Ask students able to progress further to **find out**:

- What a centrifuge is used for and name another separation process that does the same job. [O1]
- Why salt water has a higher BP than pure water. [O3]
- Why leaching is used in some copper mines and not others. [O2]

Plenary suggestions

Write the following terms on the board: 'pure',' impure', 'melting point', 'boiling point', 'filtration', 'crystallisation', 'simple distillation', 'fractional distillation', 'chromatography', 'solution', 'solute', 'solvent', 'soluble' and 'insoluble'. Allow students a few minutes to **compose** two sentences using as many of the words as possible. They can **share** their sentences with the rest of the class.

Put one student in the hot seat. The rest of the class can ask questions based on the lesson's content. Students can swap places when an answer is wrong.

Answers to Worksheet 2.1.1

1. Filtration: an insoluble substance from a solution or pure liquid/filter funnel and paper/sand from water; Evaporation: a soluble solid from a solution/evaporating dish, heating apparatus/sodium chloride from sodium chloride solution; Crystallisation: a soluble substance from its solution/crystallising dish (heating apparatus)/copper sulfate crystals from copper sulfate solution; Simple distillation: two miscible liquids/heating apparatus, flask, condenser, thermometer, collecting vessel/water and ethanol; Fractional distillation: more than two miscible liquids/flask, heating apparatus, fractionating column, collecting vessels/crude oil; Chromatography: soluble coloured solutes in a solution/chromatography paper, beaker, lid/inks

Answers to Worksheet 2.1.2

1. a) Simple distillation; b) Chromatography; c) Filtration; d) Simple distillation; e) Dissolving, filtration, simple distillation; f) Chromatography; 2. a)(i) Ethanol has a distinct m.p., impure ethanol has a lower m.p. and melts over a range of temperatures, (ii) Deionised water has a distinct b.p., seawater has a higher b.p; b) (i) C; (ii) E; (iii) D; it has a distinct m.p.

Answers to Practical sheet 2.1.1

Should include adding water to dissolve the copper sulfate; filtering to remove insoluble substances; evaporating to reduce the volume; crystallising to obtain crystals of copper sulfate

Answers to Practical sheet 2.1.2

1. Decreases; 2. Increases

Lesson 2: Relative formula mass

Lesson overview

OCR Specification reference

OCR C2.1

Learning objectives

- Review the differences between the isotopes of an element.
- Distinguish between the mass of an atom and the relative atomic mass of an element.
- Use relative atomic masses to calculate relative formula masses.

Learning outcomes

- Identify the relative atomic mass of an element from the periodic table. [O1]
- Calculate relative formula masses from relative atomic masses. [O2]
- Verify the law of conservation of mass in a balanced equation. [O3]

Skills development

- WS 1.2 Use models in explanations.
- WS 1.2 Calculate quantities based on a model.
- WS 4.1 Use scientific vocabulary, terminology and definitions.

Resources needed Equipment as listed in the Technician's notes; Worksheets 2.2.1, 2.2.2 and 2.2.3; Technician's notes 2.2

Digital resources Presentation 2.2 'Relative atomic mass'

Key vocabulary atomic mass, conservation of mass, formula mass

Teaching and learning

Engage

- Display slide 1 of Presentation 2.2 'Relative atomic mass' and challenge groups to identify the number of true statements. [O1]

Challenge and develop

- Take feedback from representatives of each group and use slides 2–5 of Presentation 2.2 'Relative atomic mass' to introduce the definition of relative atomic mass. Then explain that relative atomic masses have no units because they are all 'relative' to 1/12th of the mass of an atom of carbon-12. [O1]
- Use 'Formula mass' in Section 2.2 of the Student Book to show how relative atomic masses can be added to give relative formula masses. [O2]

Explain

Low demand

- Use the cards from Worksheet 2.2.1 to practise **calculating** relative formula masses for the compounds listed in Worksheet 2.2.2. [O1, O2]

Standard and high demand

- Use the periodic table to practise **calculating** relative formula masses for some of the compounds listed in Worksheet 2.2.2. [O1, O2]

High demand

- Challenge students to **deduce** the formulae of these compounds: lithium oxide ($M_r = 40$), beryllium hydroxide ($M_r = 43$) and an oxide of nitrogen ($M_r = 46$). [O2]

Consolidate and apply

Low demand

- Students complete Worksheet 2.2.3 to demonstrate that the total mass of reactants used is always the same as the total mass of products made. [O3]

Standard and high demand

- Students work in pairs to **choose** a chemical reaction, write a balanced symbol equation to describe it and use this equation to **explain** why the total mass of products is always the same as the total mass of reactants. [O3]

Extend

Ask students able to progress further to answer question 6 in Section 2.2 of the Student Book.

Plenary suggestion

Put these gas molecules in order of mass: CH_4, CO, CO_2, Cl_2, N_2, O_2, SO_2.

Answers to Worksheet 2.2.2

a) 28; b) 44; c) 34; d) 18; e) 34; f) 46; g) 17; h) 16; i) 64; j) 111; k) 53.5; l) 80; m) 36.5; n) 98; o) 63; p) 120; q) 96; r) 164; s) 74; t) 132

Answers to Worksheet 2.2.3

1. a) 36; b) 46; c) 64; d) 44; e) 18; 2. Atoms are never lost during chemical reactions; they rearrange to form new products; so the total mass always stays the same

 © HarperCollins*Publishers* Limited 2016

Lesson 3: Mixtures

Lesson overview

OCR Specification reference

OCR C2.1

Learning objectives

- Recognise that all substances are chemicals.
- Understand that mixtures can be separated into their components.
- Suggest suitable separation and purification techniques for mixtures.

Learning outcomes

- Know that mixtures consist of two or more elements or compounds not combined together chemically. [O1]
- Separate a mixture of copper and sodium chloride. [O2]
- Explain how filtration, crystallisation, simple distillation, fractional distillation and chromatography are used to separate mixture and select suitable methods. [O3]

Skills development

- WS 2.2 Plan experiments or devise procedures to make observations, produce or characterise a substance, test hypothesis, check data or explore phenomena.
- WS 2.3 Apply a knowledge of a range of techniques, instruments, apparatus and materials to select those appropriate to the equipment.
- WS 2.4 Carry out experiments appropriately having due regard for the correct manipulation of apparatus, the accuracy of measurements and health and safety considerations.

Resources needed Equipment as listed in the Technician's notes; Worksheets 2.3.1 and 2.3.2; Practical sheet 2.3; Technician's notes 2.3

Key vocabulary: chromatography, filtration, mixture, separation

Teaching and learning

Engage

- Remind students that table salt is sodium chloride – a compound of sodium and chlorine – and display a model of its structure. Contrast this with a model of the element copper to emphasise the difference. Add the element copper to the salt to make a mixture. Emphasise that this mixture contains an element and a compound. Add more sodium chloride to the mixture and ask students if and how the mixture has changed. [O1], [O2]
- Ask students to work in pairs and list as many properties of mixtures as they can. Draw their ideas together, establishing that mixtures contain two or more elements or compounds, not chemically combined together, in any proportions, and the components retain their individual properties. [O1]

Challenge and develop

Low demand

- Give out Practical sheet 2.3. Students work in pairs to **separate** copper from sodium chloride. [O2]

Standard demand

- Make the apparatus listed in Technician's notes 2.3 available and challenge students to **separate** the mixture of copper and sodium chloride. They work in pairs or threes to complete the task. [O2]

High demand

- Challenge students to **select** the apparatus they need to separate a mixture of copper and sodium chloride. Students work in pairs or threes to complete the task. [O2]

Explain

Low demand

- Students complete the questions on Practical sheet 2.3. [O2]

Standard demand

- Students **describe** how they separated copper and sodium chloride. [O2]

High demand

- Students use diagrams to **explain** how their method worked. [O2]

Consolidate and apply

- All students can be asked to read 'Mixtures' in Section 2.3 in the Student Book and complete Worksheet 2.3.1. [O3]

- Groups prepare answers to the questions on Worksheet 2.3.2 and appoint a spokesperson to **present** their explanations. Different groups can be selected to **explain** each technique, with others invited to **offer** additional points they haven't covered. [O3]

Extend

Ask students able to progress further to:

- Work in pairs and list ten products they would find in a kitchen or bathroom that are mixtures. Students can research the components of their mixtures. [O1]

- Find an image of a simple solar still and **identify** its advantages and disadvantages compared with the laboratory distillation apparatus illustrated in the Student Book. [O3]

Plenary suggestions

Ask students to draw a large triangle with a smaller inverted triangle that just fits inside it (so they have four triangles). In the outer three ask them to write about this lesson – something they've seen, something they've done and something they've discussed. Then add in the central triangle something they've learned.

Ask students to **discuss** in pairs what makes sodium chloride a compound and toothpaste a mixture. They can **share** their ideas with the class.

Answers to Worksheet 2.3.1

1. D, H, O; 2. E, F, L; 3. A, G, M; 4. B, J, N; 5. C, I, K

Answers to Worksheet 2.3.2

1. Filtration: filter funnel, filter paper, conical flask; the solution passes through holes in the filter paper, but the silver chloride particles are too big to pass through them; 2. a) Chromatography; b) To separate or identify colourings in foods; 3. Crystallisation; As water evaporates; the solution becomes too concentrated; for the copper sulfate to remain dissolved; 4. Fractional distillation; Ethanol evaporates more easily than water; because it has a lower boiling point

Answers to Practical sheet 2.3

Predictions: 1. Sodium chloride; 2. Copper; 3. Sodium chloride; *Evaluation*: 1. They are both soluble; 2. Sodium chloride is soluble but copper is insoluble; 3, No

Lesson 4: Formulations

Lesson overview

OCR Specification reference

OCR C2.1

Learning objectives

- Identify formulations given appropriate information.
- Explain the particular purpose of each chemical in a mixture.
- Explain how quantities are carefully measured for formulation.

Learning outcomes

- Recognise that fuels, cleaning agents, paints, medicines, alloys, fertilisers and foods are formulations. [O1]
- Explain the purpose of each ingredient in an aspirin tablet. [O2]
- Explain that the ingredients in formulations like aspirin are carefully measured so that each tablet contains exactly the same formula. [O3]

Skills development

- WS 1.4 Explain the everyday and technological applications of science; evaluate associated personal, social, economic and environmental implications; and make decisions based on the evaluation of evidence and arguments.
- WS 2.2 Plan experiments or devise procedures to make observations, produce or characterise a substance, test hypotheses, check data or explore phenomena.
- WS 2.5 Recognise when to apply a knowledge of sampling techniques to ensure that any samples collected are representative.

Maths focus Use ratios, fractions and percentages

Resources needed Shampoo bottle (or similar) showing an ingredients list; Worksheets 2.4.1 and 2.4.2

Digital resources Internet access

Key vocabulary formulation, fertilisers, active ingredient, alloys

Teaching and learning

Engage

- Show students the ingredient lists on a shampoo bottle (or similar). Tell them that shampoo is a mixture of different chemicals – but there are no chemical reactions going on in the shampoo bottle. Ask them if they think all shampoo bottles of the same type and manufacture will contain exactly the same ingredients in the same proportions. Establish that a constant composition is important and explain that shampoo is a formulated product. Tell students that a formulated product is made to an exact recipe. [O1]
- Ask students to work in pairs and make a list of other products they think are formulated. They can keep their lists until the end of the lesson. [O1]

Challenge and develop

Low demand

- Emphasise that many of the mixtures we use and rely on every day are formulated. Students can answer question 1 in Worksheet 2.4.1. [O1]

Explain

Standard demand

- Explain that each ingredient in a formulated product has a role to play. Students can work in pairs and list the different roles that chemicals must play in a bottle of shampoo (e.g., smell, colour, cleaning agents, making foam, conditioning, preserving). They can answer question 2 in Worksheet 2.4.1. They will need to look at Section 2.4 of the Student Book and have online access. You may prefer to set this question as homework. [O2]

High demand

- Ask students how manufacturers ensure that every bottle of shampoo is the same. Establish that the components have to be carefully measured. Students can read 'Measuring quantities' in Section 2.4 of the Student Book and then answer the questions in Worksheet 2.4.2. This worksheet focuses on the production of aspirin and ibuprofen tablets and the measures needed to ensure consistency of the product. Question 6 [covering skills WS 2.2 and WS 2.5] asks students to list the measurements needed by quality control to ensure that all tablets are the same. Students should include several points in their answers. [O3]

Consolidate and apply

- Students can answer question 5 in Section 2.4 of the Student Book and then answer question 3 in Worksheet 2.4.1. The question requires students to use percentage calculations. [O3]

- Students can **research** labels of ingredients from formulated foods. There are many on Google images. They can find out what each ingredient contributes to the food. [O2]

Extend

Ask students able to progress further to:

- **Discuss** the problems facing fuel producers when the composition of crude oil varies from oil field to oil field, and petrol is a formulated product. [O1]

- **Find out** what spectroscopy is and how infrared spectroscopy can be used to analyse samples.

Plenary suggestions

Students can **check** and **amend** their initial lists of formulated products and each **share** one thing they have learned with the rest of the class (or group).

Ask students to write down two ideas they have learned in this lesson. They can **share** their ideas in pairs and then double up. A class list of ideas can be compiled.

Answers to Worksheet 2.4.1

1. a) iii, iv, v, vi, vii, viii; b) Student's response; c) So that they will be the same/consistent; 2. a) Reduces rusting; makes the metal less brittle; b) Promotes shoot growth; c) Painkiller; d) Reduces use of fossil fuel; e) Reduces the density of the alloy; f) Digests biological stains; g) Preserves the fruit; h) Colour the confectionary; 3. A; G

Answers to Worksheet 2.4.2

1. To ensure consistency; 2. A sample from each batch; 3. So that the batches can be traced if there is a problem; 4. So that the amounts of components are not changed by residues left in the machinery; 5. Lactose, starch, methyl cellulose, magnesium stearate; 6. Answers should include consistency of quality of components, suitability for consumption, mass of aspirin in each tablet, masses of other ingredients in each tablet, colour of tablet, mass of tablet

 © HarperCollins*Publishers* Limited 2016

Lesson 5: Chromatography

Lesson overview

OCR Specification reference

OCR C2.1

Learning objectives

- Explain how to set up chromatography paper.
- Distinguish pure from impure substances.
- Interpret chromatograms and calculate R_f values.

Learning outcomes

- Explain how to use chromatography paper to produce a chromatogram of coloured substances. [O1]
- Know that pure substances produce one spot only on a chromatogram and that impure substances produce more than one spot. [O2]
- Be able to make appropriate measurements from a chromatogram (distance moved by substance and distance moved by solvent) and use them to calculate R_f values. [O3]

Skills development

- WS 2.2 Plan experiments or devise procedures to make observations, produce or characterise a substance, test hypothesis, check data or explore phenomena.
- WS 3.1 Present observations and other data using appropriate methods.
- WS 3.2 Translate data from one form to another.
- WS 3.3 Carry out and represent mathematical and statistical analysis.

Maths focus Recognise and use expressions in decimal form; use ratios, fractions and percentages; make estimates of the results of simple calculations; use an appropriate number of significant figures

Resources needed: Practical sheet 2.5, Worksheet 2.5, Technician's notes 2.5

Key vocabulary chromatography, R_f value, stationary phase, mobile phase

Teaching and learning

Engage

- Ask students to **recall** what chromatography is, when they have used it and what it showed. Most will have used chromatography to separate the mixtures of coloured compounds in sweets such as Smarties or M&Ms, or in inks/coloured pens. Establish that chromatography is used to separate mixtures and tell students that it is also used to help identify unknown compounds. Examples include drug testing and forensic science. [O1]

Challenge and develop

Low demand

- Introduce students to Practical sheet 2.5: Investigate how paper chromatography can be used to separate and tell the difference between coloured substances. Students should calculate R_f values.

- Depending on the ability of students you may wish to follow the whole of Practical sheet 2.5, or select parts. For those who need more guidance use steps 1–4 in the method to explain how to set up the chromatography paper. Emphasise that this is *paper* chromatography and the chromatography paper is the stationary phase and the solvent/water is the mobile phase. Explain that the lid on the beaker helps to keep the atmosphere in the beaker saturated with solvent. Allow students to complete steps 1–4. If time allows, they can do step 5 using ethanol as the solvent.

- More able students can skip the explanations and work through Practical sheet 2.5 at their own pace, using Section 2.5 of the Student Book to explain the procedure. [O1] Students can carry out step 6 on Practical sheet 2.5. [O1]

Explain

Standard demand

- Students can read the example in Section 2.5 of the Student Book to **interpret** their chromatograms. Emphasise that the individual spots are of one compound only – if an ink produces only one spot on the chromatogram, it is a pure substance. If it produces more than one spot, it is a mixture. [O2]

High demand

- Explain that how far the different compounds move along the chromatography paper depends on their molecules' attraction for the paper and for the molecules of the solvent. Either use the Student Book or explain how to calculate R_f values for each spot produced. They can use the table in the Evaluation section of Practical sheet 2.5. to record their results and R_f values. They will need to draw a different table for each chromatogram. Students can answer questions 2 and 3 in Practical sheet 2.5. [O3]

Consolidate and apply

- Students can complete Worksheet 2.5 to consolidate the procedure. [O1, O2, O3]

Extend

Ask students able to progress further to:

- **Find out** how gas chromatography differs from paper chromatography. [O1]
- **Research** other uses of chromatography. [O1, O2, O3]

Plenary suggestions

Ask students to work in pairs and **compile** a list of 'Don'ts' for producing a good chromatogram. They can **share** their lists with the rest of the class and a class list can be **compiled**.

Organise students into groups of six to eight. Hand out strips of coloured paper – each student needs two strips. They write a question about something they have learned in the lesson on one strip, and the answer on another. Shuffle and distribute the questions and answers so that each student has one question and one answer. They can read out their questions in turn and the student with the correct answer reads it out, followed by their question.

Answers to Worksheet 2.5

1. a)(i) So that all the samples can be on the same line; the pencil will not contribute to the chromatogram; (ii) Larger amounts of samples produce more distinct spots; if the sample spots are too large, they merge into others and produce a confusing chromatogram; (iii) Maintains an atmosphere saturated with solvent; (iv) Cannot measure the solvent line to calculate R_f values; b)(i) The paper; (ii) The solvent; 2. a) The attraction between molecules of the compound and the paper; and between the compound and the solvent; b)(i) One; (ii) More than one; c) Pure substances produce one spot, impure substances produce more than one spot; 3. a) The distance moved by sample divided by the distance moved by solvent; b) The distance from the solvent line to the pencil line; the distance from the centre of the spot the the pencil line; c) The attraction of different compounds to the paper and the solvent differ; d) Different compounds have different attractions to different solvents; e) Any impurity will produce an extra spot on a chromatogram

© HarperCollins*Publishers* Limited 2016

Lesson 6: Practical: Investigate how paper chromatography can be used in forensic science to identify an ink mixture used in a forgery

Lesson overview

OCR Specification reference

OCR C2.1

Learning objectives

- Describe the safe and correct manipulation of chromatography apparatus and how accurate measurements are achieved.
- Make and record measurements used in paper chromatography.
- Calculate R_f values.

Learning outcomes

- Describe how apparatus is used to produce a paper chromatogram and where measurements are made to identify the spots. [O1]
- Carry out an experiment to identify the dyes in a mixture of inks, or of food colourings. [O2]
- Use the ratio distance moved by spot/distance moved by solvent to calculate R_f values. [O3]

Skills development

- WS 2.2 Plan experiments or devise procedures to make observations, produce or characterise a substance, test hypothesis, check data or explore phenomena.
- WS 2.4 Carry out experiments appropriately having due regard for the correct manipulation of apparatus, the accuracy of measurements and health and safety concerns.
- WS 2.6 Make and record observations and measurements using a range of apparatus and methods.
- WS 2.7 Evaluate methods and suggest possible improvements and further investigations.

Maths focus Use ratios, fractions and percentages; substitute numerical values into algebraic equations using appropriate units for physical quantities

Resources needed Practical sheets 2.6.1 and 2.6.2; Technician's notes 2.6

Digital resources Computer and data projector

Key vocabulary chromatogram, solvent front, chromatography

Teaching and learning

Engage

- Set the scenario for the practical. Some forged festival tickets have appeared on the market. The ink is being analysed to determine its composition. Students are making a chromatogram of the ink used on the forgery and of known inks to see if there is a match. [O1]

Challenge and develop

Low demand

- Students can read 'Safe and correct use of apparatus' and 'Making and recording results' in Section 2.6 of the Student Book. Discuss how chromatography can be used to separate the different coloured compounds in an ink, as appropriate to the ability of the students. They can answer questions 1–6. [O1]
- Students can complete Practical sheet 2.6.1 to produce a suitable chromatogram. This gives step-by-step instructions. [O1]

Standard and high demand

- Students can use Practical sheet 2.6.2 to plan and carry out their own investigation. You will need to check their plans before they commence the practical. [O2]

35 © HarperCollins*Publishers* Limited 2016

Explain

- Students can read 'Calculating R_f values' in Section 2.6 of the Student Book and answer questions 7–10 for practice. [O3]

- Students can **complete** the Evaluation sections on Practical sheets 2.6.1 and 2.6.2. Step-by-step instructions on calculating R_f values are given in Practical sheet 2.6.1, but not in Practical sheet 2.6.2. [O3]

Consolidate and apply

- Students can **compare** their R_f values and conclusions. They could **discuss** why they obtain similar R_f values when each practical group has a different measurement for the distance moved by the solvent. Ask them why pooling their results to obtain mean R_f values is good practice. [O3]

- Ask students to **compare** the method used in this lesson to simpler methods they have used in the past, such as with single strips of chromatography paper or filter papers. Several images available from Google images showing these methods, if they need a memory jog. [O1]

Extend

Ask students able to progress further to:

- **Investigate** the coloured compounds used to make the black Sharpie pen. They will need to produce a paper chromatogram using 2-methylpropan-2-ol as the solvent. They can **compare** the black pen with other colours of Sharpie pens. [O2]

- **Find out** how gas chromatography is used to separate compounds. [O1]

Plenary suggestions

Ask students to work in pairs to **decide** on one practical tip for producing a good paper chromatogram safely. Students can double up and **share** their tips and eventually produce a class list.

Ask them to **compile** a question and a mark scheme based on the lesson's work. They can try their questions out on each other. Encourage them to write three-mark questions.

Answers to Evaluations on Practical sheets 2.6.1 and 2.6.2

Answer will depend on the inks used and student's results

Lesson 7: Maths skills: Use an appropriate number of significant figures

Lesson overview

Learning objectives

- Measure distances on chromatograms.
- Calculate R_f values.
- Record R_f values to an appropriate number of significant figures.

Learning outcomes

- Make accurate measurements. [O1]
- Use R_f values to identify dyes. [O2]
- Record R_f values to an appropriate number of significant figures. [O3]

Skills development

- WS 2.2 Apply an understanding of apparatus and techniques.
- WS 2.6 Read measurements off a scale in a practical context.
- WS 4.2 Recognise the importance of scientific quantities.

Maths focus Read scales in integers and using decimals and substitute numerical values into equations.

Resources needed rulers

Digital resources Presentation 2.7 'R_f values'

Key vocabulary chromatogram, solvent front, R_f value

Teaching and learning

Engage

- Display slide 1 of Presentation 2.7 'R_f values' to prompt groups to **discuss** how accurately distances can be measured. Then challenge groups to answer questions 1 and 2 in Section 2.7 of the Student Book as accurately as possible. The results from each group could be **collated** to check that they agree. [O1]

Challenge and develop

- Display slide 2 of Presentation 2.7 'R_f values' to prompt groups to **recall** why R_f values are used. Check that students understand that R_f values are ratios and will remain the same however far the solvent runs. [O2]

Explain

- Students read 'Calculating R_f values' in Section 2.7 of the Student Book and answer questions 3 and 4. [O2]

Consolidate and apply

- Display slide 3 of Presentation 2.7 'R_f values' to elicit students' ideas about significant figures. The rule for determining the number of significant figures in an R_f value appears on click, and examples from Section 2.7 of the Student Book can be used to reinforce it.

- Pairs complete questions 5 and 6 in Section 2.7 of the Student Book to practise **recording** the correct number of significant figures in answers. [O3]

Extend

Ask students able to progress further to to find three things that chromatography is used for apart from coloured dyes [Google search string:] Chromatography in everyday life.

Plenary suggestion

Ask students to each write down one thing about significant figures they are sure of and one thing they are unsure of. Ask them to work in groups of 5–6 to **agree** on group lists. Ask each group to say what they **decided** and **agree** as a class about what they are confident about, what they are less sure of and what things they want to know more about.

Lesson 8: Comparing metals and non-metals

Lesson overview

OCR Specification reference

OCR C2.2

Learning objectives

- Review the physical properties of metals and non-metals.
- Compare the oxides of metals and of non-metals.
- Make predictions about unknown metals and non-metals.

Learning outcomes

- Use physical properties to distinguish between metals and non-metals. [O1]
- Use the pH of solutions of their oxides to distinguish between metals and non-metals. [O2]
- Explain the differences between metals and non-metals on the basis of their characteristic physical and chemical properties. [O3]

Skills development

- WS 2.3 Carry out experiments appropriately.
- WS 2.6 Make and record observations.
- WS 3.5 Use data to make predictions.

Resources needed Equipment as listed in the Technician's notes; Worksheet 2.8; Practical sheet 2.8; Technician's notes 2.8

Digital resources Presentation 2.8.1 'Metal or non-metal'; Presentation 2.8.2 'One of these is not like the others'

Key vocabulary electrical conductor, lustrous, tensile strength, thermal conductivity

Teaching and learning

Engage

- Display a large, unlabelled lump of graphite (or display slide 1 of Presentation 2.8.1 'Metal or non-metal' if no graphite is available). Suggest that it could be a metal because it is slightly lustrous (shiny). Ask groups to **suggest** tests you could do to support or refute the idea that an element is a metal, and what results you would expect. [O1]

Challenge and develop

- Low and standard demand: Students use the examples on slide 2 of Presentation 2.8.1 'Metal or non-metal' to **practice** using physical properties to distinguish between metals and non-metals. [O1]

- Higher demand: Use a video clip to explore the atomic structure of metals and explain why they are both malleable and strong [Google search string:] BBC Bitesize – GCSE chemistry – The atomic structure of metals. [O1]

- Present the main properties of graphite using slide 3 of Presentation 2.8.1 'Metal or non-metal' and confirm that it is a non-metal despite having some metallic properties. [O1]

Explain

Low and standard demand

- Students read 'Physical properties' from Section 2.8 in the Student Book and complete 'Metal versus non-metal 1' in Worksheet 2.8. [O1]

High demand

- Students **rank** the physical properties of metals and non-metals in order of the usefulness in distinguishing which group an element belongs to. [O3]

Consolidate and apply

Low demand

- Give out Practical sheet 2.8. Students work in pairs to **compare** the pH values of metal oxides and non-metal oxides that have been mixed with water. [O2]

Low and standard demand

- Students read 'Chemical properties' from Section 2.8 in the Student Book and complete 'Metal versus non-metal 2' from Worksheet 2.8. [O1]

Standard and high demand

- Make the apparatus listed in Technician's notes 2.8 available and challenge pairs to spot a difference between the pH values of metal and non-metal oxides that have been mixed with water. [O2]

High demand

- Students read 'Chemical properties' from Section 2.8 in the Student Book and compare and contrast the chemical properties of metals and non-metals. [O3]

Extend

Ask students able to progress further to **find out** what is meant when an element is called a 'metalloid' element and list the elements in Periods 1–4 that are usually classified as metalloids.

Plenary suggestion

Use Presentation 2.8.2 'One of these is not like the others' to display three elements or compounds and ask groups to say why one of them is not like the others. Encourage students to find as many possible answers as they can, but don't allow the use of 'metal' or 'non-metal' in their answers.

Answers to Worksheet 2.8

Metal versus non-metal 1: Table as shown below

Metal versus non-metal 2: Table as shown below

Physical property	Metal	Non-metal
Lustrous	Yes	No
Hard	Yes	No
Density	High	Low
Tensile strength	High	Low
Melting or boiling point	High	Low
Conductor of heat	Yes	No
Conductor of	Yes	No

Chemical property	Metal	Non-metal
Most react with oxygen	Yes	Yes
Most react with acid	Yes	No
Their oxides are acidic	No	Yes
Their oxides are basic	Yes	No

40 © HarperCollins*Publishers* Limited 2016

Lesson 9: Electron structure

OCR Specification reference

OCR C2.2

Lesson overview

Learning objectives

- Find out how electrons are arranged in atoms.
- Use diagrams and symbols to show which energy levels they occupy.
- Use number notation to represent electronic structure

Learning outcomes

- Know that electrons are arranged in shells or energy levels around the nucleus in an atom. [O1]
- Be able to draw electronic structures for the first 20 elements. [O2]
- Be able to describe electronic structure using number notation, e.g. 2,8,1 for the first 20 elements. [O3]

Skills development

- WS 1.2 Use a variety of models such as representational, spatial, descriptive, computational and mathematical to solve problems, make predictions and to develop scientific understanding of familiar and unfamiliar facts.
- WS 2.7 Evaluate methods.
- WS 4.1 Use scientific vocabulary, terminology and definitions.

Maths focus Visualise and represent two-dimensional (2D) and 3D forms, including 2D representations of 3D objects

Resources needed Technician's notes 2.9, Worksheets 2.9.1, 2.9.2.and 2.9.3

Key vocabulary electronic structure, electron, shell, energy level

Teaching and learning

Engage

- **Display** the Rutherford–Bohr model of a helium atom used in Lesson 1.3 with two positive charges in the nucleus and two electrons in a shell around the outside. Explain that we need to improve the model to describe atoms bigger than helium because only two electrons will fit in this shell. [O1]

Challenge and develop

Low and standard demand

- Remind pupils that different elements have different numbers of electrons in their atoms. Give each pair a copy of Worksheet 2.9.1 and a set of model electrons. **Demonstrate** how these can be used to model the electron arrangements in atoms; explaining that the first shell has a maximum of two electrons, the second shell a maximum of eight electrons, eight electrons then fill the third shell. The terms energy levels or shells are both acceptable at this stage. Emphasise that electrons always fill the lower energy levels first, so shells fill in order. Pupils can read 'The 'build up' of electrons' in Section 2.9 in their Student books. [O1]
- Pupils can answer question 1 on Worksheet 2.9.1 to draw electron configurations given the number of electrons in an atom. Check their answers. [O2]

High demand

- Remind pupils that their diagrams are actually 2D representations of 3D objects (atoms). Pupils can **discuss** the advantages and disadvantages of using these 2D diagrams in pairs, then double up to share their ideas. [O2]

Explain

Standard demand

- Remind pupils that Atomic number equals the number of protons in an atom of an element, and that the number of protons also equals the number of electrons. So pupils can draw electronic structures from atomic numbers. [O2]

High demand

- Explain that we also use numbers to describe the electronic structure of an element. You could use the electron models and Worksheet 2.9.1 if necessary to describe how numbers such as 2,8,1 are used. Pupils can use Worksheet 2.9.2 to add number representations to their diagrams. Pupils can answer questions 1 and 2 on Worksheet 2.9.3. [O3]

Consolidate and apply

- Pupils can answer question 3 on Worksheet 2.9.3. [O3]

- Pupils could **compare** the use of diagrams and of numbers to describe electronic structure, listing the advantages and drawbacks of each method. [O2], [O3]

Extend

Ask students able to progress further to:

- **Find out** why some electronic structure diagrams group electrons in pairs. [O2]

- **Find out** how electron shells are filled when atoms have more than 20 electrons. [O2], [O3]

Plenary suggestions

Ask pupils to write down one thing they are sure of, one thing they are unsure of and one thing they need to know more about. Pupils can form groups of five or six to agree on a master list, which can be shared with the class. You could identify any weak areas.

Pupils can write their own question with a mark scheme and work in pairs to try them out on each other.

Answers to Worksheet 2.9.2

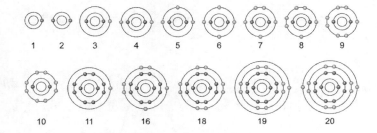

Answers to Worksheet 2.9.3

1a) 2,8,2; b) 2,8,3; c) 1; d) 2,8,8,2; e) 2,5; f) 2,4; g) 2,8,7; h) 2,8,7; i) 2,1, j) 2,8,1
3a) 2,8,1; b) 2,8,5; c) 2,8,8; d)2,8,6; e) 2,2

Lesson 10: Metals and non-metals

Lesson overview

OCR Specification reference

OCR C2.2

Learning objectives

- Explore the links between electron configurations of elements and their properties.
- Find out what happens to the outer electrons when metals react.
- Draw diagrams to show how ions form.

Learning outcomes

- Relate the electron configurations of metals and non-metals to their position in the periodic table. [O1]
- Describe how metal atoms form ions when they react with non-metals. [O2]
- Explain how the reactions of elements are related to the arrangement of electrons in their atoms. [O3]

Skills development

- WS 1.2 Recognise, draw and interpret diagrams.
- WS 3.5 Recognise and describe patterns in data.
- WS 4.1 Use scientific vocabulary, terminology and definitions.

Resources needed Equipment as listed in the Technician's notes; Worksheets 2.10.1, 2.10.2 and 2.10.3; Technician's notes 2.10

Key vocabulary ions, atomic structure, metalloids

Teaching and learning

Engage

- Point out element number 34 – selenium. Ask students whether the element is a metal or a non-metal and how we can tell. [O1]
- Ask students to draw a metal–non-metal dividing line between Be and B, Al and Si, Ge and As, Sb and Te, and Po and At. [O1]

Challenge and develop

- Use a video clip to demonstrate the reaction between sodium and chlorine. [Google search string:] Practical chemistry videos: reaction of sodium and chlorine. Then use a model to show the structure of the sodium chloride produced.

Low and standard demand

- Give each student pair a copy of Worksheet 2.10.1 and a set of model electrons. Ask alternate groups to **produce** the electron arrangements in sodium and chlorine, and then join another pair with the other atoms. Explain that when a sodium atom reacts with a chlorine atom its outer electron is transferred to the non-metal and fills the gap in its outer shell (highest energy level). This converts both atoms into charged ions, which attract each other to form a crystal. [O1, O2]

High demand

- Explain that when sodium reacts with chlorine its outer electron is transferred to the non-metal and fills the gap in its outer shell (highest energy level). Challenge pairs to **explain** why this makes sodium chloride form a regular crystal structure. They should then join with another group to **compare** ideas. [O1, O2]

43

Explain

Low and standard demand

- Students complete Worksheet 2.10.2. [O1, O2, O3]

High demand

- Students **explain** how atoms of sodium and chlorine are changed when they react with each other. [O1, O2]

Consolidate and apply

- Use a video clip to review the formation of ions. [Google search string:] BBC Bitesize – GCSE chemistry – Ionic compounds and the periodic table. [O1, O2, O3]

- Ask students to **annotate** their periodic table to show the charges on the ions formed by elements in Groups 1–3 and 5–7. [O3]

Low and standard demand

- Students read Section 2.10 in the Student Book and answer questions 1–4. [O1, O2, O3]

High demand

- Students read Section 2.10 in the Student Book and **explain** how the reactions of elements are related to the arrangement of electrons in their atoms.[O3]

Extend

Ask students able to progress further to complete Worksheet 2.10.3. [O2]

Plenary suggestion

Ask students to draw a large triangle with a smaller inverted triangle that just fits inside it (so they have four triangles). In the outer three ask them to write something they've seen; something they've done; and something they've discussed. Then add something they've **learned** in the central triangle.

Answers to Worksheet 2.10.2

1. Na^+, Mg^{2+}; Al^{3+}; 2. O^{2-}; F^-; 3. a) Li^+; b) Cl^-; c) Ca^{2+}; d) S^{2-}; e) Br^-; f) K^+

Answers to Worksheet 2.10.3

1.

2.

3.

4.

 © HarperCollins*Publishers* Limited 2016

Lesson 11: Chemical bonds

Lesson overview

OCR Specification reference

OCR C2.2

Learning objectives

- Describe the three main types of bonding.
- Explain how electrons are used in the three types of bonding.
- Explain how bonding and properties are linked.

Learning outcomes

- Know that bonding may be ionic, covalent or metallic. [O1]
- Describe how electrons are transferred in ionic bonds, shared in covalent bonds and delocalised in metallic bonds. [O2]
- Know that substances with ionic bonding have crystalline structures and high melting points, substances with covalent bonding form molecules with low melting points, and substances with metallic bonding are metals with high melting points and are good conductors of electricity. [O3]

Skills development

- WS 1.2 Use a variety of models, such as representational, spatial, descriptive, computational and mathematical, to solve problems, make predictions and develop scientific explanations and understanding of familiar and unfamiliar facts.
- WS 4.1 Use scientific vocabulary, terminology and definitions.

Resources needed Samples of sodium chloride crystals (large if possible), water in a beaker and copper metal; Worksheets 2.11.1 and 2.11.2

Digital resources YouTube video (optional)

Key vocabulary covalent, delocalised, ionic, metallic

Teaching and learning

Engage

- Show students samples of sodium chloride (large crystals if possible), water and copper metal. Ask them to (1) list the types of atom contained in each; (2) put the substances in order of ascending melting point (sodium: 801 °C, copper 1085 °C); (3) identify which will conduct electricity. Explain that melting point and electrical conductivity are dependent on the type of chemical bonds in the substance. [O1]

Challenge and develop

- Introduce students to the terms 'ionic', 'covalent' and 'metallic'. Explain that these are three different types of chemical bonds. Sodium chloride, water and copper have different properties because sodium chloride has ionic bonding, water has covalent bonding and copper has metallic bonding. [O1]
- Students form groups of three. Within each group, one student **researches** ionic bonding, one **researches** covalent bonding and one **researches** metallic bonding. They can use Chapter 2 in the Student Book for information and question 1 on Worksheet 2.11.1 to outline their research and record information. This should take no longer than 10 minutes. [O1]
- Students can complete question 2 on Worksheet 2.11.1 to show which elements in the periodic table form which types of bonds. [O1]

Explain

- Explain that in forming ionic bonds, electrons are transferred from one atom to another. Atoms of elements with one, two or three electrons in their outer shells can transfer electrons to atoms of elements with five, six or seven electrons in their outer shells. Tell students that the resulting particles are no longer atoms, but ions that carry an electric charge – either positive or negative. Explain that these opposite charges attract and are called *electrostatic forces*. There are several animations of ionic bonding on YouTube, but the details of ionic bonding are dealt with in the next lesson. Students can answer question 1 on Worksheet 2.11.2. [O2]

- Explain that covalent bonds involve sharing electrons between atoms. Discuss how covalent bonds differ from ionic bonds. Students can complete question 2 on Worksheet 2.11.2. [O2]

- Explain metallic bonding as one or more of the outer electrons moving away from their original atoms and moving throughout the metal. Explain the term 'delocalised' and ask students to suggest what attracts metal atoms to each other. They can **share** their theories with the rest of the class. Students can answer question 3 on Worksheet 2.11.2. [O2]

- Ask students to use the Student Book to find out how the type of bonding in a substance determines its properties. They can work in their original groups of three and **devise** a table to record their findings and those of the rest of the group. [O3]

Consolidate and apply

- Students can answer question 4 on Worksheet 2.11.2. They identify statements as true or false. [O1, O2]

- Students can then **devise** their own lists of true or false statements and quiz each other. [O1, O2, O3]

Extend

Ask students able to progress further to **find out** how diagrams that show the electronic structure of an element are used to illustrate ionic bonding and covalent bonding. [O2]

Plenary suggestions

Students can work in groups of three to **produce** a 30-second drama sketch to illustrate ionic bonding, covalent bonding and metallic bonding. They can **perform** their sketches.

Students can draw a large triangle on a sheet of paper with a smaller inverted triangle inside. They can write something they have done, something they have discussed and something they have learned in the three outer triangles. They can write something they need to know in the inner triangle **identifying** any weak areas.

Divide the class into three groups representing ionic bonding, covalent bonding and metallic bonding. Ask questions to which the answers are 'covalent', 'ionic' or 'metallic' bonding. Students can 'claim' their questions.

Answers to Worksheet 2.11.1

1. a) *Ionic*: e.g. sodium chloride, metals with non-metals, electrons are transferred; b) *Covalent*: e.g. water, non-metals with non-metals, electrons are shared; c) *Metallic*: e.g. copper, metallic, electrons are delocalised; 2. a) Groups 1, 2, 3, 5, 6 and 7 should be shaded; b) All non-metals should be shaded; c) All metals should be shaded

Answers to Worksheet 2.11.2

1. a) One or more outer electrons move from one atom to another; b) Sodium, etc.; c) Chlorine, etc.; d) Ions; e) Electrostatic forces; 2. a) Electrons are shared; b) Atoms with shared electrons; c) Ionic substances have charged particles but covalent ones do not; electrons are transerred in ionic bonds and shared in covalent bonds; ionic bonds are between metal and non-metal atoms; but covalent bonds are between non-metal atoms only; d) B, D, E; 3. a) Outer electrons that move away from their original atom; and are shared by the structure; b) Metal ions surrounded by delocalised electrons; c) Attraction between positive metal ions and negative delocalised electrons; 4. a) False; b) True; c) False; d) False; e) True; f) False; g) True; h) False; i) False; j) True

© HarperCollins*Publishers* Limited 2016

Lesson 12: Ionic bonding

Lesson overview

OCR Specification reference

OCR C2.2

Learning objectives

- Represent an ionic bond with a diagram.
- Draw dot-and-cross diagrams for ionic compounds.
- Work out the charge on the ions of metals and non-metals from the group number of the element (1, 2, 6 & 7).

Learning outcomes

- Be able to draw diagrams of electronic structures and use them to show how ionic bonds form. [O1]
- Draw dot-and-cross diagrams for ionic compounds formed from Groups 1 and 2 metals and Groups 6 and 7 non-metals. [O2]
- Work out the charge on the ions of metals and non-metals from the group number of the element – limited to the metals in Groups 1 and 2, and non-metals in Groups 6 and 7. [O3]

Skills development

- WS 1.2 Use a variety of models, such as representational, spatial, descriptive, computational and mathematical, to solve problems, make predictions and develop scientific explanations and understanding of familiar and unfamiliar facts.
- WS 4.1 Use scientific vocabulary, terminology and definitions.

Maths focus Visualise and represent two-dimensional (2D) and 3D forms, including 2D representations of 3D objects

Resources needed Copies of a periodic table; Practical sheet 2.12; Worksheet 2.12; Technician's notes 2.12

Digital resources TES Ionic bonding PowerPoint or similar

Key vocabulary dot-and-cross, electron transfer, electrostatic attraction, square bracket

Teaching and learning

Engage

- Write the words 'magnesium', 'oxygen' and 'magnesium oxide' on the board. Ask students to identify the type of bonding they would expect in each substance. Check their answers. Now ask them to **list** the changes that occur in the arrangements of electrons when magnesium reacts with oxygen to form magnesium oxide. Alternatively, carry out the exercise using sodium and chlorine. [O1]
- Demonstrate the burning of magnesium in oxygen and explain that this rearrangement of electrons involves several energy changes. [O1]

Challenge and develop

Low demand

- Ask students to draw the electronic structure of a sodium atom and, alongside, of a chlorine atom. You could use white boards for this exercise. Remind students that atoms are more stable if they have a full outer shell of electrons and ask what is needed for sodium and chlorine to achieve this. Students can add arrows to their diagrams to show the electron transfer. Emphasise the terminology of electron transfer. [O1]
- Students can answer question 1 on Worksheet 2.12. [O1]

Explain

Standard demand

- Explain the dot-and-cross notation used to show the formation of an ionic compound, in which dots and crosses are used to represent electrons and to show where they originated. You could work through the dot-and-cross diagram for the formation of sodium chloride or students could use Section 2.12 in the Student Book. Ask students to **compile** their own set of rules for drawing dot-and-cross diagrams. Check that these include showing the outer electron shell only, using either dots or crosses for the metal and non-metal reactants and using square brackets around the ions with the charge as a superscript. [O2]

- Students can answer question 2 on Worksheet 2.12. Check their diagrams. Some may need help when the electron transfer involves two electrons. [O2]

High demand

- Allow students to work in pairs and suggest a shortcut for calculating the charge on an ion using the periodic table. Deduce that the charges on the ions produced by metals in Groups 1 and 2 and non-metals in Groups 6 and 7 relate to the group number (this specification is not using the new IUPAC numbering for groups in the periodic table at GCSE). So Group 1 atoms produce ions with a +1 charge, Group 2 with a +2 charge, Group 6 with a −2 charge and Group 7 with a −1 charge. Students can answer question 3 on Worksheet 2.12. [O3]

Consolidate and apply

- Allow students to work through the TES Ionic bonding presentation (selected slides only). If suitable, students can **supply** their own commentary. [O1, O2, O3]

Extend

Ask students able to progress further to:

- Draw dot-and-cross diagrams to show the bonding in aluminium chloride and in aluminium oxide. [O2]

- **Explain** the charges on the ions in aluminium chloride and aluminium oxide.

- **Discuss** in pairs whether or not electrons shown as dots are different from electrons shown as crosses.

Plenary suggestions

Ask students to work in pairs and **compile** a list of points about ionic bonding. The pairs can join into fours and merge their lists. One student from each group can **report** back to the class. Or ask students to work in pairs to draw a dot-and-cross diagram for rubidium fluoride. Students can **peer assess** diagrams with other groups.

Answers to Worksheet 2.12

1. a)

b)

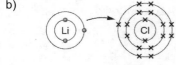

c)

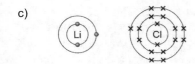

2. $Li^{\bullet} + {}^{\times}_{\times}\overset{\times\times}{\underset{\times\times}{Cl}}{}^{\times} \rightarrow \left[Li\right]^{+}\left[{}^{\times}_{\times}\overset{\times\times}{\underset{\times\times}{Cl}}{}^{\times}\right]^{-}$

$Na^{\bullet} + {}_{\times}\overset{\times\times}{\underset{\times\times}{F}}{}^{\times} \rightarrow \left[Na\right]^{+}\left[{}_{\times}\overset{\times\times}{\underset{\times\times}{F}}{}^{\times}\right]^{-}$

$Mg{}^{\times}_{\times} + 2{}_{\times}\overset{\times\times}{\underset{\times\times}{Cl}}{}^{\times} \rightarrow \left[Mg\right]^{2+}\left[{}^{\times}_{\times}\overset{\times\times}{\underset{\times\times}{Cl}}{}^{\times}\right]^{-}\left[{}^{\times}_{\times}\overset{\times\times}{\underset{\times\times}{Cl}}{}^{\times}\right]^{-}$

$Mg{}^{\times}_{\times} + \overset{\times\times}{\underset{\times\times}{O}}{}^{\times} \rightarrow \left[Mg\right]^{2+}\left[{}^{\times}_{\times}\overset{\times\times}{\underset{\times\times}{O}}{}^{\times}\right]^{2-}$

$2Na^{\bullet} + \overset{\times\times}{\underset{\times\times}{O}}{}^{\times} \rightarrow \left[Na\right]^{+}\left[Na\right]^{+}\left[{}^{\times}_{\times}\overset{\times\times}{\underset{\times\times}{O}}{}^{\times}\right]^{2-}$

$2Li^{\bullet} + \overset{\times\times}{\underset{\times\times}{O}}{}^{\times} \rightarrow \left[Li\right]^{+}\left[Li\right]^{+}\left[{}^{\times}_{\times}\overset{\times\times}{\underset{\times\times}{O}}{}^{\times}\right]^{2-}$

$K^{\bullet} + {}_{\times}\overset{\times\times}{\underset{\times\times}{Cl}}{}^{\times} \rightarrow \left[K\right]^{+}\left[{}^{\times}_{\times}\overset{\times\times}{\underset{\times\times}{Cl}}{}^{\times}\right]^{-}$

$2K^{\bullet} + \overset{\times\times}{\underset{\times\times}{O}}{}^{\times} \rightarrow \left[K\right]^{+}\left[K\right]^{+}\left[{}^{\times}_{\times}\overset{\times\times}{\underset{\times\times}{O}}{}^{\times}\right]^{-}$

3. a) Both sodium and chlorine atoms have equal numbers of protons and positive charges; and electrons and negative charges; b) Charges in order are +11, −10, +1, +17, −18, −1; c) Magnesium ion has lost two electrons/negative charges, so has an overall charge of +2; an oxide ion has gained two negative charges, so has an overall charge of −2. d)(i) +1; (ii) +2; (iii) −2; (iv) −1

Lesson 13: Ionic compounds

Lesson overview

OCR Specification reference

OCR C2.2

Learning objectives

- Identify ionic compounds from structures.
- Explain the limitations of diagrams and models.
- Work out the empirical formula of an ionic compound.

Learning outcomes

- Recognise ball-and-stick diagrams, close-packed diagrams and two-dimensional (2D) and 3D representations showing giant structures having ionic bonding. [O1]
- Explain the usefulness and limitations of ball-and-stick diagrams and close-packed diagrams showing giant structures of ionic compounds. [O2]
- Work out the empirical formula of an ionic compound from a given model or diagram that shows the ions in the structure. [O3]

Skills development

- WS 1.2 Use a variety of models, such as representational, spatial, descriptive, computational and mathematical, to solve problems, make predictions and develop scientific explanations and understanding of familiar and unfamiliar facts.
- WS 3.3 Carry out and represent mathematical and statistical analysis.
- WS 4.1 Use scientific vocabulary, terminology and definitions.

Maths focus Visualise and represent two-dimensional (2D) and 3D forms, including 2D representations of 3D objects

Resources needed Data projector (optional), ball-and-stick and close-packed models of a sodium chloride lattice; Practical sheet 2.13; Worksheets 2.13.1, 2.13.2 and 2.13.3; Technician's notes 2.13

Key vocabulary 3D representation, electrostatic empirical formula, giant lattice

Teaching and learning

Engage

- Show students some large sodium chloride crystals. Ask them to suggest why they have such a regular shape. They can keep their ideas to the end of the lesson. Alternatively, students could use Practical sheet 2.13 and observe sodium chloride crystals under the microscope, or set up some sodium chloride crystals under the microscope and use the data projector.

Challenge and develop

Low demand

- Remind students that sodium ions are attracted to chloride ions by strong electrostatic attractions. Show them a model of a sodium chloride lattice, either ball-and-stick or close-packed. If these are not available, they can use the diagrams in Section 2.13 in the Student Book. Explain that a sodium ion is not just attracted to one chloride ion, but to as many as will fit around it, and vice versa. The result is that the ions are arranged in a large giant structure called a 'lattice'. You may need to explain that these are measured on the nano scale. [O1]
- Students can complete the questions on Worksheet 2.13.1. [O1]

Explain

Standard demand

- Ask students what the stick represents in a ball-and-stick model. Establish that it represents the electrostatic attraction between the ions but, in reality, this force acts in all directions round the ion. Emphasise that electrostatic forces act in all directions in a lattice. [O2]

- Ask students to identify the different ways used to represent the giant structure of a sodium chloride lattice. Name these as ball-and-stick models, close-packed models, 2D diagrams and 3D diagrams (2D diagrams show one side of the lattice only; 3D diagrams show three sides) and dot-and-cross diagrams. Allow students to work in pairs or in small groups to **discuss** how useful these models and diagrams are. You may wish to escalate this into a class **discussion** and input on the idea of space filling shown by close-packed models versus ball-and-stick models. They can use Worksheet 2.13.2 to **record** their ideas. [O2]

High demand

- Ask students why we write the formula of sodium chloride as 'NaCl' when, in fact, a lattice of sodium chloride can contain trillions of sodium ions and chloride ions. Establish that *NaCl* represents the *ratio* of sodium ions to chloride ions and is called the *empirical formula*. Ask students to **identify** the information needed to work out the empirical formula of an ionic compound and **deduce** that they need to know the charges on the ions. Work through some examples of **calculating** empirical formula (e.g. magnesium oxide, lithium oxide, magnesium fluoride and barium chloride). Alternatively, students can read Section 2.13 in the Student Book. [O3]

- Students complete Worksheet 2.13.3 – explain that a sulfide ion is just S^{2-}. [O3]

Consolidate and apply

- Ask students to revisit their initial ideas about why sodium chloride crystals have such a regular shape and amend their ideas if necessary. Discuss their ideas and establish that the regular arrangement of ions in a lattice produces a regular shaped crystal. [O1]

- Ask students to **decide** which type of model is best to show the giant structure of an ionic lattice and to give reasons for their choice. [O2]

Extend

- The empirical formula for caesium chloride is CsCl. In a caesium chloride lattice, each caesium ion is surrounded by eight chloride ions and vice versa. Ask students to use their periodic tables to **suggest** why the ions in caesium chloride are arranged differently to the ions in sodium chloride. [O3]

Plenary suggestions

- List the 'Key vocabulary' words on the board and ask students to choose one of the words and compile a question to which this word is the answer. Students can ask their questions around the group/class.

- Ask students to write down one thing they are sure of, one thing they are unsure of and one thing they need to know more about. They can share their lists in pairs, double up and devise a group list to report back to the class. Note any areas that need revisiting.

Answers to Worksheet 2.13.1

1. a) Electrostatic forces, b) Electrostatic forces act in all directions; so a positive ion is surrounded by negative ions; and vice versa; c) Lattice; 2. In all directions; 3. B

Answers to Worksheet 2.13.2

These depend on the responses from students

Answers to Worksheet 2.13.3

1. a) KCl; b) CaO; c) $CaCl_2$; d) K_2O; e) MgS; f) $MgCl_2$; g) LiF; h) Li_2O; j) Na_2S; 2. a)(i) NaCl, (ii) MgO, (iii) LiCl; b) CaO

Lesson 14: Properties of ionic compounds

Lesson overview

OCR Specification reference

OCR C2.2

Learning objectives

- Describe the properties of ionic compounds.
- Relate their melting points to forces between ions.
- Explain when ionic compounds can conduct electricity.

Learning outcomes

- Know that ionic compounds have high melting points and boiling points and conduct electricity when melted or dissolved in water. [O1]
- Explain that ionic compounds have high melting points because large amounts of energy are needed to overcome the strong electrostatic forces between the oppositely charged ions. [O2]
- Explain that ionic compounds conduct electricity when molten or dissolved in water because the ions are free to move. [O3]

Skills development

- WS 2.1 Use scientific theories and explanations to develop hypotheses.
- WS 2.2 Plan experiments or devise procedures to make observations, produce or characterise a substance, test hypotheses, check data or explore phenomena.
- WS 2.4 Carry out experiments appropriately having due regard for the correct manipulation of apparatus, the accuracy of measurements and health and safety considerations.
- WS 2.6 Make and record observations and measurements using a range of apparatus and measurements.

Resources needed Ball-and-stick model of sodium chloride lattice (optional); Practical sheets 2.14.1 and 2.14.2; Worksheet 2.14; Technician's notes 2.14.1 and 2.14.2

Key vocabulary conducting electricity, energy, high melting point, ions free to move

Teaching and learning

Engage

- Write the following on the board: 'evidence', 'fair test', 'independent variable', 'dependent variable', 'repeatable', 'reproducible', 'valid'. Allow students a few minutes in pairs to write definitions for the terms when applied to experiments. Alternatively, present a hypothetical experiment, such as an investigation to find the school's favourite pizza, and ask students to apply the terms. Discuss their definitions. [O1, O2, O3]

Challenge and develop

Low demand

- Introduce Practical sheet 2.14.1. Students are **devising** their own experiments to investigate electrical conductivity in ionic and covalent substances. Note that the content of this investigation overlaps with Lesson 2.15. Students are provided with a range of ionic and covalent substances and must first propose a hypothesis. Tell students that their hypothesis must relate to the type of bonding present in the substance. Remind them that the terms in the 'Engage' section must apply to their experiment. Allow them to work in pairs to **plan** their experiments. Depending on the students' prior experiences, you may need to provide information about electric circuits. Check their plans and allow them to **carry out** their experiments. They need to **devise** a table to record their results and **draw** conclusions. [O1]

- Discuss students' results and establish that solid ionic substances and covalent substances do not conduct electricity, but solutions of ionic compounds do. [O1]

- Use Practical sheet 2.14.2 to demonstrate that molten zinc chloride (ionic compound) conducts electricity. This demonstration must be carried out in a fume cupboard. Note that chlorine gas is given off during the electrolysis. You could show students a ball-and-stick model of a sodium chloride lattice and then model melting to show that the ions can now move away from their fixed positions. [O1]

Explain

Standard demand

- Remind students that the zinc chloride used in the demonstration has a high melting point. Explain that this is typical of ionic compounds. Students can **suggest** reasons for the high melting points. They can **check** their ideas in Section 2.14 in the Student Book. [O2]

High demand

- Tell students that in this lesson they are concentrating on ionic compounds. Students can read Section 2.14 in the Student Book to find out why solid ionic compounds do not conduct electricity, but molten ionic compounds and those dissolved in water do. They can answer question 1 on Worksheet 2.14. [O3]

- Students can answer question 2 on Worksheet 2.14. [O2]

Consolidate and apply

- Tell students that pure water does not conduct electricity, but tap water does. Ask them to **suggest** why. [O3]

- Remind students of the initial list of words in the 'Engage' section. Ask them to work in their practical groups and write down how each word relates to their experiment. [O1]

Extend

Ask students able to progress further to:

- Prepare for the next lesson by **suggesting** why covalent substances do not conduct electricity. [O1]

- Carry out an online **search** to find the melting points and boiling points of more ionic (and covalent) substances to provide further evidence for the ideas in this lesson.

Plenary suggestions

Ask students to work in pairs and **identify** the two main ideas encountered in this lesson. They can share their ideas with another pair and, eventually, the whole class. Students can work in groups of three or four and **create** a 'freeze frame' to illustrate one point they have learned in this lesson. Other students can **suggest** what each freeze frame represents.

Answers to Worksheet 2.14

1. a)(i) No; the ions are not free to move/held in place by strong electrostatic forces; (ii) When sodium chloride melts, the ions move away from their fixed position; and are free to carry charge; (iii) The sodium and chloride ions are free; to move between the water molecules; and carry the charge; b) B, C, E; 2. a) B; C; b) Ionic bonds are strong; so a lot of energy is needed to overcome them

Lesson 15: Properties of small molecules

Lesson overview

OCR Specification reference

OCR C2.2

Learning objectives

- Identify small molecules from formulae.
- Explain the strength of covalent bonds.
- Relate the intermolecular forces to the bulk properties of a substance.

Learning outcomes

- Recognise that formulae of small molecules have no charge and contain two or more atoms. They are usually gases or liquids at room temperature, have relatively low melting points and boiling points and do not conduct electricity. [O1]
- Explain that covalent bonds are strong bonds and are not broken when a covalent substance melts or boils. [O2]
- Explain how intermolecular forces increase with the size of the molecules and larger molecules have higher melting points and boiling points because more energy is needed to overcome the intermolecular forces. [O3]

Skills development

- WS 1.2 Use a variety of models, such as representational, spatial, descriptive, computational and mathematical, to solve problems, make predictions and develop scientific explanations and understanding of familiar and unfamiliar facts.
- WS 4.1 Use scientific vocabulary, terminology and definitions.

Maths focus Visualise and represent two-dimensional (2D) and 3D forms, including 2D representations of 3D objects

Resources needed Class sets of Molymods or similar; Worksheets 2.15.1 and 2.15.2

Key vocabulary covalent bond, free electrons, intermolecular force, molecule

Teaching and learning

Engage

- Remind students of the experiments they devised in the previous lesson and ask them to **recall** whether or not the covalent substances conduct electricity. Remind students that covalent substances can have small molecules or have giant structures, like diamond. Ask students to work in pairs to name and write the formulae of five covalent substances that have small molecules. You may need to remind them that covalent bonds only form between non-metals. Compile a class list of the small molecules suggested. [O1]

Challenge and develop

Low demand

- Ask students to identify what the formulae for their five small molecules have in common. Establish that they contain symbols for non-metals only, have more than one atom and do not have a charge. Students can read Section 2.15 in the Student Book to find out/recall that small molecule compounds are usually gases or liquids at room temperature because they have relatively low melting points and boiling points, and they do not conduct electricity because they have no charged particles. Students can answer question 1 in Worksheet 2.15.1. [O1]

Explain

Standard demand

- Ask students to work in pairs and make as many water molecules from the Molymod kits as possible. They can join up with other pairs to form larger groups. Ask students to follow the instructions in question 2(a) in Worksheet 2.15.1 and use their model water molecules to show how they are arranged in liquid water and what happens to them when water boils. They could **perform** their demonstrations to the rest of the class. Students can answer question 2(b) in Worksheet 2.15.1. Ensure that they understand that covalent bonds are strong bonds and do not break when small molecules melt or boil. You may wish to contrast this with the melting and boiling of ionic compounds, but avoid the pitfall of thinking that ionic bonds are weaker than covalent bonds — they have about the same strength, so a lot of energy is needed to overcome them. [O2]

High demand

- Ask students to suggest how their water molecules might be arranged in ice. Establish that they will have a fixed position and tell them that they are held together by intermolecular forces. They can read about these in Section 2.15 in the Student Book. Explain that these are between one-tenth and one-hundredth of the strength of a covalent bond or an ionic bond and are weak forces, but these are the forces that have to be overcome before the substance can melt or boil. Ask students why substances with small molecules have relatively low melting points and boiling points and are usually gases or liquids at room temperature. Students can answer questions 2 and 3 in Section 2.15 of the Student Book and answer question 3 in Worksheet 2.15.1. [O3]

Consolidate and apply

- Students can answer the questions in Worksheet 2.15.2. This worksheet applies the lesson's content to ammonia. [O1, O2, O3]

Extend

Ask students able to progress further to:

- **Find out** what causes intermolecular forces and why larger molecules have stronger intermolecular forces.

- **Find out** if all covalent bonds have the same strength.

Plenary suggestions

Ask students to each write down one thing they are sure of, one thing they are unsure of and one thing they need to know more about. They can **agree** on a group list in groups of five or six, and then as a whole class.

Ask students to write 'Intermolecular forces' in the centre of a sheet of paper. They can use arrows to add all the facts they have **learned** about intermolecular forces in this lesson.

Answers to Worksheet 2.15.1

1. a)(i) H_2O, Cl_2, NH_3, NO_2; (ii) They contain non-metal atoms only; (iii) They contain non-metal atoms only and have more than one atom; b)(i) Melting and boiling points are relatively lower; (ii) Most are gases or liquids at room temperature; (iii) They do not conduct electricity; 2. a) Depends on class activity; b) Nothing; 3. a) Force/attraction between two adjacent molecules; b) The larger the molecule; the stronger the intermolecular force; c) Intermolecular forces have to be overcome; d) They have weaker intermolecular forces; that do not require large amounts of energy to overcome; e) HBr; Br is larger than Cl; so HBr should have stronger intermolecular forces between its molecules; f) Melting point; boiling point

Answers to Worksheet 2.15.2

1. a) The atoms are all non-metals; there are only four atoms in the molecule; b) Relatively low melting point; and boiling points; non-conductor of electricity; 2. There are no charged particles; to carry the current; 3. Gas; 4. a) Should show three covalent bonds between N and H in each molecule; and wavy lines to show intermolecular forces between molecules; b) Nothing; c) They are overcome; d) Intermolecular forces are weak; and easily overcome

 © HarperCollins*Publishers* Limited 2016

Lesson 16: Covalent bonding

Lesson overview

OCR Specification reference

OCR C2.2

Learning objectives

- Identify single bonds in molecules and structures.
- Draw dot- and-cross diagrams for small molecules.
- Deduce molecular formulae from models and diagrams.

Learning outcomes

- Identify single lines in diagrams of small molecules, repeating units of polymers and giant covalent structures as representing single covalent bonds. [O1]
- Draw dot-and-cross diagrams for the molecules H_2, Cl_2, O_2, N_2, HCl, H_2O, NH_3 and CH_4. [O2]
- Deduce the molecular formula of a substance from a given model showing the atoms and bonds in a molecule. [O3]

Skills development

- WS 1.2 Use a variety of models, such as representational, spatial, descriptive, computational and mathematical, to solve problems, make predictions and develop scientific understanding of familiar and unfamiliar facts.
- WS 4.1 Use scientific vocabulary, terminology and definitions.

Maths focus Visualise and represent two-dimensional (2D) and 3D forms, including 2D representations of 3D objects

Resources needed Molecular models of diamond/silicon(IV) oxide/graphite (as available), Molymods (optional); Worksheets 2.16.1 and 2.16.2

Key vocabulary covalent, giant covalent structure, polymer, single bond

Teaching and learning

Engage

- Write the following formulae on the board and ask students which of the compounds have ionic bonding: NaCl, H_2O, MgO, Li_2O, CO_2 and CH_4. Remind them that the ionic compounds they have learned about contain a Group 1 or 2 element and a Group 6 or 7 element. Deduce that H_2O, CO_2 and CH_4 do not fit the pattern of ionic compounds studied so far. Introduce students to covalent bonding and explain that it is present in a wide range of substances from those with small molecules to those with large molecules. [O1]

Challenge and develop

Low demand

- Challenge students to **suggest** the difference between molecules and the large ionic structures they have learned about in previous lessons. They can write their ideas in their note books and refer to them at the end of the lesson. You could remind students about Lesson 2.10.

- Ask students to draw the electronic structure of a hydrogen atom, plus a second hydrogen atom alongside. Show them how electrons are shared so that each hydrogen atom has a complete outer shell. Use a diagram with overlapping outer electron shells. Tell students this is a covalent bond and that two hydrogen atoms covalently bonded together make a hydrogen molecule. Show them alternative ways of representing the hydrogen–hydrogen covalent bond: a dot-and-cross diagram, H–H and a ball-and-stick model (see Worksheet 2.16.1). Explain that these bonds are single covalent bonds and ask students to **describe** how they are shown in each diagram. [O1, O2]

- Students can answer question 1 on Worksheet 2.16.1. [O1]

Explain

Standard demand

- Students can answer question 2 on Worksheet 2.16.1, drawing dot-and-cross diagrams for small molecules. Question 2(a) involves single covalent bonds only. Question 2(b) includes the double bond in O_2 and question 2(c) includes the triple bond in N_2. These are explained in Worksheet 2.16.1, but some may need additional assistance. Emphasise the difference between single, double and triple covalent bonds and the way they are represented. [O2]

- Explain that some covalently bonded substances have very large molecules, such as polymers. Others have giant structures, such as silicon dioxide and diamond. Students can answer question 1 on Worksheet 2.16.2. [O1]

High demand

- Make the connection between the dot-and-cross, overlapping shell diagrams, displayed formulae and ball-and-stick models and the formula. Explain that because these are molecules, they are molecular formulae. Compare molecular formulae with the empirical formulae used to describe ionic compounds. Students can answer question 2 on Worksheet 2.16.2. [O3]

Consolidate and apply

- Students can **answer** question 3 on Worksheet 2.16.2. Check their answers. [O3]

Extend

Ask students able to progress further to:

- Draw dot-and-cross diagrams of H_2S, PH_3, F_2, C_2H_6, C_2H_4 and C_2H_2 for extra practice (these are not in the specification). [O2]

- **Discuss** what formula they should use for poly(ethene) and diamond. [O3]

Plenary suggestions

Ask students to refer back to their initial lists of substances and **update** their ideas if necessary. Organise a 'hot seat' whereby each student thinks up a question based on this lesson and one student sits in the hot seat. Tell them to ask their questions and say at the end whether it is correct or incorrect.

Answers to Worksheet 2.16.1

1. a) As a single line; b) Depends on student response;
2. a) H:H   b) O:O (c) :N:N

:Cl:Cl:

H:Cl:

H:O:H

H:N:H
   H

   H
H:C:H
   H

Answers to Worksheet 2.16.2

1. a) A single covalent bond; b) *Similarities*: Both have a large number of particles; both have strong bonds, etc.; *Differences*: A giant covalent structure has covalent bonding/shared electron pairs; an ionic structure has ionic bonding/ions attracted by electrostatic forces; c) *Similarities*: Both have single covalent bonds, and form molecules; *Difference*: Poly(ethene) molecules contain a large number of atoms; CH_4 molecules have just five; 2. a) *Covalent substance*: molecular formula; the type and number of particles in a molecule; molecules; e.g. water; *Ionic substance*: empirical formula; the type of atoms that make up the substance and the ratio they are present in; ions; e.g. NaCl;

3. a) $H^+(aq) + Cl^-(aq)$; (H Cl) b) Ions; molecules;

c)(i) Empirical formula; (ii) molecular formula

Lesson 17: Giant covalent structures

Lesson overview

OCR Specification reference

OCR C2.2

Learning objectives

- Recognise giant covalent structures from bonding and structure diagrams.
- Explain the properties of giant covalent structures.
- Recognise the differences in different forms of carbon.

Learning outcomes

- Recognise structural diagrams of diamond, graphite and silicon dioxide as giant covalent structures where the atoms are linked with strong covalent bonds. [O1]
- Explain that properties of hardness, melting point, solubility and electrical conductivity of giant covalent structures are the result of the many strong covalent bonds linking the atoms together. [O2]
- Know that diamond and graphite are forms of carbon with giant covalent structures, but different properties. [O3]

Skills development

- WS 1.2 Use a variety of models, such as representational, spatial, descriptive, computational and mathematical, to solve problems, make predictions and develop scientific explanations and understanding of familiar and unfamiliar facts.
- WS 4.1 Use scientific vocabulary, terminology and definitions.

Maths focus Visualise and represent two dimensional (2D) and 3D forms, including 2D representations of 3D objects

Resources needed Molecular models of diamond, graphite and silicon dioxide if available, class sets of Molymods, samples of graphite; Practical sheet 2.17; Worksheets 2.17.1 and 2.17.2; Technician's notes 2.17

Digital resources Digital microscope, data projector

Key vocabulary diamond, giant covalent structure, graphite, silicon dioxide

Teaching and learning

Engage

- Remind students that both small molecule compounds and polymers are covalent substances. Show students photos of diamonds and other gemstones (there are many on Google images) and/or use Section 2.17 in the Student Book and the photo of amethysts. Use Practical sheet 2.17 to view sand grains under the microscope. This uses a digital microscope and data projector. If these are not available, a few microscopes could be set up prior to the lesson for students to use. Tell students that diamonds, amethysts and sand (quartz or silicon dioxide) all have giant covalent structures. Ask what else they have in common. [O1]

Challenge and develop

Low demand

- Students can use Section 2.17 in the Student Book to **find out** about the structure of diamond and silicon dioxide. If models are available, use these too. Tell students that silicon dioxide is the second most abundant mineral in the Earth's crust. It makes up sand, clay, quartz and many other gemstones, such as amethysts and citrine. Students can use the Molymods and work in groups to make a model of diamond. These could be combined to make a class model. Students can answer question 1 in Worksheet 2.17.1. [O1]

Explain

Standard demand

- Check students' answers to question 1(d). They were asked to predict some bulk properties of giant covalent structures. The term 'bulk properties' may need explaining. They can use the table in Section 2.17 in the Student Book to **compare** some properties of diamond and silicon dioxide. Discuss how these properties are a result of the giant covalent structure, emphasising that when giant covalent substances melt, some covalent bonds have to be broken so that the atoms can move away from the structure. You could contrast this with small molecule substances melting. Students can answer question 2 in Worksheet 2.17.1. [O2]

High demand

- Explain that carbon has more than one giant covalent structure – it also forms graphite. Show students some samples of graphite. They could try writing with them. They can use Section 2.17 in the Student Book and/or models of diamond and graphite and/or the diagrams on Worksheet 2.17.2 to **find out** how the structure of graphite is different from that of diamond. Discuss how these differences affect the bulk properties. Students can complete Worksheet 2.17.2. [O3]

Consolidate and apply

- Ask students to work in pairs and **compare** the properties of substances with small molecules, such as water and carbon dioxide, with the properties of substances with giant covalent structures, such as diamond. [O2]

- Ask students to **predict** what structure makes a large flawless diamond. They could suggest the conditions required for large flawless diamonds to form. You could give them the hint that many diamonds are found in South Africa in the cores of old, inactive volcanoes. Students could **research** their ideas online. [O1]

Extend

Ask students able to progress further to **find** out why graphite can conduct electricity and why it is used as a lubricant. [O3]

Plenary suggestions

Ask students to work in pairs and make a list of giant covalent structure facts. They can double up with another pair and **merge** their lists. A class list can be **compiled**. They can work in groups of three or four and **create** a 'freeze frame' of an idea they have **learned** this lesson. Other students can **suggest** what it represents.

Answers to Worksheet 2.17.1

1. a) Covalent bonds; b) Have many atoms in their structures; structures have bonds in all directions; c) Small molecule substances contain few atoms; giant covalent structures contain many; d) Depends on predictions; 2. a)(i) Covalent bonds; (ii) Intermolecular forces; (iii) Covalent bonds are strong bonds; lots of energy is needed to break them; b)(i) Atoms are arranged in a regular pattern; (ii) Strong covalent bonding with bonds in all directions; (iii) There are no charged particles; to carry the current

Answers to Worksheet 2.17.2

1. a) A single straight line should be labelled; 2. *Diamond*: Tetrahedral structure with many carbon atoms joined with single covalent bonds; Both are forms of carbon, have covalent bonds, have many atoms in their structure; Tetrahedral structure, each carbon atom makes four covalent bonds; Strong structure means high melting point, hard substance, does not conduct electricity; *Graphite*: Carbon atoms bond in hexagons and flat sheets; Both are forms of carbon, joined with single covalent bonds, have many atoms in their structure; Layers of carbon atoms bonded in hexagonal pattern, each carbon atom makes three covalent bonds; Conducts electricity; layers rub off

 © HarperCollins*Publishers* Limited 2016

Lesson 18: Polymer structures

Lesson overview

OCR Specification reference

OCR C2.2

Learning objectives

- Recognise polymers from their unit formulae.
- Explain why some polymers can stretch.
- Explain why some plastics do not soften on heating.

Learning outcomes

- Know that polymers are very large molecules and recognise them from their repeating unit formulae. [O1]
- Explain that the intermolecular forces between some polymer chains can be overcome by applying a force and the polymer can stretch. [O2]
- Explain that the weaker the intermolecular forces, the lower the melting point, and that some plastics have strong covalent bonds linking adjacent polymer chains together. These do not soften easily. [O3]

Skills development

- WS 1.2 Use a variety of models, such as representational, spatial, descriptive, computational and mathematical to solve problems, make predictions and develop scientific explanations and understanding of familiar and unfamiliar facts.
- WS 2.1 Use scientific theories and explanations to develop hypotheses.
- WS 2.2 Plan experiments or devise procedures to make observations, produce or characterise a substance, test hypotheses, check data or explore phenomena.

Maths focus Substitute numerical values into algebraic equations using appropriate units for physical quantities.

Resources needed Class sets of Molymods, or similar; Practical sheet 2.18; Worksheet 2.18; Technician's notes 2.18

Digital resources BBC Bitesize (images of HDPE and LDPE structure required)

Key vocabulary intermolecular force, monomer, polymer, repeating unit

Teaching and learning

Engage

Ask students to work in pairs and use the Molymod kits to join a string of 'carbons' together, and then add two 'hydrogens' to every carbon. They can join their models together until they have one long molecule. Explain that if this was about 10 000 carbons long, it would be a model of poly(ethene), commonly called polythene and used to make poly bags. You could explain the correct way of writing poly(ethene), but this is covered later in the Chapter 6 lessons. Tell students that poly(ethene) is an example of a polymer. [O1]

Challenge and develop

Low demand

- Discuss the type of bonding between the carbon and hydrogen atoms in poly(ethene) and how we represent these bonds in a displayed formula. Explain that the number of carbon atoms in a poly(ethene) molecule is not exactly 10 000, but varies, and that writing a molecular formula is not possible. Show students how formulae for repeating units are used to represent a polymer. Explain that all polymers have a carbon chain, but that different atoms or groups of atoms can be attached to them, making different types of polymers. Note that at this stage, they just need to be aware that polymers are very long molecules that can be represented using repeating units. [O1]

- Students can answer question 1 on Worksheet 2.18. [O1]

Explain

Standard demand

- Remind students that the properties of substances with small molecules are linked to their structure. Ask students to **predict** the properties of polymers from their structure. Establish that in a polymer the chains usually lie side–by–side and that intermolecular forces exist between polymer chains. Explain that these intermolecular forces will be stronger in polymers than in smaller molecules and will need more energy to overcome them. So polymers are solids at room temperature and have higher melting and boiling points than substances with small molecules. Students can answer question 2 on Worksheet 2.18. [O2]

- Students can use Practical sheet 2.18 to **explore** how polymers stretch. This is an observation and planning exercise only. Check that students appreciate that polymers like poly(ethene) stretch because, although the intermolecular forces are stronger in polymers than in smaller molecules, they are still easily overcome and the polymer chains can slide over each other. Students could **peer assess** each other's plans. [O2]

High demand

- Ask students to write a sentence to **summarise** how melting points and boiling points vary with the strength of the intermolecular forces. Check their answers and explain that some polymers have strong covalent bonds between their polymer chains. These do not break easily and the polymer does not soften when heated. Students could **predict** whether these polymers stretch or not. Students can answer question 3 on Worksheet 2.18. [O3]

Consolidate and apply

- Explain that there are two types of poly(ethene) – high density poly(ethene) (HDPE) and low density poly(ethene) (LDPE). There are suitable images of their structure on Bitesize. Students can **discuss** which type of poly(ethene) has the higher melting point and why. [O2, O3]

Extend

Ask students able to progress further to:

- **Explain** why poly(ethene) does not have a definite melting point, but melts over a range of temperatures. [O2]

- **Research** the properties and the uses of HDPE and LDPE.

- **Find out** why some polymers have plasticisers added. [O3]

Plenary suggestions

Ask students to work in pairs to list what they know about polymers. They can **merge** their lists with other pairs and keep doubling up until a class list is achieved. Students can each write a question about polymers and **devise** a mark scheme, trying to include at least two points and two marks. They can try their questions out on each other.

Answers to Worksheet 2.18

1.a) A long/large molecule/carbon chain; b) Poly(ethene), etc.; c)(i) Length of carbon chain varies in a sample of poly(ethene); (ii) Section of the polymer chain that is repeated many times, (iii) Many; (iv) Covalent bonds; (v) 5000; (vi) $C_{10000}H_{20002}$; 2. a) Covalent; b) Polymer chains are larger/longer than small molecules; c) Stronger intermolecular forces; mean more energy is needed to overcome them; 3. a) Strong covalent bonds; hold the polymer chains in a fixed position; b) Diagram should have no cross-links; and intermolecular forces should be shown

 © HarperCollins*Publishers* Limited 2016

Lesson 19: Metallic bonding

Lesson overview

OCR Specification reference

OCR C2.2

Learning objectives

- Describe that metals form giant structures.
- Explain how metal ions are held together.
- Explain delocalisation of electrons.

Learning outcomes

- Describe that metals consist of giant structures of atoms arranged in a regular pattern. [O1]
- Explain that the metal ions are held together by the strong electrostatic forces between the positive metal ions and the negative sea of delocalised electrons. [O2]
- Explain that electrons are delocalised when they leave the outer shell of metal atoms and move throughout the metal structure. [O3]

Skills development

- WS 1.2 Use a variety of models, such as representational, spatial, descriptive, computational and mathematical, to solve and develop scientific explanations and to understand familiar and unfamiliar facts.
- WS 3.5 Interpret observations and other data, including drawing conclusions.
- WS 4.1 Use scientific vocabulary, terminology and definitions.

Maths focus Visualise and represent two-dimensional (2D) and 3D forms, including 2D representations of 3D objects

Resources needed Poly(styrene) spheres (optional); Practical sheet 2.19; Worksheets 2.19.1 and 2.19.2; Technician's notes 2.19

Digital resources Digital microscope, data projector, YouTube video

Key vocabulary delocalised electrons, metal ions, metallic bonds, sea of electrons

Teaching and learning

Engage

- Remind students that ionic bonds form between metals and non-metals, and that covalent bonds form between non-metals. Neither of these types of bonding occurs between just metal atoms. Ask students to **recall** some of the properties of metals and establish that metals are strong, conduct electricity, can be shaped and have high melting points. Establish that metals must have strong bonds between their atoms. [O1]

- Carry out the demonstration in Practical sheet 2.19 to grow silver crystals. Students should be able to observe silver crystals growing on the cathode. Discuss the regular shape of the crystals and establish that the particles making this shape must also have a regular structure. If a digital microscope and data projector are not available, there is a suitable video on YouTube that shows the reaction. Alternatively, show students some photos of metal crystals. There are many good photos in Google images. Ask students to suggest what structure could produce these crystals.

Challenge and develop

Low demand

- Tell students that metals also form giant structures with the metal atoms arranged in a regular pattern. Remind them that ionic compounds and some covalent substances also form giant structures. You could use poly(styrene) spheres to model a metal structure in two dimensions. Students can answer question 1 in Worksheet 2.19.1. [O1]

Standard demand

- Students can read the relevant section in Section 2.19 in the Student Book to find out how metal atoms are held together. They can answer question 2 in Worksheet 2.19.1. Check their answers. They need to appreciate that strong electrostatic forces between the positive metal ions and negative 'sea' of electrons hold the metal atoms together. [O2]

Explain

High demand

- Ask students to draw the electronic structures of a sodium atom and a magnesium atom, and a sodium ion and a magnesium ion. Establish that a sodium atom loses one electron when a sodium ion forms and that a magnesium atom loses two electrons when a magnesium ion forms. Explain that these are the electrons that form the 'sea' of electrons in a metal structure. These electrons are then free to move throughout the metal; they do not stay with any one metal ion. Tell students that these are called 'delocalised electrons' and that the sharing of delocalised electrons produces strong metallic bonds. Students can use Section 2.19 in the Student Book to reinforce these ideas. [O3]

- Students can answer Worksheet 2.19.2 to reinforce their understanding of delocalised electrons and to explore metal properties that result from this structure. [O3]

Consolidate and apply

- Students can **consolidate** their knowledge of metallic bonding by **comparing** it with the other types of bonding they have **learnt** about – ionic and covalent. Question 3 in Worksheet 2.19.2 has an outline of a suitable table. Students could add extra rows to their tables to answer their own questions. This table may be useful later as a revision tool. [O1, O2, O3]

Extend

Ask students able to progress further to:

- **Explain** why a text book or website may refer to 'ionic compounds', but 'covalent substances'. Then ask them to **decide** what is the best equivalent for metals. [O2]

- **Predict** how the strength of aluminium's metallic bonds and its properties compare with those of sodium and magnesium.

Plenary suggestions

Divide the class into three groups. Label the groups 'ionic bonds', 'covalent bonds' and 'metallic bonds'. Ask questions to which ionic bonds, covalent bonds or metallic bonds are the answers. They can **identify** which questions are theirs.

Start a clock ticking. Students have 30 seconds to list as many points they have learned today as they can. Students can **share** their lists with the class.

Answers to Worksheet 2.19.1

1. **A** false; **B** true **C** true; **D** false; **E** true; **F** false; 2. a) Metals can be shaped, bent and twisted; b) Regular arrangement of atoms produces a regular shape; 3. a) The outer electrons of the metal atoms; b) Negative/−1; c) Metal atoms lost electrons; d) Strong electrostatic attractions; between the positive ions and the sea of electrons

Answers to Worksheet 2.19.2

1. a) Sentences should include: formed by electrons lost from outer shells of metal atoms; move throughout the structure; are not confined to any one metal ion; have negative charge; b) Strong attraction between the positive ions; and the delocalised electrons; 2. a) Delocalised electrons are free to move; and carry the current; b) It is disrupted; c) A lot of energy is needed; to overcome; the strong metallic bonds; d)(i) One; (ii) Two; (iii) Magnesium; (iv) Stronger metallic bonds need more energy; to be overcome;, so melting point is higher; e) Lithium and potassium are in Group 1; and each atom releases one delocalised electron; barium and calcium are in Group 2; and each atom releases two delocalised electrons; so barium and calcium have stronger metallic bonds; and more energy is needed to disrupt them; 3. *Ionic*: Group 1 or 2 with Group 6 or 7; electrostatic forces between ions; giant structures; high melting points, do not conduct electricity when solid, etc.; *Covalent*: non-metals; shared electron pairs; molecules or giant structures; lower melting points, do not conduct electricity; *Metallic:* metals; electrostatic attraction between metal ions and delocalised electrons; giant structures; high melting points, conduct electricity, can be shaped

Lesson 20: Properties of metals and alloys

Lesson overview

OCR Specification reference

OCR C2.2

Learning objectives

- Identify metal elements and their properties, and metal alloys.
- Describe the purpose of a tin–lead alloy.
- Explain why alloys have different properties to those of elements.

Learning outcomes

- Identify metals as having giant structures of atoms with strong metallic bonding. [O1]
- Know that pure metals, such as gold, copper, iron and aluminium, do not always have the properties required for everyday use, so are mixed with other metals to make alloys with the desired properties. [O2]
- Explain that different sized atoms in alloys distort the layers of metal atoms, making it more difficult for the layers to slide over each other and changing the properties of the pure metal. [O3]

Skills development

- WS 1.2 Use a wide variety of models, such as representational, spatial, descriptive, computational and mathematical, to solve problems, make predictions and develop scientific explanations and understanding of familiar and unfamiliar facts.
- WS 1.4 Explain everyday and technological applications of science.
- WS 2.4 Carry out experiments appropriately having due regard for the correct manipulation of apparatus, the accuracy of measurements and health and safety considerations.

Maths focus Use ratios, fractions and percentages.

Resources needed Poly(styrene) spheres (optional); Practical sheet 2.20; Worksheets 2.20.1 and 2.20.2; Technician's notes 2.20

Key vocabulary alloy, distort, ductile, malleable

Teaching and learning

Engage

- Show students a large image of a piece of gold jewellery, or a gold artefact. Ask them to **identify** the metal used and allow students to complete Worksheet 2.20.1. In this students meet the idea that pure metals do not always have the required properties, and can be mixed with other metals to produce the desired properties. They find out about the carat rating of gold alloys and **calculate** the percentages of gold in different gold alloys. [O1]

Challenge and develop

Low demand

- Ask students to recall the type of bonding in pure metals. Establish that metals have giant structures with strong metallic bonds. The giant structures contain metal atoms and delocalised electrons. Recall that this means most metals have high melting and boiling points. Remind students that metals are good conductors of electricity because the delocalised electrons can carry the electric charge. The delocalised electrons also transfer heat energy, so metals are good conductors of thermal energy.
- Students can use Section 2.20 in the Student Book to **identify** which metals are used to make bronze, brass and steel. They can answer question 1 in Worksheet 2.20.2. [O1]

Standard demand

- Students can read about the tin–lead alloy in Section 2.20 in the Student Book. Introduce Practical sheet 2.20 and explain the hazards of handling very hot objects. Students can carry out the experiment in small groups. Students are **comparing** the melting points of tin, lead and solder by heating them on a crucible lid and **observing** the order in which they melt. They must wear safety glasses. Discuss their results. [O2]

Explain

High demand

- Remind students that metals have atoms arranged in regular patterns and explain that the layers of metal atoms are able to slide over each other. This is why metals can be shaped – hammered (malleable), bent, twisted and pulled into wire (ductile). You could use poly(styrene) spheres to model this. Explain that adding another metal with different sized atoms, to make an alloy, disrupts the regular pattern of the metal atoms. This means the layers can no longer easily slide over each other and the alloy is harder. Students can answer question 2 on Worksheet 2.20.2. [O3]

Consolidate and apply

- Students can answer question 3 on Worksheet 2.20.2. This revisits ideas met in this lesson and applies them to the metals used in smart-phone cases. [O1, O2, O3]

Extend

Ask students able to progress further to read about nitinol in Section 2.20 in the Student Book and **find out** how smart alloys such as nitinol are able to revert to their original shape when heated. (The crystalline structure in nitinol is temperature sensitive. When it is heated, the crystalline structure reverts to the original.) [O3]

Plenary suggestions

Ask each student to **compile** a question and a mark scheme on the properties of metals and alloys. They must try to include two points at least, so a two-mark answer at least. They can pair up and try their questions out on each other.

Ask students to draw a large triangle and then a smaller inverted triangle inside it. In the three outer triangles, they write something they have done, something they have discussed and something they have seen. In the inner triangle, they write something they have learned.

Answers to Worksheet 2.20.1

1. Gold alloys are harder; 2. a) 75%; b) 58.3%; c) 50%; d) 37.5%

Answers to Worksheet 2.20.2

1. a) Giant structure; b) Metallic bonding; c) Metallic bonding is strong; a lot of energy is needed to overcome it; d) Delocalised electrons; carry the charge through the metal; (e) Delocalised electrons; transfer thermal energy; f)(i) Copper and tin; (ii) Copper and zinc; (iii) Iron and carbon; 2. a) Regular sized particles; arranged in rows; b) The layers of metal atoms; can slide over each other; c) Different sized metal atoms; distort the regular arrangement; and prevent the layers sliding over each other; 3. a) It is a light metal; b) Layers of metal atoms; can slide over each other; c) Layers of metal atoms are distorted in an alloy; they do not bend as easily

Answers to Practical 2.20

1. Lowest melting point to highest: solder, tin, lead; 2. The melting point of the alloy is lower than the melting point of the constituent metals

Lesson 21: Key concept: The outer electrons

Lesson overview

OCR Specification reference

OCR C2.2

Learning objectives

- Review the patterns in the periodic table.
- Compare the trends in Group 1 and Group 7.
- Relate these trends to the number of outer electrons and the sizes of atoms.

Learning outcomes

- Predict the reactivity of elements from their position in the periodic table. [O1]
- Predict whether a compound will contain molecules or ions. [O2]
- Explain why the trends down the group in Group 1 and in Group 7 are different. [O3]

Skills development

- WS 1.2 Use models in explanations.
- WS 1.2 Make predictions based on a model.
- WS 3.5 Recognise and describe patterns or trends.

Resources needed Periodic tables, equipment as listed in the Technician's notes; Worksheet 2.21; Technician's notes 2.21

Digital resources Presentation 2.21 'Outer electrons'

Key vocabulary transfer, share, electrons, outer shell

Teaching and learning

Engage

- Challenge groups to list as many things as possible that they can tell about an element from its position in the periodic table. [O1, O2, O3]

Challenge and develop

Low and standard demand

- Ask pairs to sort cards cut from Worksheet 2.21 into stable and unstable atoms based on their electron configurations. They should recognise that only He, Ne and Ar are stable because these have full outer shells. [O1]

- Display slide 1 of Presentation 2.21 'Outer electrons' and ask pairs to **divide** the reactive elements into those that definitely transfer outer electrons to other atoms to form positive ions, those that definitely do not, and those they are not sure about. They should **appreciate** that the metals in Groups 1 and 2 transfer their outer electrons to other atoms when they react and form positive ions. Although boron and aluminium can form ions in solution, their small size makes it difficult for them to transfer electrons to other atoms. [O1, O2]

- Check that students **appreciate** that reactive elements that don't form positive ions can form molecules, and the most reactive non-metals can also accept electrons to form negative ions. [O1, O2]

High demand

- Display slide 2 of Presentation 2.21 'Outer electrons' and ask pairs to copy the diagram and then use the electron configuration cards to **place** the first 18 elements in the correct groups. [O1, O2]

Explain

Low and standard demand

- Students copy the diagram on slide 2 of Presentation 2.21 'Outer electrons' and use their electron configuration cards to find an example for each group. [O2]

High demand

- Ask pairs to **explain** how an element's outer electrons determine whether ions or molecules form when it reacts. Use sodium chloride and silicon chloride as examples. [O2]

Consolidate and apply

- Use a video clip to remind students of the trends in reactivity in Groups 1 and 7 [Google search string:] Science Bank Patterns of Reactivity. [O3]

- Ask pairs of students to use the electron configuration cards to **explain** why reactivity increases as you go down Group 1, and as you go up Group 7. Pairs should then join up into groups of four to **compare** ideas and **agree** on an explanation. The symbols on these cards reflect the sizes of the atoms to prompt students to **consider** the distance between the outer electrons and the nucleus. [O3].

- If appropriate, use a video clip mentioned in Lesson 1.17 to summarise the reasons for the trends in the reactivity of Group 1 and Group 7 elements. [Google search string:] Bitesize – GCSE Chemistry – Reactivity of Group 1 and 7 elements. [O2]

Extend

Ask students able to progress further to **research** boron and aluminium. Do they form ions or molecules when they react? [O3]

Plenary suggestion

Ask students to write one thing about the outer electrons they are sure of, one thing they are unsure of and one thing they need to know more about. Then let groups of five or six students **formulate** group lists to **feed back** to the rest of the class.

 © HarperCollins*Publishers* Limited 2016

Lesson 22: The periodic table

Lesson overview

OCR Specification reference

OCR C2.2

Learning objectives

- Explain how the electronic structure of atoms follows a pattern.
- Recognise that the number of electrons in an element's atoms outer shell corresponds to the element's group number.
- Use the periodic table to make predictions.

Learning outcomes

- Relate the number of electrons and energy levels to an element's position in the periodic table. [O1]
- Know that Group 1 elements all have one electron in their outer shell, Group 2 elements have two electrons and so on, and that this corresponds to the Group number. [O2]
- Explain that since chemical reactions involve rearranging the outer electrons, elements in the same Group can be expected to have similar chemical reactions. [O3]

Skills development

- WS 1.2 Use a variety of models such as representational, spatial, descriptive, computational and mathematical to solve problems, make predictions and to develop scientific understanding of familiar and unfamiliar facts.
- WS 3.5 Interpret observations and other data (presented in verbal, diagrammatic, graphical, symbolic or numerical form), including identifying patterns and trends, making inferences and drawing conclusions.
- WS 4.1 Use scientific vocabulary, terminology and definitions.

Maths focus Visualise and represent two-dimensional (2D) and 3D forms, including 2D representations of 3D objects

Resources needed Student copies of the periodic table, a large poster of the periodic table (or copy on data projector) Worksheets 2.22.1, 2.22.2 and 2.22.3

Digital resources Class Internet access (optional)

Key vocabulary electron shells, energy levels, group, period

Teaching and learning

Engage

- Explain that as new elements were being discovered through history, chemists made many attempts to put them in some sort of order. The outcome of this is the periodic table. Give out class copies of the periodic table and tell students that in this lesson, they are looking for links and patterns between the periodic table and the electronic structures of atoms they drew last lesson. [O1]

Challenge and develop

Low demand

- Remind students that the number of protons and electrons in an atom increases by one across each row of the periodic table. Tell students that the vertical columns in the periodic table are called 'groups' and the horizontal rows, 'periods'. Recap on the electronic structures of atoms covered last lesson, if necessary. Students can complete Worksheet 2.22.1 to draw electronic structure diagrams on a blank of the periodic table. [O1]

- Ask students to work in pairs and **list** the links/patterns between the periodic table and the electronic structure of the atoms. They can double up, then **share** their ideas as a class to develop a class list. This should include: a new shell corresponds with a new period; all elements in the same period have the same number of shells; all elements in the same group have the same number of electrons in the outer shell; shells increase by one as you down a group. [O1]

Explain

Standard demand

- Remind students that the groups in the periodic table are numbered 1 to 7 and 0. (Note that the new IUPAC numbering system is not being used at GCSE.) Ask students to identify the number of electrons in the outer shell of each group. Students can complete question 1 on Worksheet 2.22.2. [O2]

High demand

- Explain that when chemical reactions happen, the outer electrons in the atoms are rearranged. Students could **hypothesise** what this means for the chemical reactions of a group of elements. If time allows, students could do an Internet search to check their hypothesis; to find out whether all elements in Group 1 (e.g.) have similar reactions. Students can answer question 2 on Worksheet 2.22.2. They will learn about Group 1, 7 and 0 elements later in Chapter 4. [O3]

Consolidate and apply

- Students can use Worksheet 2.22.3 to **sort** the cards into groups and **identify** the connections. [O1], [O2].

Extend

Ask students able to progress further to:

- **Find out** why new elements are being added to the periodic table today. [O2]

- **Find out** how electron shells are filled when atoms have more than 20 electrons. [O2], [O3]

Plenary suggestions

Ask students to write down the three 'big ideas' they **learnt** this lesson. They can join up in pairs, then double up again, each time **deciding** on the three big ideas. These can be **shared** with the class to **agree** a class list.

Students can write their own question with a mark scheme and work in pairs to try them out on each other.

Answers to Worksheet 2.22.1

The following should be added: H 1; He 2; Li 2,1; Be 2,2; B 2,3; C 2,4; N 2,5; O 2,6; F 2,7; Ne 2,8; Na 2,8,1; Mg 2,8,2; Al 2,8,3; Si 2,8,4; P 2,8,5; S 2,8,6; Cl 2,8,7; Ar 2,8,8; K 2,8,8,1; Ca 2,8,8,2

Answers to Worksheet 2.22.2

1. a) Group 1: 1, e.g. sodium, 2,8,1; Group 2: 2, e.g. magnesium, 2,8,2; Group 3: 3, e.g. aluminium, 2,8,3; Group 4: 4, e.g. silicon, 2,8,4; Group 5: 5, e.g. phosphorus, 2,8,5; Group 6: 6, e.g. sulfur, 2,8,6; Group 7: 7, e.g. chlorine, 2,8,7; Group 0: 8, e.g. argon, 2,8,8; b) The group number is the same as the number of electrons in the outer shell; 2. Elements in the same group are: argon, helium and neon; lithium, sodium and potassium; fluorine, chlorine and bromine; The observations do support the statement

Answers to Worksheet 2.22.3

Groups are: Lithium; 2,1; $^{7}_{3}Li$; electron configuration diagram for 2,1. Magnesium; 2,8,2; $^{24}_{12}Mg$; electron configuration diagram for 2,8,2. Chlorine; 2,8,7; $^{35}_{17}Cl$; electron configuration diagram for 2,8,7. Neon, 2,8. $^{20}_{10}Ne$; electron configuration diagram for 2,8. Oxygen; 2,6; $^{16}_{8}O$; electron configuration diagram for 2,6

Lesson 23: Developing the periodic table

Lesson overview

OCR Specification reference

OCR C2.2

Learning objectives

- Find out how the periodic table has changed over the years.
- Explore Mendeleev's role in its development.
- Consider the accuracy of Mendeleev's predictions.

Learning outcomes

- Describe the main steps in the development of the periodic table. [O1]
- Explain why the periodic table has changed throughout the years. [O2]
- Describe and explain how testing a prediction can support or refute a new scientific idea. [O3]

Skills development

- WS 1.2 Match features of the periodic table to data and observations.
- WS 1.2 Make predictions based on the periodic table.
- WS 1.2 Recall Mendeleev's predictions and explain why they boosted support for his periodic table.

Resources needed Periodic tables; equipment as listed in the Technician's notes; Worksheets 2.23.1 and 2.23.2; Technician's notes 2.23

Digital resources Presentation 2.23 'Mendeleev'

Key vocabulary periodicity, predictions, properties, patterns

Teaching and learning

Engage

- Scientists are coming up with new ideas all the time. Ask groups to **suggest** how we can tell whether their ideas are correct or not. [O3]

Challenge and develop

- Give out sets of cards cut from Worksheet 2.23.1 and use Presentation 2.23 'Mendeleev' to guide students through the main steps in the development of the periodic table. [O1, O2]
- Use a video clip to review Mendeleev's work [Google search string:] Mendeleev's dream video – BIG QUESTIONS Mendeleev's dream. [O1, O2, O3]

Explain

Low and standard demand

- Use Section 2.23 in the Student Book to draw a timeline for the development of the periodic table. Then answer questions 1–4. [O1, O2]

Standard and high demand

- **Explain** why the periodic table has changed throughout the years. [O2]

Consolidate and apply

- Use a video clip to review and extend students' understanding of Mendeleev's contribution. [Google search string:] The genius of Mendeleev's periodic table. [O1, O2, O3]

Standard and high demand

- Complete Worksheet 2.23.2. [O3

Extend

Ask students able to progress further to **explain** why the discovery of isotopes made it possible to show why an element order based on atomic weights was not always correct.

Plenary suggestion

Ask students to draw a large triangle with a smaller inverted triangle that just fits inside it (so they have four triangles). In the outer three ask them to write something they've seen; something they've done; something they've discussed. Then add in the central triangle something they've learned.

Answers to Worksheet 2.23.2

1. The results did not match predictions based on the model; 2. Germanium's properties matched Mendeleev's predictions; 3. The elements in each group have the same number of outer electrons; 4. Ar and K; or Te and I; (or Hs and Mt); 5. Atomic number or proton number

Lesson 24: Diamond

Lesson overview

OCR Specification reference

OCR C2.3

Learning objectives

- Identify why diamonds are so hard.
- Explain how the properties relate to the bonding structure in diamond.
- Explain why diamond differs from graphite.

Learning outcomes

- Know that each carbon atom in diamond forms four covalent bonds with other carbon atoms in a giant covalent structure, and this makes diamond very hard. [O1]
- Explain that diamond has a very high melting point and does not conduct electricity because of its structure and bonding. [O2]
- Explain that diamond, unlike graphite, does not have free electrons to conduct electricity. [O3]

Skills development

- WS 1.2 Use a variety of models, such as representational, spatial, descriptive, computational and mathematical, to solve problems, make predictions and develop scientific understanding of familiar and unfamiliar facts.
- WS 1.4 Explain everyday and technological applications of science.

Maths focus Visualise and represent two-dimensional (2D) and 3D forms, including 2D representations of 3D objects

Resources needed Ball-and-stick model of diamond; Worksheet 2.24

Digital resources Google images of diamond structure

Key vocabulary diamond, directional bonds, electrical non-conductor, tetrahedral bonding

Teaching and learning

Engage

- Show students a ball-and-stick model of diamond. Ask them to work in pairs to **identify** the type of bonding and the type of structure. Deduce that the 'sticks' represent covalent bonds and the model shows a giant covalent structure. [O1]

Challenge and develop

Low demand

- Explain that this is a model of diamond. Ask students to draw one carbon atom from the model and add its four bonds. Alternatively, students could use Molymod kits and join four bond links to a carbon atom. Explain that the shape made by the four bonds is called a *tetrahedron* and that each carbon atom in diamond bonds tetrahedrally. Students can use Figures 2.70 and 2.71 in the Student Book to see diagrams of carbon's tetrahedral bonding and how the atoms link together to make diamond. Several images on Google images show the diamond structure from different angles. Ask students to predict whether the giant covalent structure will produce a hard, soft, rigid or flexible substance and establish that diamond is hard because of its giant covalent structure and strong covalent bonds. In fact, diamond scores top marks on the scale used by geologists to measure the hardness of rocks and minerals (Mohr scale). [O1]
- Students can answer question 1 in Worksheet 2.24. [O1]

Explain

Standard demand

- Students can work in pairs to **discuss** question 2 in Worksheet 2.24 and then answer the questions. They are **agreeing** on reasons for diamond's relatively high melting point and that diamond does not conduct electricity. These are ideas they have met before and you may wish to combine this lesson with Lesson 2.13 for more able students. [O2]

- Students can work in pairs again. They need to **recall** Section 2.17 of their Student Book on giant covalent structures and the structure of graphite. They can **discuss** question 3 in Worksheet 2.24, and then answer it in their note book. Check their answers. [O3]

Consolidate and apply

- List the uses of diamond on the board: 'as a gemstone in jewellery', 'in cutting tools'. Students can **explain** which properties of diamond make it suitable for these uses. [O1, O2, O3]

- Explain that diamond is also unreactive chemically. It is not attacked by common acids or alkalis. Students could **suggest** reasons for the lack of reactivity of diamonds and why medical research is currently finding out if diamond is a suitable substance to use in replacement joints and surgical tools. [O2]

High demand

- Tell students that, although diamonds are usually clear, coloured diamonds (grey, red, yellow, orange and blue green) also occur. The colour is sometimes caused by impurities, with different impurities making differently coloured diamonds. Students can **suggest** what effect impurities have on the giant covalent structure of diamond. [O3]

Extend

Ask students able to progress further to:

- **Find out** how synthetic (man-made) diamonds are made, how they are different from natural diamonds and what they are used for. [O1]

- **Find out** why diamonds are sparkly. [O3]

Plenary suggestions

Ask students to work in pairs and **compile** a list of diamond facts. They can double up and **merge** their lists. A class list can be **compiled**.

Write the properties 'hardness', 'high melting point' and 'does not conduct electricity' on the board. Ask students to work in pairs and explain the properties in terms of the structure and bonding in diamond.

Answers to Worksheet 2.24

1. a) Single covalent bonds; b) Carbon atoms; c) Student diagram; d) Tetrahedral bonding; e) Each carbon atom forms four strong covalent bonds; holding each atom firmly in place; 2. b)(i) Many strong covalent bonds; need a lot of energy to break; (ii) There are no free electrons; electricity is a flow of electrons; 3. a) 2,4; b) Four; c) Four; d) Three; e) They move between the layers of carbon atoms/are delocalised; f) Free electrons; can move through the structure

Lesson 25: Graphite

Lesson overview

OCR Specification reference

OCR C2.3

Learning objectives

- Describe the structure and bonding of graphite.
- Explain the properties of graphite.
- Explain the similarity to metals.

Learning outcomes

- Know that carbon atoms are arranged in a giant covalent structure in graphite: carbon atoms make three single covalent bonds and are arranged in layers of hexagonal rings with delocalised electrons between the layers. [O1]
- Explain that graphite has a high melting point because a lot of energy is needed to break the strong covalent bonds. It conducts electricity because the delocalised electrons are free to move and it is soft because the weak bonds between the layers of hexagonal rings are easily overcome. [O2]
- Explain that both metals and graphite have delocalised electrons and can conduct electricity. [O3]

Skills development

- WS 1.2 Use a variety of models, such as representational, spatial, descriptive, computational and mathematical, to solve problems, make predictions and develop scientific explanations and understanding of familiar and unfamiliar facts.
- WS 1.4 Explain everyday and technological applications of science.
- WS 4.1 Use scientific vocabulary, terminology and definitions.

Maths focus Visualise and represent two-dimensional (2D) and 3D forms, including 2D representations of 3D objects

Resources needed Tracing paper, scissors, molecular model of graphite, samples of graphite; Worksheets 2.25.1 and 2.25.2

Key vocabulary delocalised electrons, graphite, hexagonal rings, weak bonds

Teaching and learning

Engage

- Students need tracing paper, a pencil and a ruler. Ask them to use Worksheet 2.25.1 to copy the tessellations of regular hexagons onto tracing paper. They need at least three copies and could work in groups to achieve this. Students can follow Worksheet 2.25.1 to **construct** a model of graphite.

Challenge and develop

Low demand

- Relate their models to the structure of graphite. The sides of the hexagons represent single covalent bonds and carbon atoms are at the corners, and the hexagonal rings form layers that stack up on each other. Explain that weak bonds exist between the layers. Show students a ball-and-stick model of graphite, if available, or use Figure 2.74 in the Student Book to illustrate the structure of graphite. Students can answer questions 1 and 2 in Worksheet 2.25.1. [O1]

Explain

Standard demand

Allow students to use samples of graphite to write with. Ask them to **explain/discuss** why some of the graphite giant covalent structure can be left on paper. Deduce that the bonds between the layers of hexagonal rings are much weaker than the strong covalent bonds between the carbon atoms in the layers. The weaker bonds are easily broken, so you can write with graphite. Emphasise that the stronger covalent bonds are not broken, only the weaker bonds between the layers. **Explain** that pencil 'lead' is graphite. Students can **find out** how it got this name from Section 2.19 in the Student Book. [O2]

- Explain that melting graphite requires some of the strong covalent bonds holding the carbon atoms together to be broken. This requires a lot of energy and graphite has a high melting point (almost 4000 °C). Use the models and Figure 2.75 in the Student Book to tell students that the electrons between the layers of hexagonal rings are delocalised and free to move throughout the structure. This means that graphite can conduct electricity. Emphasise that graphite is an exception to the rule that non-metals do not conduct electricity. Explain that because the delocalised electrons are free to move between the layers, graphite is also a good conductor of heat – the delocalised electrons transfer the thermal energy. Students can answer question 1 in Worksheet 2.25.2. Check their answers. [O2]

High demand

- Remind students that the electrons between the layers of hexagonal rings are delocalised. Students can answer question 2 in Worksheet 2.25.2 to explore the source of the delocalised electrons. They may recall the term 'delocalised electrons' used in descriptions of metallic bonding. Allow them a few minutes to revisit metallic bonding using either their notes or Section 2.19 in the Student Book. They can complete the table in question 3 in Worksheet 2.25.2 to **compare** the structure of graphite with metals. They need to **realise** that graphite and metals are similar because they both have delocalised electrons. [O3]

Consolidate and apply

- Ask students to work in pairs and **evaluate** how good the paper model and the ball-and-stick model are at representing graphite. They will find the Student Book useful in explaining the relative distances between the layers of hexagonal rings in graphite. They can make a list of points and then double up and **merge** their lists with another group. A class discussion can follow as to the usefulness of the models. [O1]

- Remind students that diamond and graphite are both forms of carbon. They can **devise** a suitable table to compare them. They need to **decide** on points to compare and subtitles in their table. [O1, O2, O3]

Extend

Ask students able to progress further to **find out** how graphite is used in nuclear reactors, such as the one in the EDF power station at Dungeness in Kent. [O2]

Plenary suggestions

Ask students to work in pairs and **compile** a list of graphite facts. They can double up and merge their lists, eventually **compiling** a class list. This could be **compared** to the class list compiled in Lesson 24 on Diamond.

Divide the class into two groups – one group is 'diamond', the other 'graphite'. Ask questions to which the answer is either 'graphite' or 'diamond'. Students need to **claim** their questions. The winning group is the one that makes no mistakes.

Answers to Worksheet 2.25.1

1. a) Single covalent bonds; b) Carbon atoms; c) The area containing delocalised electrons; 2. a) 120°; b) Carbon atoms in diamond form four bonds; so less than 120° (109°)

Answers to Worksheet 2.25.2

1. a) Strong covalent bonds between carbon atoms; covalent bonds are broken when graphite melts; requires a lot of energy so a high melting point; b)(i) As lead in pencils; (ii) The weak bonds between the layers of hexagonal rings; are easily broken; and the layers can rub off; c) Graphite is soft and slippery; because the layers can rub off; d)(i) Electrons that move away from their original atom; (ii) One; (iii) Between the layers of hexagonal rings; (iv) The delocalised electrons; can carry thermal energy ;and/or electric charge; (v) Most non-metals do not conduct electricity; (vi) The material has to be a conductor of electricity; needs a high melting point; and must not be very reactive; 2. a) 2,4; b) Seven; c) It is delocalised; 3. a) *Bonding*: Layers of hexagonal rings with delocalised electrons between the layers (graphite); and regular arrangement of metal ions with delocalised electrons moving in structure (metals); *Structure*: Giant covalent structure (graphite) and giant metallic tructure (metals); *Properties*: Soft and slippery, conducts electricity and heat, high melting point (graphite); hard, can be shaped, conducts electricity and heat, high melting point (metals); b) Both are good conductors of electricity; and heat

Lesson 26: Graphene and fullerenes

Lesson overview

OCR Specification reference

OCR C2.3

Learning objectives

- Describe the structure of graphene.
- Explain the structure and uses of the fullerenes.
- Explain the structure of nanotubes.

Learning outcomes

- Know that graphene is made of hexagonal rings of carbon atoms bonded by strong covalent bonds and is one layer thick. [O1]
- Explain that fullerenes are molecules of carbon atoms with hollow shapes, such as C_{60}. [O2]
- Explain that nanotubes are fullerenes with cylindrical shapes. [O3]

Skills development

- WS 1.2 Use a variety of models, such as representational, spatial, descriptive, computational and mathematical, to solve problems, make predictions and develop scientific explanations and understanding of familiar and unfamiliar facts.
- WS 1.4 Explain everyday and technological applications of science, evaluate the associated personal, social, economic and environmental implications and make decisions based on the evaluations of evidence and arguments.
- WS 1.6 Recognise the importance of the peer review of results and of communicating results to a range of audiences.

Maths focus Visualise and represent two-dimensional (2D) and 3D forms, including 2D representations of 3D objects; recognise and use expressions in standard form

Resources needed Ball-and-stick models of graphene, C_{60} or other fullerene, nanotubes (if available), a football showing the Buckminsterfullerene structure; Worksheets 2.26.1 and 2.26.2

Digital resources Images of fullerenes

Key vocabulary cylindrical, fullerene, graphene, nanotube

Teaching and learning

Engage

Ask students to imagine a material 200 times stronger than steel that conducts electricity and is transparent. It is described as a 2D material because it is only one atom thick. It promises to transform everything from smart phones and computers to cars, buildings and satellites. Explain that this material exists and is called *graphene* and is a single layer of graphite. Ask students to describe the structure of graphene. [O1]

Challenge and develop

Low demand

- Students can complete questions 1 and 2 in Worksheet 2.26.1. Check their answers. [O1]

Standard demand

- Tell students that a sheet of graphene does not contain a set number of carbon atoms, and remind them that it, like diamond and graphite, has a giant structure. But carbon does form molecules called *fullerenes*. Students can complete Worksheet 2.26.2 to read about the discovery of Buckminsterfullerene. This worksheet encourages students to think about the importance of peer review and communication of scientific discoveries. [O2]

Explain

- Explain that Buckminsterfullerene is just one of many fullerenes and that fullerenes are carbon molecules with hollow shapes – C_{60} is spherical, but C_{70} is the shape of a rugby ball. Use models or diagrams of C_{60} or other fullerenes to ask students to **identify** the number of carbon atoms that make up the carbon rings in fullerenes. Establish that the structure is based on hexagons containing six carbon atoms, but pentagons with five carbon atoms and heptagons with seven carbon atoms also occur. If a model is available, students can **determine** the number of hexagons and pentagons in C_{60}. [O2]

- Students can read about the uses of fullerenes, such as Buckminsterfullerene, in Section 2.26 of the Student Book. They can answer question 3 in Worksheet 2.26.1. [O2]

High demand

- Explain that some fullerenes are tubes that are open ended – these are called *carbon nanotubes*. Explain that nanotubes are like rolled-up sections of graphite and the carbon atoms are arranged in hexagonal rings. There are several images of nanotubes online that students can view. Students can use Section 2.26 in the Student Book to find out the properties of nanotubes and how they are used. They can answer question 4 in Worksheet 2.26.1. [O3]

Consolidate and apply

- Ask students to explain why Buckminsterfullerene is classed as a small molecule. [O2]

- Students can make a table to **compare** graphene, carbon nanotubes and spherical fullerenes like Buckminsterfullerene. They can **peer assess** their comparisons. [O1, O2, O3]

Extend

Ask students able to progress further to suggest a possible structure for the carbon rings in Buckminsterfullerene, given that the rings contain single and double bonds. Students could draw a small section of the fullerene, adding single or double lines to show the single bonds and double bonds. [O2]

Plenary suggestions

Ask students to imagine life 10 years on. They can work in pairs or small groups to list how they **predict** graphene and fullerenes could change their lives. Ask students to write the word 'Carbon' in the centre of a sheet of paper. They can use arrows to add the different forms of carbon and some key facts about them.

Answers to Worksheet 2.26.1

1. a) Because pencil lead is graphite; b) The carbon atoms have strong covalent bonds between them; c) Three; d) It has delocalised electrons; 2. Diagram should show a single layer of tessellated hexagons; 3. a)(i) Delivering a drug or medicine to a target part of the body only; (ii) They have a hollow structure in which the drug can fit; b) Buckminsterfullerene is spherical and can roll, so provide lubrication between two surfaces; c)(i) 1×10^{-7}; (ii) 1×10^{7}; 4. a) A hollow cylindrical fullerene; b) A layer of carbon atoms in hexagonal rings rolled up into a tube structure; c) For example, to strengthen tennis rackets and golf clubs; d) They have a large surface area, and other atoms can be bonded to the nanotube

Answers to Worksheet 2.26.2

1. A new form of carbon was discovered; 2. To inform other scientists; so that their results could be reproduced by other scientists; and to claim the discovery as theirs; 3. Other scientists reproduced their results; providing more data; and Kroto and colleagues continued to gather more evidence; 4. To check their results and conclusions were valid; 5. To see whether the results were reproducible; by different scientists in different labs; if not, there may have been a flaw in Kroto's experiments/apparatus; 6. The initial discovery; publication of the results; more evidence gathered; and acceptance of its discovery

76 © HarperCollins*Publishers* Limited 2016

Lesson 27: Nanoparticles, their properties and uses

Lesson overview

OCR Specification reference

OCR C2.3

Learning objectives

- Relate the sizes of nanoparticles to atoms and molecules.
- Explain that there may be risks associated with nanoparticles.
- Evaluate the use of nanoparticles for specific purposes.

Learning outcomes

- Identify sizes of nanoparticles in nanometres and know some applications of nanoparticles. [O1]
- Explain the safe use of nanoparticles to prevent health problems. [O2]
- Explain the changes in surface area to volume ratio as particles become smaller and how this affects nanoparticle properties compared to those of the bulk substance. [O3]

Skills development

- WS 1.2 Use a variety of models such as representational, spatial, descriptive, computational and mathematical to solve problems, make predictions and to develop scientific understanding of familiar and unfamiliar facts.
- WS 1.3 Appreciate the power and limitation of science and consider any ethical issues that may arise.
- WS 1.4 Explain everyday and technological applications of science; evaluate associated personal, social, economic and environmental implications; and make decisions based on the evaluation of evidence and argument.
- WS 4.1 Use scientific vocabulary, terminology and definitions.
- WS 4.3 Use SI units (e.g. kg, g, mg, km, m, mm, kJ, J) and IUPAC nomenclature unless inappropriate.
- WS 4.4 Use prefixes and powers of 10 for orders of magnitude (e.g. tera, giga, mega, kilo, centi, micro and nano).
- WS 4.5 Interconvert units.

Maths focus Make orders of magnitude calculations; calculate areas of triangles and rectangles, surface areas and volumes of cubes.

Resources needed Objects as listed on Technician's notes; Worksheets 2.27.1, 2.27.2 and 2.27.3; Technician's notes 2.27

Digital resources Class computer access, YouTube video: GCSE Nanoparticles – data projector

Key vocabulary nanoparticle, surface area to volume ratio

Teaching and learning

Engage

- Ask students how big they think the average atom is. Explain that we use nanometres to measure atoms and other small particles. A nanometre (nm) is 1×10^{-9} m (one billionth of a metre or 0.000000001 m). Explain that we have known about small particles for a long time, but have recently found out they are far more useful than we thought and have developed new uses for them. Explain that this is called *nanotechnology*. Show students examples of products containing nanoparticles, as in Technician's notes 2.27. [O1]

Challenge and develop

Low and standard demand

- Use Figure 2.81 in Section 2.27 in the Student Book to illustrate the size of nanoparticles compared with those of other particles. You could reference these sizes to that of C_{60}, one nanometre. Students can complete Worksheet 2.27.1 to practice using measurements for small particles. [O1]

Standard demand

- Students can complete Worksheet 2.27.2 to find out about some uses and disadvantages of nanoparticles. They will need computer access to **research** the problems associated with nanoparticle use. They will find 'Using nanoparticles safely' in Section 2.27 of the Student Book useful. For Worksheet 2.27.2, students write a short paragraph to **explain** some problems associated with the use of nanoparticles. [O1, O2]

Explain

High demand

- Explain that nanoparticles may have different properties from those of the same material in bulk because of their large surface area. Students can complete Worksheet 2.27.3 to investigate how the surface area to volume ratio changes as particles get smaller. Check that students understand that a larger surface area to volume ratio is one reason nanoparticles have different properties to those of their bulk materials. They also need to know that as the side of a cube decreases by a factor of 10, the surface area to volume ratio increases by a factor of 10. This is covered in question 2 of Worksheet 2.27.3. [O3]

Consolidate and apply

- Students can watch the video: Nano particles GCSE Chemistry (search 'Nanoparticles, GCSE' on the YouTube website). [O1]

- Ask students why scientists are developing surgical masks and wound dressings that contain nanoparticles of silver. [O2]

Extend

Ask students able to progress further to:

- **Investigate** whether or not the surface area to volume ratio always increases by a factor of 10 when the side of a cube decreases by a factor of 10. They can select different sized cubes. [O3]

- **Research** further uses of nanoparticles. [O2]

Plenary suggestions

Ask students to work in groups of three or four and **create** a freeze frame of an idea in this lesson. They can **present** their ideas to the class in turn and other students can **suggest** what it represents.

Students can work in pairs and write three sentences to **summarise** the content of this lesson. They can read their sentences to the class in turn.

Answers to Worksheet 2.27.1

1. a) Scale should read from 1×10^{-9} to 1×10^{-2}; b)(i) 1 to 100 nm; (ii) 1×10^{-9} to 1×10^{-7} m; c) 1×10^{-7} to 2.5×10^{-6}; d) 10 000 nm and 2500 nm; 2. a) Particles are 10 micrometres; b) 10 000 nm; c) PM_{10}

Answers to Worksheet 2.27.2

Should include the harmful effects of inhaling nanoparticles; nanoparticles being absorbed by the skin from sun screens and cosmetics; nanoparticles being washed into the environment from clothing.

Answers to Worksheet 2.27.3

1. a) 64 cm^3; b) 96 cm^2; c) 64 cm^3; d) 192 cm^2; e) Smaller cubes; f) It increases; g)(i) 384 cm^2; (ii) 64 cm^3; h) It increases; 2. a)(i) 600 cm^2; (ii) 1000 cm^3; b)(i) 6000 cm^2; (ii) 1000 cm^3; c) It has increased by a factor of 10

© HarperCollins*Publishers* Limited 2016

Lesson 28: Maths skills: Use ratios, fractions and percentages

Lesson overview

OCR Specification reference

OCR C2.3

Learning objectives

- Consider ways of comparing the amounts of gases in the atmosphere.
- Review what balanced symbol equations show.
- Compare the yields in chemical reactions.

Learning outcomes

- Use fractions and percentages to describe the compositions of mixtures. [O1]
- Use ratios to determine the mass of products expected. [O2]
- Calculate percentage yields in chemical reactions. [O3]

Skills development

- WS 1.4 Recognise the importance of scientific quantities and understand how they are determined, Interconvert units

Maths focus M1 Use ratios, fractions and percentages.

Resources needed Equipment as listed in the Technician's notes; Worksheet 2.28.1 and 2.28.2; Technician's notes 2.28

Digital resources Presentation 2.28 'Changing ratios'

Key vocabulary ratio, empirical formula

Teaching and learning

Engage

- Ask pairs to sort cards cut from Worksheet 2.28.1 into three sets of five. Introduce parts per million (ppm) as a way of comparing the amounts of gases that form a very small percentage of the atmosphere. [O1]

Challenge and develop

- Low demand: Students use a video clip to demonstrate how to **interconvert** fractions, decimals and percentages [Google search string:] Learning Fractions, Decimals, and Percents | abcteach. [O1]
- Standard and high demand: Ask students to read 'Proportion of substances' in Section 2.28 of the Student Book and answer questions 1 and 2. [O1]

Explain

- Students explain how to **convert** numbers from fractions into percentages and vice versa. [O1]

Consolidate and apply

- Use 'Empirical formulae' in Section 2.28 in the Student Book to introduce the use of ratios to compare the numbers of molecules that take part in chemical reactions. [O2]

- Low and standard demand: **Use** Presentation 2.28 'Changing ratios' to remind students that the amount of oxygen available determines the products formed when hydrocarbons burn. Students complete worksheet 2.28.2. [O2]

- High demand: Students complete questions 3–5 in Section 2.28 of the Student Book. [O2]

- Standard and high demand: Students read 'Ratios of substances' in Section 2.28 of the Student Book and answer questions 6–9. [O3]

Extend

Ask students able to progress further to use their results to **calculate** the percentage yield of carbon dioxide. They can be given the theoretical yield for this reaction, which is 0.36 grams of carbon dioxide per gram of copper carbonate. Their yields are likely to be less that 100% and they could be challenged to **suggest** why. [O3]

Plenary suggestion

Challenge groups to **suggest** pairs of reactants that combine in each of these ratios: 1 : 1, 1 : 2 and 1 : 3.

Answers to Worksheet 9.12.1

The three sets are ABGIM; CDFKL; EHJNO

Answers to Worksheet 9.12.2

1. a) 1 : 2, carbon + water; b) 1 : 3, carbon + carbon monoxide + water; c) 1 : 4, carbon dioxide + carbon monoxide + water; d) carbon dioxide + water. 2. As the amount of oxygen in the mixture increases; the carbon (soot) produced is replaced by carbon monoxide; and then by carbon dioxide

When and how to use these pages: Check your progress, Worked example and End of chapter test

Check your progress

Check your progress is a summary of what students should know and be able to do when they have completed the chapter. Check your progress is organised in three columns to show how ideas and skills progress in sophistication. Students aiming for top grades need to have mastered all the skills and ideas articulated in the final column (shaded pink in the Student book).

Check your progress can be used for individual or class revision using any combination of the suggestions below:

- Ask students to construct a mind map linking the points in Check your progress
- Work through Check your progress as a class and note the points that need further discussion
- Ask the students to tick the boxes on the Check your progress worksheet (Teacher Pack CD). Any points they have not been confident to tick they should revisit in the Student Book.
- Ask students to do further research on the different points listed in Check your progress
- Students work in pairs and ask each other what points they think they can do and why they think they can do those, and not others

Worked example

The worked example talks students through a series of exam-style questions. Sample student answers are provided, which are annotated to show how they could be improved.

- Give students the Worked example worksheet (Teacher Pack CD). The annotation boxes on this are blank. Ask students to discuss and write their own improvements before reviewing the annotated Worked example in the Student Book. This can be done as an individual, group or class activity.

End of chapter test

The End of chapter test gives students the opportunity to practice answering the different types of questions that they will encounter in their final exams. You can use the Marking grid provided in this Teacher Pack or on the CD Rom to analyse results. This shows the Assessment Objective for each question, so you can review trends and see individual student and class performance in answering questions for the different Assessment Objectives and to highlight areas for improvement.

- Questions could be used as a test once you have completed the chapter
- Questions could be worked through as part of a revision lesson
- Ask Students to mark each other's work and then talk through the mark scheme provided
- As a class, make a list of questions that most students did not get right. Work through these as a class.

© HarperCollins*Publishers* Limited 2016

Chapter 2: Elements, compounds and mixtures

| Student Name | Getting Started [Foundation Tier] | | | | Going Further [Foundation and Higher Tiers] | | | | | More Challenging [Higher Tier] | | | | Most demanding [Higher Tier] | | | | Total marks | Percentage |
	Q. 1 (AO1) 1 mark	Q. 2 (AO1) 1 mark	Q. 3 (AO2) 2 marks	Q. 4 (AO1) 1 mark	Q. 5 (AO2) 2 marks	Q. 6 (AO2) 2 marks	Q. 7 (AO1) 2 marks	Q. 8 (AO1) 1 mark	Q. 9 (AO2) 2 marks	Q. 10 (AO1) 1 mark	Q. 11 (AO3) 4 marks	Q. 12 (AO2) 2 marks	Q. 13 (AO1) 1 mark	Q. 14 (AO1) 1 mark	Q. 15 (AO2) 2 marks	Q. 16 (AO3) 2 marks	Q. 17 (AO2) 2 marks	Q. 18 (AO2) 2 marks	Q. 19 (AO2) 4 marks	Q. 20 (AO3) 4 marks	

Check your progress

You should be able to:

describe how to separate mixtures of elements and compounds →	use word equations to describe chemical reactions →	use balanced equations to describe reactions
be able to calculate a relative formula mass from the sum of the relative atomic masses →	calculate the sum of the relative formula masses of reactants and products →	show how the relative formula masses of reactants are equal to the relative formula masses of products
describe, explain and identify examples of processes of separation such as filtration, crystallisation and distillation →	suggest separation and purification techniques for mixtures →	distinguish pure and impure substances using melting point and boiling point data
describe how to set up paper chromatography →	distinguish pure from impure substances →	interpret chromatograms and determine R_f values
describe how Mendeleev was able to leave spaces for elements that had not yet been discovered →	explain why the modern periodic table has the elements in order of atomic number →	explain how Mendeleev was able to make predictions of as yet undiscovered elements such as eka-silicon
describe the pattern of the electrons in shells for the first 20 elements' →	explain how the electronic arrangement of atoms follows a pattern up to the atomic number 20 →	explain how the electronic arrangement of transition metal atoms put them into a period
describe a number of physical properties of metals and non-metals →	explain that atoms of metals have 1, 2 or 3 electrons in their outer shell →	explain that non-metals need to gain or share electrons during reactions and that metals need to lose electrons during reactions.
describe three main types of bonding →	explain how electrons are used in the three types of bonding →	explain how bonding and properties are linked
represent an ionic bond with a diagram →	draw a dot and cross diagrams for ionic compounds →	work out the charge on the ions of metal and non-metals from the group number of the element
identify ionic compounds from structures →	explain the limitations of diagrams and models →	work out the empirical formula of an ionic compound
describe that metals form giant structures →	explain how metal ions are held together →	explain how metallic bonding is enabled by the delocalisation of electrons
identify small molecules from formulae →	identify polymers from their unit formula →	relate the intermolecular forces to the bulk properties of a substance
explain how the properties relate to the bonding in diamond →	explain why diamond differs from graphite →	explain the similarity of graphite to metals
describe the structure of graphene →	explain the structures and uses of fullerenes →	compare 'nano' dimensions to dimensions of atoms and molecules

83

© HarperCollinsPublishers Limited 2016

Worked example

1 **Suggest which piece of apparatus you would use to measure out exactly 25 cm³ of acid or alkali solution each time.**

a pipette

> The answer is correct.

2 **Suggest the difference between the melting points of a pure and an impure substance.**

An impure substance has a lower melting point than a pure one.

> The answer is correct.

3 **Draw the spots that will appear on the chromatogram if the mixed dye contains colours Q, R and T.**

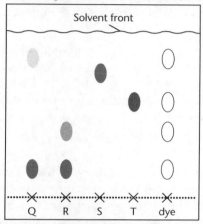

Solvent front

Q   R   S   T   dye

> This answer is correct as the distances of all the spots are the same as the original and the red spot is not there.

4 **Calculate the R_f value for the red spot S.**

R_f value = solvent distance/spot distance
= 14/10.5 = 1.3 mm

> This answer is incorrect as the R_f value is spot distance / solvent distance. Also there are no units to an R_f value. The answer should have been R_f = 10/14.5 = 0.69

5 **An element from Group 1, X, bonds with an element from Group 7, Y.**

a Identify the type of bonding.

a) metallic   b) (ionic)   c) covalent   d) giant

> The answer is correct.

b Draw the dot and cross diagram for the resulting compound.

Use X to represent the metal and Y to represent the non-metal. Show the outer shell only.

> X has lost an electron so the 1+ charge is missing. Y has gained an electron so the 1– charge is missing. Both charges should be at the top right outside the brackets. Electrons should be kept in pairs in a circle.

6 **Explain why both diamond and silicon dioxide are hard with high melting points.**

They have covalent bonds that act in all directions.

> This answer explains what the bonding is like but needs to be linked to the energy required to break the bonds for melting to happen.

7 **Fill in the missing data in the table.**

Substance	Metal	Small molecule	Giant covalent	Ionic
Melting point	High	Low	High	High
Conducts electricity	Yes	No	Yes	Yes when melted, no when solid

> The metal, small molecule and ionic columns are all correct. The student has correctly stated the difference in conductivity between an ionic solid and liquid. Giant covalent structures do not normally conduct electricity. Graphite is an exception and this should be explained.

84

© HarperCollins*Publishers* Limited 2016

Chemical reactions: Introduction

When and how to use these pages

This unit builds on ideas from Unit 1 and 2 about ionic, covalent and metallic substances and the particles they are composed of.

Overview of the unit

As they work through the unit students will improve their ability to work with symbols and equations. They will learn to calculate relative formula masses and will be introduced to moles. They will use moles to calculate reacting masses and to balance equations.

Students will identify exothermic and endothermic changes, use reaction profiles to describe them, calculate theoretical energy transfers using bond energies and investigate the variables that affect the temperature changes in solutions.

Students will explain the oxidation and reduction of metals in terms of loss or gain of oxygen or electrons. They will deduce a reactivity series for metals based on experimental results, and relate this to the tendency of metals to form positive ions and the extraction method used to extract each metal. They will also make soluble salts by neutralising acids with metals, metal oxides, carbonates or alkalis and write equations for these reactions. They will distinguish between strong acids and concentrated acids, and explain what happens during neutralisation. Students will also identify the products formed when molten or dissolved binary compounds are electrolysed, and write equations for the reactions at each electrode.

As part of the practical work students will identify independent, dependent and control variables; identify the main hazards in practical contexts; plan experiments to test hypotheses; carry out experiments appropriately; describe techniques; read measurements from scales; make and record observations; present data appropriately; recognise and describe patterns and trends; use models in explanations; use data to make predictions; and communicate findings and reasoned conclusions.

There are a number of opportunities for students to use mathematics. They will calculate the mean of sets of experimental values; translate information between graphical and numeric forms; use ratios to write simple formulae, recognise and use fractions and percentages, use expressions in decimal form and in standard form, use an appropriate number of significant figures, and substitute numerical values into chemical equations to balance them.

Obstacles to learning

Many students find the concept of the mole difficult, confusing the 'mass' of a substance with the 'amount' of a substance. Many will not realise that the mole is a counting unit. The word 'mole' is similar to the terms 'molecule', 'molar' and 'molecular', which can also confuse students. The term 'amount' can also presents confusion since because is commonly used to describe mass and volume in everyday life.

Individuals may also struggle to appreciate that the overall energy change during a reaction depends on both the energy taken in to break bonds and the energy released when new bonds form.

Students need good laboratory practice to complete the titration. They will need to work in an organised, tidy and clean manner, taking care not to contaminate their solutions.

Students may struggle to recognise that reduction takes place when positive metal ions gain negative electrons and are discharged from solution as neutral atoms. Their work with symbol equations needs to be carefully structured so that they experience early success and are not overwhelmed by the process.

Practicals in this unit

In this unit students will do the following practical work:

- Measure the mass of carbon dioxide released per gram of copper carbonate decomposed

- Measure the volume of carbon dioxide produced when varying amounts of hydrochloric acid are reacted with excess marble chips

- Witness exothermic and endothermic changes

- Make models to show the bonds in methane, oxygen, carbon dioxide and water

- Measure the mass of fuel needed to raise the temperature of 100g of water by about 10°C

- Investigate displacement reactions.

- React a metal oxide with charcoal to remove oxygen.

- Prepare sodium chloride by neutralising sodium hydroxide with hydrochloric acid.

- Prepare copper sulfate by reacting copper oxide or copper carbonate with sulfuric acid.

- Prepare zinc sulfate by reacting zinc with sulfuric acid.

- Prepare magnesium sulfate by reacting magnesium carbonate with sulfuric acid.

- Use universal indicator to classify solutions as weak or strong acids or alkalis and explore the pH changes that take place when a strong acid and a strong alkali neutralise each other.

- Find the reacting volumes of solutions of acids and alkalis by titration.

- Explore the pH changes that take place when a strong acid and a strong alkali are diluted by factors of 10.

- Test their own hypothesis about a variable that affects the temperature change produced by a reaction.

- Explore the effect of passing an electric current through a solution of potassium manganate(VII).

- Investigate the products formed when an electric current is passed through a solution of copper(II) sulfate.

- Investigate the products formed at the cathode when an electric current is passed through different salt solutions.

Students will also observe:

- How iron and sulfur react to form a compound

- How magnesium burns

- Molar quantities of a variety of substances

- The conservation of mass when copper(II) sulfate reacts with sodium hydroxide

- The use of an electric spark to provide the activation energy for an exothermic reaction

- Magnesium ribbon reacting with a strong acid and with a weak acid.

 © HarperCollins*Publishers* Limited 2016

	Lesson title	Overarching objectives
1	Elements and compounds	Use chemical symbols and formulae to describe elements and compounds.
2	Atoms, formulae and equations	Know that the atoms in an element are all the same and use symbols and formulae to represent elements and compounds and simple chemical reactions.
3	Moles	Know that chemical amounts are measured in moles, and be able to use the relative formula mass of a substance to calculate the number of moles and vice versa.
4	Key concept: Conservation of mass and balanced equations	Explain the law of conservation of mass and apply it to balance symbol equations.
5	Key concept: Amounts in chemistry	Use formula masses to convert grams into miles and vice versa
5	Mass changes when gases are in reactions	Explain observed changes in mass during reactions in terms of loss or gain of gases from the atmosphere.
7	Using moles to balance equations	Use moles to balance an equation given the masses of the reactants and products.
8	Key concept: Limiting reactants and molar masses	Explain how limiting quantities of a reactant affect the amount of products produced.
9	Amounts of substances in equations	Calculate the masses of reactants and products from balanced symbol equations and the mass of a given reactant or product.
10	Key concept: Endothermic and exothermic reactions	Identify exothermic and endothermic reactions and investigate the variables that affect temperature changes in reacting solutions.
11	Reaction profiles	Use reaction profiles to identify reactions as exothermic or endothermic and describe the activation energy of a reaction.
12	Energy change of reactions	Use bond energies to describe the energy changes in bond breaking and bond making and explain how a reaction is endothermic or exothermic overall.
13	Maths skills: Recognise and use expressions in decimal form	Measure temperature changes accurately and use them to compare the energy released by different fuels.
14	Oxidation and reduction in terms of electrons	Explain reduction as a gain of electrons and oxidation as loss of electrons and write ionic equations for displacement reactions.
15	Key concept: Electron transfer, oxidation and reduction	Explain oxidation and reduction by electron transfer and relate the ease of losing electrons to reactivity.
16	Neutralisation of acids and salt production	Describe ways that salts can be made and predict the products formed from given reactants.
17	Soluble salts	Describe how to make a pure, dry sample of a soluble salt and derive its formula.
18	Reaction of metals with acids	Describe how to make salts by reacting metals with acids and write equations for the reactions.
19	Practical: Preparing a pure, dry sample of a soluble salt from an insoluble oxide or	Prepare a pure, dry sample of a soluble salt from an insoluble oxide or carbonate.

20	pH and neutralisation	Use the pH scale to identify acidic and alkaline solutions and recognise how the pH changes when a strong acid neutralises a strong alkali.
21	Strong and weak acids	Explain the difference between strong acids and concentrated acids, and explain what happens during neutralisation.
22	Maths skills: Make order of magnitude calculations	Calculate acid concentrations and deduce the effect of hydrogen ion concentration on the numerical value of pH.
23	Practical: Investigate the variables that affect temperature changes in reacting solutions such as acid plus metals, acid plus carbonates, neutralisations, displacement of metals	Investigate the variables that affect temperature changes in reacting solutions such as acid plus metals, acid plus carbonates, neutralisations, displacement of metals.
24	The process of electrolysis	Describe electrolysis and write half equations for the reactions at each electrode.
25	Electrolysis of molten ionic compounds	Predict the products of the electrolysis of molten binary compounds and explain how ions are discharged at each electrode.
26	Electrolysis of aqueous solutions	Explain the products of electrolysis of copper sulfate using inert electrodes and predict the products that other aqueous solutions will form.
27	Practical: Investigating what happens when aqueous solutions are electrolysed using inert electrodes	Investigate what happens when aqueous solutions are electrolysed using inert electrodes

88   © HarperCollins*Publishers* Limited 2016

Lesson 1: Elements and compounds

Lesson overview

OCR Specification reference

OCR C3.1

Learning objectives

- Identify symbols of elements from the periodic table.
- Recognise the properties of elements and compounds.
- Identify the elements in a compound.

Learning outcomes

- Recognise that the periodic table contains all known elements and be able to use names and symbols of the first 20 elements. [O1]
- Describe how a compound such as iron sulfide is made from elements in a chemical reaction. [O2]
- Distinguish between formulae for elements and compounds and name the elements in a compound. [O3]

Skills development

- WS 1.2 Use a variety of models such as representational, spatial, descriptive, computational and mathematical to solve problems, make predictions and to develop scientific explanations and understanding of familiar and unfamiliar facts.
- WS 3.5 Interpret observations and other data (presented in verbal, diagrammatic, graphical, symbolic or numerical form), including identifying patterns and trends, making inferences and drawing conclusions.
- WS 4.1 Use scientific vocabulary, terminology and definitions.

Resources needed Student copies of the periodic table; class poster of the periodic table; circles of coloured card or modelling clay to model atoms (optional); Worksheets 3.1.1 and 3.1.2; Practical sheet 3.1; Technician's notes 3.1

Digital resources Internet access is required for the 'Extend' section

Key vocabulary balanced, compound, element, equation, symbol

Teaching and learning

Engage

- Display 12 kg of barbecue charcoal and identify it as the average mass of the element carbon in a 15-year-old girl. Introduce the idea that we get carbon into our bodies by eating. [O1]
- Ask for examples of foods that contain carbon and discuss the fact that that the element carbon cannot be seen because it is chemically joined with other elements in compounds. Bread can be toasted to make the compounds (carbohydrates and proteins) in it react and leave a layer of carbon on its surface. [O1]
- Remind students that our bodies excrete some carbon when we breathe out. We cannot see the particles of carbon because it is now combined with oxygen as carbon dioxide. Establish that carbon is an element and carbon dioxide is a compound.

Challenge and develop

Low demand

- Explain that carbon is one of 92 naturally occurring elements and there are 26 more that are made synthetically. Show students how we organise these elements in the periodic table. They can practice finding symbols for named elements to familiarise themselves with the table. Explain that all the compounds in the known Universe are made from these elements. Students can answer questions 1 and 2 on Worksheet 3.1.1. [O1]

Standard demand

- Use Practical sheet 3.1 to demonstrate the reaction between iron and sulfur to make iron sulfide. This should be carried out in a fume cupboard. [O2]

Explain

Standard demand

- Students can answer the questions on the practical sheet. Check their answers and establish that the elements iron and sulfur have combined in a chemical reaction to make the compound iron sulfide. The reaction can be modelled using circles of coloured card or different coloured balls of plasticine to show that one iron atom combines with one sulfur atom. Note that because iron sulfide has an ionic structure, this model has its limitations, but is useful to show the 1:1 ratio of iron sulfide and the idea of elements combining in fixed proportions. [O2]

- Students can read the section 'Elements and compounds' in Section 3.1 in the Student Book and answer question 3 on Worksheet 3.1.1. [O2]

High demand

- Explain that we use the chemical symbols for elements to write formulae for compounds. Students can answer the questions on Worksheet 3.1.2 to practise **identifying** the elements present in different compounds and how they are named. [O3]

Consolidate and apply

- Students can **discuss** what is meant by the statement 'the same compound always has the same formula, no matter where and how it is made or found'. [O3]

- Students can answer questions 1 to 7 in Section 3.1 of the Student Book. [O1, O2, O3].

Extend

Ask students able to progress further to **research** the discovery of the elements with atomic numbers 113, 115, 117 and 118 and why this discovery has completed the seventh row of the periodic table. They will need internet access. Students can **present** their finding to the class. [O1]

Plenary suggestions

Students need two strips of paper. They can write a question on today's lesson on one strip and the answer on the other. In groups of five or six, the questions and answers are shuffled and redistributed so that each student has a question and an answer. One student asks their question and the student with the correct answer responds and asks their question in turn.

Answers to Worksheet 3.1.1

1. a) C; b) O; c) Ca; d) Cu; e) Na; f) H; g) K; h) Fe; 2. a) S is the symbol for sulfur; b) e.g. sodium Na; lead Pb; potassium K; iron Fe; 3. (a) Sodium; chlorine; b) Sodium chloride; c) Na and Cl; d) Energy is given out; a new substance is made; e) A white powder is produced; f) Another chemical reaction

Answers to Worksheet 3.1.2

1. Magnesium and oxygen, copper and sulfur, sodium and bromine, potassium and iodine, lithium and oxygen, aluminium and chlorine, copper and nitrogen, copper and sulfur; 2. a) Magnesium sulfide, magnesium and sulfur; b) Iron sulfide, iron and sulfur; c) Copper oxide, copper and oxygen; d) Potassium iodide, potassium and iodine; e) Sodium chloride, sodium and chlorine; f) Calcium oxide, calcium and oxygen; g) Lithium bromide, lithium and bromine; h) Magnesium oxide, magnesium and oxygen

Answers to Practical sheet 3.1

1. Fe, S; 2. Use a magnet to attract the iron particles; 3. Heat and light given out; 4. Mixture is grey/yellow powder; iron sulfide is black and solid; 5. Iron sulfide; 6. No

Lesson 2: Atoms, formulae and equations

Lesson overview

OCR Specification reference

OCR C3.1

Learning objectives

- Learn the symbols of the first 20 elements in the periodic table.
- Use symbols to describe elements and compounds.
- Use formulae to write equations.

Learning outcomes

- Recognise elements and compounds of the first 20 elements from their formulae. [O1]
- Write formulae correctly using elements from Groups 1, 2, 6 and 7. [O2]
- Write equations for simple reactions. [O3]

Skills development

- WS 1.2 Use models in explanations.
- WS 2.4 Carry out experiments appropriately.
- WS 4.1 Use scientific terminology.

Maths focus Use ratios

Resources needed Equipment as listed in the Technician's notes; a periodic table for each pupil; a large wall-mounted periodic table; a printed copy of each slide in Presentation 3.2 displayed on a notice board visible to the students; Worksheets 3.2.1, 3.2.2 and 3.2.3; Practical sheet 3.2; Technician's notes 3.2

Digital resources Presentation 3.2 'Name that compound'

Key vocabulary compound, element, molecule

Teaching and learning

Engage

- Display unlabelled samples of magnesium powder and calcium oxide. Ask students to **deduce** which is an element and which is a compound. [O1, O2, O3]
- Present the formula of each sample and discuss how symbols can make it easier to identify elements and compounds. Make sure every student has a periodic table. [O1, O2, O3]

Challenge and develop

- Introduce the idea that each compound has a set formula and that this contains the symbols of the elements that reacted to make it. Students **complete** the quiz on Presentation 3.2 to practice identifying symbols. [O1, O2]

Low demand

- Give out Practical sheet 3.2. Students work in pairs to burn a sample of magnesium ribbon and answer the questions. [O1, O2]
- Ask students to read 'Elements and compounds' from Section 3.2 of the Student Book and complete questions 1 and 2. [O1]

Standard and high demand

- Challenge students to burn magnesium and **describe** the product formed using words and symbols. **CARE!** Students must hold the magnesium with tongs at arm's length and must not look directly at the flame. [O1, O2]

- Challenge groups to identify patterns in the formulae of the compounds in Presentation 3.2 and ask each group in turn to offer one new point until every group 'passes'.

Explain

Low demand

- Read 'Formulae' from Section 3.2 of the Student Book and then **complete** questions 3 and 4, and those in part 1 of Worksheet 3.2.1. [O1, O2]

Standard and high demand

- Read 'Formulae' from Section 3.2 of the Student Book and then **complete** questions 3 and 4, and those in part 1 of Worksheet 3.2.2. [O1, O2]

Consolidate and apply

Low demand

- Students **construct** the formulae and equations in Worksheets 3.2.1 and 3.2.2. [O2, O3]

Standard and high demand

- Students **construct** the formulae and equations in Worksheet 3.2.3. [O2, O3]

Extend

Ask students able to progress further to **devise** a help sheet to support students who don't know how to balance equations. [O3]

Plenary suggestion

Ask students to write down three things they **learned** about writing equations. Then ask them to **share** their facts in groups and **compile** a master list, with the most important facts at the top.

Answers to Worksheet 3.2.1

a) CaO; b) $CaCl_2$; c) NaF, d) BeS; e) BeF_2; f) LiF; g) MgS; h) KCl; i) MgF_2; j) Li_2O; k) Na_2O; l) Li_2S

Answers to Worksheet 3.2.2

a) $2Ca + O_2 \rightarrow 2CaO$; b) $Ca + Cl_2 \rightarrow CaCl_2$; c) $2Na + F_2 \rightarrow 2NaF$; d) $Be + S \rightarrow BeS$; e) $Be + F_2 \rightarrow BeF_2$; f) $2Li + F_2 \rightarrow 2LiF$; g) $Mg + S \rightarrow MgS$; h) $2K + Cl_2 \rightarrow 2KCl$; i) $Mg + F_2 \rightarrow MgF_2$; j) $4Li + O_2 \rightarrow 2Li_2O$; k) $4Na + O_2 \rightarrow 2Na_2O$; l) $2Li + S \rightarrow Li_2S$

Answers to Worksheet 3.2.3

1. a) CaO; b) $CaCl_2$; c) NaF; d) BeS; e) BeF_2; f) LiF; g) MgS; h) KCl; i) MgF_2; j) Li_2O; k) Na_2O; l) Li_2S; 2. a) $2Ca + O_2 \rightarrow 2CaO$; b) $Ca + Cl_2 \rightarrow CaCl_2$; c) $2Na + F_2 \rightarrow 2NaF$; d) $Be + S \rightarrow BeS$; e) $Be + F_2 \rightarrow BeF_2$; f) $2Li + F_2 \rightarrow 2LiF$; g) $Mg + S \rightarrow MgS$; h) $2K + Cl_2 \rightarrow 2KCl$; i) $Mg + F_2 \rightarrow MgF_2$; j) $4Li + O_2 \rightarrow 2Li_2O$; k) $4Na + O_2 \rightarrow 2Na_2O$; l) $2Li + S \rightarrow Li_2S$

Answers to Practical sheet 3.2

1. Any two differences; 2. Magnesium oxide; 3. A; 4. B; 5. D; 6. Magnesium + oxygen → magnesium oxide; 7. The oxygen atoms are bonded/joined together

 © HarperCollins*Publishers* Limited 2016

Lesson 3: Moles

Lesson overview

OCR Specification reference

OCR C3.1

Learning objectives

- Describe the measurements of amounts of substances in moles.
- Calculate the amount of moles in a given mass of a substance.
- Calculate the mass of a given number of moles of a substance.

Learning outcomes

- Know that the amount of a substance is measured in moles and is the relative formula mass in grams containing 6.02×10^{23} particles, Avogadro's constant. [O1]
- Calculate the amount of moles of a substance from its mass in grams. [O2]
- Calculate the mass of a substance, given the amount in moles. [O3]

Skills development

- WS 4.1 Use scientific vocabulary, terminology and definitions.
- WS 4.2 Recognise the importance of scientific quantities and understand how they are determined.
- WS 4.3 Use SI units (e.g. kg, g, mg, km, m, mm, kJ, J) and IUPAC chemical nomenclature unless inappropriate.
- WS 4.5 Interconvert units.
- WS 4.6 Use an appropriate number of significant figures in calculations.

Maths focus Recognise and use expressions in decimal form; recognise and use expressions in standard form use an appropriate number of significant figures; understand and use the symbols $=$, $<$, $\ll$, $\gg$, $>$, $\propto$, $\sim$; change the subject of an equation

Resources needed Calculators; Worksheets 3.3.1 and 3.3.2; Technician's notes 3.3

Key vocabulary mole, chemical amount, relative formula mass, Avogadro number

Teaching and learning

Engage

- Explain that chemists have a real problem when it comes to measuring numbers of particles like atoms and molecules because these are so small. A chlorine atom is about 0.1 nm across, which makes counting chlorine atoms impossible. Explain that chemists have solved the problem by measuring particles like atoms and molecules in moles, where 1 mole is a set number of particles. This allows them to compare the amount of substance in different chemicals. [O1]

- You can make analogies. A 'bit' on a computer stores one piece of information, 0 or 1. Bits are organised into bytes – a byte holds enough information for one letter of the alphabet. Students can **suggest** other units in common use that measure amounts. [O1]

- Show students the containers that show 1 mole each of copper(II) sulfate crystals, water, sodium chloride, zinc and carbon (see Technician's notes 3.4). Explain that these all contain the same number of particles. Discuss that they look different because some particles are larger/heavier that others – they are arranged differently. [O1]

Challenge and develop

High demand

- Tell students that moles can be measured conveniently because the mass of 1 mole of a substance is its relative formula mass in grams. Students can read 'Measurement of amounts' in Section 3.3 in the Student Book and then **suggest** the mass of each substance in the containers. Discuss their answers. [O1]

- Tell students that moles are not just used to measure out atoms and molecules – they are also used to measure ions, electrons, formulae and apply to equations, but the type of particle being referred to must be named. They can answer question 1 in Worksheet 3.3.1. [O1]

Explain

High demand

- Explain that the number of particles in 1 mole is 6.02×10^{23} and that this is called the *Avogadro constant* to honour the work he did on quantitative chemistry. Writing this number out in full may help students appreciate its magnitude. They can answer question 2 in Worksheet 3.3.1. They will need calculators and some may need help with standard form calculations. [O1]

- Use 'Calculating molar mass' in Section 3.3 in the Student Book to explain how the mass of 1 mole is calculated. Students can answer question 3 in Worksheet 3.3.1. Check their answers. [O1]

- Explain how to calculate the number of moles in a given mass of a substance. Students can use the formula:

$$\text{amount in moles} = \frac{\text{mass (g)}}{A_r \text{ or relative formula mass}}$$. They can answer question 1 in Worksheet 3.3.2.

Explain how to calculate the mass of a substance from the amount in moles. Students can use the formula: mass (g) = moles $\times A_r$ (relative formula mass). They can answer question 2 in Worksheet 3.3.2.

Consolidate and apply

- Ask students why 100 g of atoms will contain different numbers of atoms, depending on which atoms are present. Pick-and-mix sweets make a good analogy to help them understand this. [O1]

- Ask students to **compile** a list of step-by-step instructions to convert mass in grams into moles and vice versa. They can test out each other's instructions and **peer assess** them. [O2, O3]

Extend

Ask students able to progress further to **find out**:

- How moles help chemists to work out what masses of reactants are needed in a chemical reaction so that there is no waste. [O1, O2, O3]

- Who Avogadro was. [O1]

Plenary suggestions

Students can work in pairs to **compile** a list of 'Mole facts'. They can double up and **agree** on the most important five facts. These can be **shared** with the rest of the class.

They can **work** in groups of five or six. Each needs two strips of paper. They write a question on one strip and the answer on the other. The questions must be calculations to convert mass into moles or vice versa. **Shuffle** the strips and redistribute them so that each student has one question and one answer. They can ask their question and the student with the correct answer responds and ask the next question.

Answers to Worksheet 3.3.1

1. a) Particles like atoms and molecules are too small to be counted; b) mol; c) 6.02×10^{23}; d) Avogadro number; e) 1 mol copper, 1 mol water, 1 mol hydrogen gas; 2. a)(i) 3.01×10^{23}; (ii) 1.204×10^{24}; (iii) 6.02×10^{24}; (iv) 6.02×10^{22}; (v) 3.01×10^{24}; 3. a)(i) 23 g/mol; (ii) 31 g/mol; (iii) 52 g/mol; (iv) 7 g/mol; (b)(i) 135 g/mol; (ii) 64 g/mol; (iii) 16 g/mol; (iv) 74 g/mol; (v) 250 g/mol

Answers to Worksheet 3.3.2

1. a) 2 mol; b) 0.25 mol; c) 0.1 mol; d) 0.2 mol; e) 10 mol; f) 0.2 mol; g) 2 mol; h) 4 mol; i) 1000 mol; j) 500 mol; 2. a) 80 g; b) 32 g; c) 230 g; d) 10 g; e) 18.5 g; f) 0.98 g; g) 42 g; h) 84 g; i) 35.5 g; j) 2070 g

Lesson 4: Key concept: Conservation of mass and balanced equations

Lesson overview

OCR Specification reference

OCR C3.1

Learning objectives

- Explore ideas about the conservation of mass.
- Consider what the numbers in equations stand for.
- Write balanced symbol equations.

Learning outcomes

- Explain the law of conservation of mass. [O1]
- Explain why a multiplier appears as a subscript in a formula. [O2]
- Explain why a multiplier appears in equations before a formula. [O3]

Skills development

- WS 1.2 Use models in explanations.
- WS 4.1 Use scientific vocabulary, terminology and definitions.
- WS 4.2 Recognise the importance of scientific quantities.

Maths focus Substitute numerical values into equations

Resources needed Equipment as listed in the Technician's notes; Worksheet 3.4; Technician's notes 3.4

Digital resources Presentation 3.4 'Rearranging atoms'

Key vocabulary balanced equation, conservation of mass, products, reactants

Teaching and learning

Engage

- Demonstrate that there is no change of mass when the atoms in copper(II) sulfate and sodium hydroxide rearrange to form products. Technician's notes 3.4 outline the method. Then elicit students' ideas about why we should expect this result. [O1]

Challenge and develop

- Display slide 1 of Presentation 3.4 'Rearranging atoms' to present the word equation. Use slide 2 to show the symbol equation and check that they understand what the subscripts and the numbers in front of formulae mean. [O2, O3]

Low demand

- Students read 'The law of conservation of mass' and 'Balanced equations and formulae' in Section 3.4 of the Student Book and answer questions 1–4. [O2]

Standard and high demand

- Give students a quick quiz to check that they can interpret subscripts and brackets in formulae, e.g. how many oxygen atoms in $Al_2(SO_4)_3$; how many nitrogen atoms in $(NH_4)_2SO_4$; how many nitrogen atoms in $Al(NO_3)_3$? [O2]

- Use a video to reinforce the idea that the reactants and the products of a reaction must have the same total mass because they contain the same atoms [Google search string:] BBC Bitesize Conservation of Mass in Chemical Reactions or [Google search string:] The Law of Conservation of Mass – Todd Ramsey. [O1]

Explain

- Ask pairs to choose a chemical equation and use it to **explain** why the mass doesn't change when a reaction takes place. [O1]

Consolidate and apply

- Use an animation to demonstrate how to balance symbol equations [Google search string:] Chemical Reaction Equation | Balancing Animation. [O2, O3]

Low demand

- Students complete Worksheet 3.4 to **practise** balancing symbol equations. [O2, O3]

Standard demand

- Students read 'Balancing equations' in Section 3.4 of the Student Book and answer question 5. [O2, O3]

High demand

- Students work in pairs to **create** a teaching resource to help other students write balanced symbol equations. [O2, O3]

Extend

Ask students able to progress further to use an online interactive balancing equations quiz [Google search string:] It's Elemental – Balancing Act!

Plenary suggestion

Show a video clip demonstrating the conservation of mass without sound and ask students to **produce** a 'voice over' for it [Google search string:] Science Fix Video Demo: Conservation of Mass.

Answers to Worksheet 3.4

1. $2H_2 + Br_2 \rightarrow 2HBr$; 2. $2Mg + O_2 \rightarrow 2MgO$; 3. $2H_2 + O_2 \rightarrow 2H_2O$; 4. $2C + O_2 \rightarrow 2CO$; 5. $2Cu + O_2 \rightarrow 2CuO$; 6. $2Na + 2H_2O \rightarrow 2NaOH + H_2$; 7. $2Na + Cl_2 \rightarrow 2NaCl$; 8. $Br_2 + 2KI \rightarrow I_2 + 2KBr$; 9. $H_2 + Cl_2 \rightarrow 2HCl$; 10. $Mg + 2HCl \rightarrow MgCl_2 + H_2$; 11. $2K + Br_2 \rightarrow 2KBr$; 12, $N_2 + 3H_2 \rightarrow 2NH_3$

 © HarperCollins*Publishers* Limited 2016

Lesson 5: Key concept: Amounts in chemistry

Lesson overview

OCR Specification reference

OCR C3.1

Learning objectives

- Use atomic masses to calculate formula masses.
- Explain how formula mass relates to the number of moles.
- Explain how the number of moles relates to other quantities.

Learning outcomes

- Calculate the formula mass of elements and compounds, including formulae with brackets. [O1]
- Explain how a number of moles relates to formula mass, gas volume, concentration of solutions and chemical equations. [O2]
- Use balanced chemical equations to calculate masses of reactants and products from a given mass or amount of a reactant or product. [O3]

Skills development

- WS 3.3 Carry out and represent mathematical and statistical analysis.
- WS 3.5 Interpret observations and other data (presented in verbal, diagrammatic, graphical, symbolic or numerical form), including identifying patterns and trends, making inferences and drawing conclusions.
- WS 4.2 Recognise the importance of scientific quantities and understand how they are determined.

Maths focus Recognise and use expressions in decimal form; use ratios, fractions and percentages.

Resources needed Class copies of the periodic table; 1 mol of water (18 g) in a beaker; 1 mol of ethanol (46 g) in a beaker and/or 1 mol of sodium chloride (58.5 g) in a beaker; and 1 mol of a lead compound in a beaker; Worksheets 3.5.1 and 3.5.2

Key vocabulary mole, molar mass, Avogadro number, molar volume

Teaching and learning

Engage

- Write the words 'atomic mass', 'formula mass', 'moles', 'balanced equations' and 'gas volume' on the board. Ask students to write as many sentences as needed to link the words together. Students can **amend** and use their sentences as the lesson progresses. [O1, O2, O3]
- Alternatively, use the terms as a discussion focus and assess students' current understanding. [O1, O2, O3]

Challenge and develop

Low demand

- Discuss what is meant by 'atomic mass' and remind students that atomic masses are shown in the periodic table. Remind them how atomic masses are used to calculate the formula mass of a substance, given its chemical formula. Include examples involving brackets in their formulae. [O1]
- Students can answer question 1 on Worksheet 3.5.1. [O1]

Explain

Standard demand

- Ask students what they understand by 'moles'. Explain the links (as necessary) to:
  - o molar mass in grams
  - o the Avogadro number

- o concentration of solutions

- o gas volume

- o balanced chemical equations.

- Revisit these topics as required. Students can **complete** the spider diagram in Worksheet 3.5.2 to show the relationship between moles and these topics. They may wish to use and/or amend their sentences from the beginning of the lesson. [O2]

- They can answer question 2 on Worksheet 3.5.1. Check their answers. [O2]

High demand

- Students can read through the higher tier example in Section 3.5 of the Student Book and answer questions 7 and 8. They can answer question 3 on Worksheet 3.5.1. [O3]

Consolidate and apply

- Ask students to imagine they are working as a chemist in a chemical plant that manufactures ammonia. They can work in pairs and **decide** how moles would be useful to them. [O2]

- Show them molar amounts of different substances in the same state at room temperature, such as 1 mol of water and 1 mol of ethanol, or 1 mol of sodium chloride and 1 mol of a lead compound. Ask them why there seems to be more ethanol than water, or more lead compound than sodium chloride. [O2]

Extend

Ask students able to progress further to:

- **Calculate** the concentration of a solution of hydrochloric acid, given that 15 cm^3 of the acid are neutralised exactly by 25 cm^3 of 0.1 mol/dm^3 of potassium hydroxide solution. [O3]

- Suggest, following on from the appearance of moles of different substances in the Consolidate and apply section, why similar observations would not apply to 1 mol of hydrogen gas and 1 mol of ammonia gas. [O2]

Plenary suggestions

Students can **compose** their own questions, with a mark scheme, and test them out on each other. They can refer back to their earlier sentences about 'atomic mass', 'formula mass', 'moles', 'balanced equations' and 'gas volumes', and either **assess** their own sentences, or work in pairs to **assess** each other's. They can **identify** what they have learned this lesson.

Answers to Worksheet 3.5.1

1. a) 63; b) 256; c) 30; d) 60; e) 162.5; f) 213; g) 132; h) 180; i) 158; j) 342; 2. a)(i) 58.5 g; (ii) 20 g; (iii) 500 g; (iv) 0.84 g; b)(i) 2 mol; (ii) 0.25 mol; (iii) 3 mol; (iv) 0.1 mol; c)(i): 1.204 × 10^{24}; (ii) 3.01 × 10^{23}; (iii) 6.02 × 10^{22}; (iv) 6.02 × 10^{21}; d)(i) 12 g; (ii) 44 g; (iii) 32 g; (iv) 32 g; e)(i) 40 g; (ii) 20 g; (iii) 4 g; (iv) 1 g; f)(i) 24 dm^3; (ii) 120 dm^3; (iii) 2.4 dm^3; (iv) 240 dm^3; 3. a) 8.0 g; b) 2.4 dm^3

Answers to Worksheet 3.5.2

These are student dependent, but diagrams should include: Molar mass = the mass in grams of 1 mole of a substance; Avogadro number = the number of particles in 1 mole of a substance; Mole = an amount of a substance that contains 6.0 × 10^{23} particles/the formula mass in grams; Balanced chemical equations = show the formulae of the substances involved in a chemical reaction and track their changes; Gas volume = 1 mole of gas occupies 24 dm^3 at rtp; Concentration of solutions = can be measured in g/dm^3 and mol/dm^3

Lesson 6: Mass changes when gases are in reactions

Lesson overview

OCR Specification reference

OCR C3.1

Learning objectives

- Find out how mass can be gained or lost during a reaction.
- Find the mass of carbon dioxide released per gram of copper carbonate decomposed.
- Assess the accuracy of our measurements.

Learning outcomes

- Explain any observed changes in mass in a chemical reaction. [O1]
- Identify the mass changes using a balanced symbol equation. [O2]
- Explain these changes in terms of the particle model. [O3]

Skills development

- WS 2.3 Carry out experiments appropriately.
- WS 2.6 Make and record observations.
- WS 2.7 Evaluate methods and suggest possible improvements.

Resources needed Equipment as listed in the Technician's notes; Worksheet 3.6; Practical sheet 3.6; Technician's notes 3.6

Digital resources Presentation 3.6.1 'Using equations'; Presentation 3.6.2 'Changing mass'; Graph Plotter 3.6
Key vocabulary gas, mass, particles, thermal decomposition

Teaching and learning

Engage

- Display the prediction on slide 1 of Presentation 3.6.1 'Using equations' – scientists estimate that cement-making adds over 1 billion tonnes of carbon dioxide to the atmosphere each year. Ask students to **suggest** how scientists came up with this value.

Challenge and develop

- Check that students **recognise** that the mass of carbon dioxide released can be deduced from the mass of calcium carbonate decomposed or the mass of calcium oxide produced. Then explain that copper carbonate reacts in a similar way and it is easier to tell when decomposition has taken place because it changes from green to black. [O1]

- Groups decompose copper carbonate following the guidance on Practical sheet 3.6 and **calculate** how much carbon dioxide they release per gram of carbonate decomposed. [O2]

Explain

Low and standard demand

- Students use the data on Worksheet 3.6 to display and **describe** the relationship between the mass of copper carbonate heated and the mass of carbon dioxide released as a line graph. They could plot their graphs by hand or use Graph Plotter 3.6. [O2]

High demand

- Students use relative atomic mass values to **calculate** how much carbon dioxide 1 gram of copper carbonate should release and **compare** that with their experimental results. [O2, O3]

Consolidate and apply

Low demand

- Students read 'Losing mass' and 'Gaining mass' in Section 3.6 in the Student Book and answer questions 1–5. [O1, O2]

Standard and high demand

- Students read 'Other reactions and limiting factors' in Section 3.6 in the Student Book and answer questions 6 and 7. [O2, O3]

Extend

Ask students able to progress further to **calculate** how many grams of dissolved HCl the $MgCO_3$ used in question 7 would react with. [O2]

Plenary suggestion

Use Presentation 3.6.2 'Changing mass' to display three different reactions and ask groups to **identify** whether the mass increases or decreases, and **explain** what causes the change.

Answers to Worksheet 3.6

1. See below; 2. The mass of carbon dioxide released; is directly proportional; to the mass of copper carbonate decomposed; so 0.35 g of carbon dioxide is released for every gram of copper carbonate decomposed; 3. The copper carbonate was not heated for long enough; and did not decompose completely; 4. Student results are likely to be lower than the theoretical values if the copper carbonate has not been heated for long enough

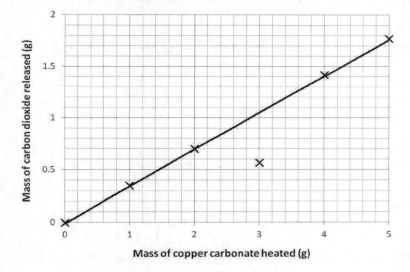

Lesson 7: Using moles to balance equations

Lesson overview

OCR Specification reference

OCR C3.1

Learning objectives

- Convert masses in grams to amounts in moles.
- Balance an equation given the masses of reactants and products.
- Change the subject of a mathematical equation.

Learning outcomes

- Use number of moles = mass of substance ÷ molar mass to calculate an amount in moles from the mass in grams. [O1]
- Convert the masses of reactants and products in a chemical equation into numbers of moles, convert numbers of moles into whole number ratios and use this to balance the equation [O2]

  Change the subject of the equation mass of chemical = molar mass × number of moles [O3]

Skills development

- WS 3.2 Translate data from one form to another.
- WS 4.2 Recognise the importance of scientific quantities and understand how they are determined.
- WS 4.3 Use SI units (e.g. kg, g, mg, km, m, mm, kJ, J) and IUPAC nomenclature unless inappropriate.
- WS 4.5 Interconvert units.

Maths focus Change the subject of an equation; substitute numerical values into algebraic equations using appropriate units for physical quantities

Resources needed Worksheet 3.7

Key vocabulary balanced equation, moles, thermal decomposition, molar mass

Teaching and learning

Engage

- Write this sequence on the board:

  balancing numbers in equations → amount in moles → masses of reactants and products

  Explain that this is the route used in the previous lesson to calculate reacting masses and products. Explain that if the process is reversed, the balancing numbers in an equation can be worked out from the masses of reactants and products in the chemical reaction. [O1, O2, O3]

Challenge and develop

High demand

- Turn the route round to read:

  masses of reactants and products → amounts in moles → balancing numbers in equations

  Explain that the first step in this process converts masses of reactants and products in grams to amounts in moles. Students carried out similar calculations in Lesson 5. (You may wish to skip this step with more able students.) Remind students how to carry out these calculations using:

  $$\text{number of moles} = \frac{\text{mass (g)}}{\text{molar mass}}$$

  Students can answer question 1 in Worksheet 3.7. Students can read the examples in Section 3.7 of the Student Book. [O1]

- Check that they can change the subject of a mathematical equation by asking them to complete question 2 in Worksheet 3.7. You may wish to make this a quick verbal exercise with more able students. [O3]

Explain

High demand

Write this information on the board: '112 g of iron reacts with 213 g of chlorine to produce 325 g of iron(III) chloride'. Calculate the number of moles of each substance (2 mol of iron, 3 mol of chlorine, 2 mol of iron(III) chloride); write the equation and insert the balancing numbers. Check students understand it is chlorine gas (Cl_2), not chlorine atoms (Cl). [O2]

- Now give students the information:

    12.0 g of ethane react with 44.8 g of oxygen to produce 35.2 g of carbon dioxide and 21.6 g of water

    (a more complicated example). Work through the calculation to obtain the ratio of the balancing numbers. You will need to show students how to obtain a whole number ratio by dividing through by the lowest number. Write out the balanced equation. [O2]

- Students can read through the example in Section 3.7 of the Student Book and do the calculation. [O2]

- They can answer question 3 in Worksheet 3.7. Check their answers. [O2]

Consolidate and apply

- Students can write their own step-by-step guide to writing balanced symbol equations from masses of reactants and products. [O2]

- Ask them to write the word 'moles' in the centre of a sheet of paper. They can use arrows to annotate the diagram to show links to concepts they have learned in this lesson. [O1, O2, O3]

Extend

Ask students able to progress further to **compose** their own examples of writing balanced symbol equations from masses of reactants and products. They can try them out on each other. [O2] They can answer question 4 in Worksheet 3.7. [O2]

Plenary suggestions

Ask students to write down the most important idea they have learned this lesson. They can **share** their ideas in pairs and double up to **agree** the main idea. These can be **shared** by the class.

Ask students to **decide** whether they are confident about this lesson or unsure. Group students so that each group contains at least one confident student and one unsure student. Confident students can answer questions/ **explain** the topic to those who are unsure.

Answers to Worksheet 3.7

1. a) 0.1 mol; b) 0.2 mol; c) 1.5 mol; d) 0.01 mol; e) 0.001 mol; 2. a) molar mass = $\dfrac{\text{mass (g)}}{\text{number of moles}}$; b) number of moles =

$\dfrac{\text{mass (g)}}{\text{molar mass}}$; 3. a) $C_3H_8 + 5O_2 \rightarrow 3CO_2 + 4H_2O$; b) $2C_2H_5OH + 6O_2 \rightarrow 4CO_2 + 6H_2O$; c) $C_{14}H_{30} \rightarrow C_7H_{16} + C_3H_6 + 2C_2H_4$; d)

$Na_2S_2O_3 + 2HCl \rightarrow 2NaCl + S + SO_2 + H_2O$; e) $2HBr + H_2SO_4 \rightarrow 2H_2O + SO_2 + Br_2$; 4. $3Cl_2 + 6NaOH \rightarrow NaClO_3 + 5NaCl + 3H_2O$

Lesson 8: Key concept: Limiting reactants and molar masses

Lesson overview

OCR Specification reference

OCR C3.1

Learning objectives

- Recognise when one reactant is in excess.
- Consider how this affects the amount of product made.
- Explore ways of increasing the amount of product.

Learning outcomes

- Identify which reactant is in excess. [O1]
- Explain the effect of a limiting quantity of a reactant on the amount of products. [O2]
- Calculate amount of products in moles or masses in grams. [O3]

Skills development

- WS 1.2 Use models in explanations.
- WS 1.2 Use models to make predictions.
- WS 4.1 Use scientific vocabulary, terminology and definitions.

Resources needed Equipment as listed in Technician's notes 3.8; Practical sheet 3.8; Worksheets 3.8.1 and 3.8.2

Digital resources Presentation 3.8 'Limiting reactants'

Key vocabulary limiting, excess, mole, directly proportional

Teaching and learning

Engage

- Display a large container of coloured balls. Add a few more of a different colour. Explain that the balls represent reactants. They react to make a new substance. Ask what decides how much of the new compound forms. If students aren't sure, ask volunteers to 'react' the two colours together and remove them from the container until one colour is used up. Describe this as the 'limiting reactant' and the other as being 'in excess'. [O1]

Challenge and develop

- Use Presentation 3.8 'Limiting reactants' to introduce the practical. [O1]
- Groups of three or four students use the method on Practical sheet 3.8 to **measure** the total volume of carbon dioxide produced when varying amounts of acid are reacted with excess marble chips. [O2]

Explain

Low and standard demand

- Students **analyse** experimental results using Worksheet 3.8.1. [O2]

High demand

- Ask students to **describe** the relationship between the volume of the limiting reactant and the volume of carbon dioxide produced. [O2]

Consolidate and apply

Low demand

- Students could use Worksheet 3.8.2 to **reinforce** their understanding of limiting reactants. [O2]

Standard and high demand

- Students read Section 3.8 in the Student Book and complete questions 1–8. [O1, O2, O3]

Extend

- Ask students (in pairs) able to progress further to **explain** why an excess of one reactant is used when we investigate factors that change the reaction rate. [O2]

Plenary suggestion

Sketch graphs Present a graph, suggest a change to the reaction conditions and ask pairs of students to **predict** what the new graph would look like and sketch it on a mini whiteboard, e.g. a plot of gas collected against time when 20 cm^3 of an acid is added to excess marble chips instead of 10 cm^3; a plot of gas collected against time when 1 cm of magnesium ribbon is added to an excess of 1 mol/dm^3 acid instead of 2 mol/dm^3 acid.

Answers to Worksheet 3.8.1

1. There are more marble chips than needed to react with the acid; so the reaction will stop when all the acid has been used up; 2. The graph should show a straight line from 0,0 to 20,100; 3. It doubles; 4. B

Answers to Worksheet 3.8.2

1. The acid particles; there are fewer of them so the reaction will stop when they have all been used up; 2. The diagram should have 12 magnesium atoms and 8 acid particles; 3. There is more than enough magnesium available; so all the extra acid particles will react and release hydrogen gas; 4. There would not be enough magnesium to react with them all; the magnesium would be the limiting reactant

Lesson 9: Amounts of substances in equations

Lesson overview

OCR Specification reference

OCR 3.1

Learning objectives

- Calculate the masses of substances in a balanced symbol equation.
- Calculate the masses of reactants and products from balanced symbol equations.
- Calculate the mass of a given reactant or product.

Learning outcomes

- Use relative atomic masses and relative formula masses to calculate the masses of all the substances in a balanced chemical equation. [O1]
- Use molar masses to calculate the masses of reactants and products from balanced symbol equations, given the mass of one reactant or product. [O2]
- Use molar masses to calculate the mass of one given reactant or product from a balanced symbol equation. [O3]

Skills development

- WS 4.1 Use scientific vocabulary, terminology and definitions.
- WS 4.3 Use SI units (e.g. kg, g, mg, km, m, mm, kJ, J) and IUPAC chemical nomenclature unless inappropriate.
- WS 4.5 Interconvert units.

Maths focus Recognise and use expressions in decimal form; use ratios, fractions and percentages; change the subject of an equation; substitute numerical values into algebraic equations using appropriate units for physical quantities

Resources needed Calculators; Worksheet 3.9

Key vocabulary mole, reactant mass, product mass, molar mass

Teaching and learning

Engage

- Explain that moles are an important concept for chemists – they enable chemists in chemical industries to measure precisely the amounts of chemicals that will react exactly, with no waste. They also enable chemists to calculate how much product they can make. Revise the skills students learnt in the previous lesson by asking them how many moles are in 88 g carbon dioxide (2 mol) and to **calculate** the mass of 0.5 mol calcium carbonate (50 g).

Challenge and develop

High demand

- Write the equation 'Mg + 2HCl → MgCl$_2$ + H$_2$' on the board. Ask students what the numbers in front of the symbols mean and establish that one atom of magnesium reacts with two formula units of hydrochloric acid to produce one formula unit of magnesium chloride and one molecule of water. Tell students that because 1 mole of any substance contains the same number of particles, the equation also tells us that 1 mole of magnesium reacts with 2 moles of hydrochloric acid to produce 1 mole of magnesium chloride and 1 mole of hydrogen molecules. Students can answer question 1 in Worksheet 3.9. You may wish to make this a quick verbal exercise with the more able students. [O1]

Explain

High demand

- Explain that because we can calculate the mass of a mole, we can calculate the mass of each reactant and product in a chemical reaction. Ask students to **calculate** the mass of each reactant and product in $Mg + 2HCl \rightarrow MgCl_2 + H_2$. Remind them that 2 mol HCl and 2 mol H_2 are involved. Explain that this means that 24 g magnesium reacts with 73 g hydrochloric acid to produce 95 g magnesium chloride and 2 g hydrogen. Students can check that the mass of the reactants is the same as the mass of the products. [O1]

- Explain how balanced symbol equations can be used to calculate masses of reactants and products from a given mass of a substance. You can refer back to the equation $Mg + 2HCl \rightarrow MgCl_2 + H_2$ and ask students to **calculate** the mass of magnesium chloride formed from 48 g magnesium. Students can read 'Masses of substance from equations' in Section 3.9 in the Student Book and answer questions 1 and 2. They can answer question 2 in Worksheet 3.9. [O2]

- Explain how moles can be used to predict masses in chemical reactions. Students can read 'Predicting masses' in Section 3.9 in the Student Book and answer questions 4 and 5. They can answer question 3 in Worksheet 3.9. Check their answers. [O3]

Consolidate and apply

- Ask students to write a list of step-by-step instructions that can be used to calculate the mass of a product or reactant from the given mass of a substance, using a balanced chemical equation. They can pair up and try out their instructions on each other. [O2]

- Ask students to list what balanced symbol equations tell us. [O1, O2, O3]

Extend

Provide students with the equations $TiO_2 + 2Cl_2 + 2C \rightarrow TiCl_4 + 2CO$ and $TiCl_4 + 2Mg \rightarrow Ti + 2MgCl_2$. Explain that these represent two stages in the extraction of titanium from titanium ore.

Ask students able to progress further to:

- **Calculate** how much titanium dioxide is required to produce 480 g titanium. [O2, O3]

- **Suggest** or **find out** why the masses of products calculated from balanced symbol equations are often not achieved in reality. [O1, O2, O3]

Plenary suggestions

Ask students to **devise** a question involving calculations from a balanced equation and work out the answer. They can pair up and **test** their questions on each other.

Ask students to write down one thing they are sure of from the lesson, one thing they are unsure of and one thing they need to know more about. Ask them to work in groups of five or six and **agree** on a group list. These can be **combined** into a class list. Note areas that need revisiting.

Answers to Worksheet 3.9

1. a)(i) 2; (ii) 2 mol; b)(i) 8 mol; (ii) 12.5 mol; 2. a) 560 g; b)(i) 27.75 g; (ii) 11 g; c)(i) 16 g; (ii) 4.4 g; d)(i) 240 g; (ii) 396 g; 3. a) 184 g; b) 18 g; c) 108 g; d) 22 g

106 © HarperCollins*Publishers* Limited 2016

Lesson 10: Key concept: Endothermic and exothermic reactions

Lesson overview

OCR Specification reference

OCR C3.2

Learning objectives

- Explore the temperature changes produced by chemical reactions.
- Consider how reactions are used to heat or cool their surroundings.
- Investigate how these temperature changes can be controlled.

Learning outcomes

- Use temperature changes to distinguish between exothermic and endothermic reactions. [O1]
- Evaluate the uses and applications of exothermic and endothermic reactions. [O2]
- Investigate the variables that affect temperature changes in reacting solutions. [O3]

Skills development

- WS 1.4 Describe and explain technological applications of science.
- WS 2.2 Plan experiments to test hypotheses.
- WS 2.6 Read measurements off a scale in a practical context.

Maths focus Translate information between graphical and numeric form.

Resources needed For the first part of the lesson: equipment as listed in Technician's notes 3.10 and Practical sheet 3.10. For the 'Consolidate and apply' section: access to computers with Excel and Word; Worksheet 3.10

Digital resources Graph plotters 3.10.1 and 3.10.2

Key vocabulary endothermic, energy transfer, exothermic, surroundings

Teaching and learning

Engage

- Show a video of a chemical explosion [Google search string:] Tianjin blasts video, and ask small groups to **look for** as many pieces of evidence as possible that this reaction released energy from the chemicals that exploded. [O1]
- Demonstrate the endothermic reaction between solid hydrated barium hydroxide and solid ammonium chloride following the method given in Technician's notes 3.10. Then ask for evidence that this reaction takes in energy. [O1]

Challenge and develop

Low demand

- Student pairs measure temperature changes during four different reactions and **record** them on Practical sheet 3.10. Stress the need to keep the polystyrene cup in the beaker to stop it tipping, and demonstrate how to hold it at an angle when measuring temperatures to keep the end of the temperature sensor, or thermometer, completely covered. [O1]

Standard and higher demand

- Each student pair could measure the temperature change for one of the reactions listed on Practical sheet 3.10. Then they **contribute** their results to a whole class table to **assess** the reproducibility of their results. [O1]

Explain

Low demand

- Students read 'Exothermic and endothermic reactions' in Section 3.10 of the Student Book and answer questions 1 and 2. They then **classify** the reactions they tested earlier as exothermic or endothermic and **suggest** one use for each type of reaction. [O2]

- Use a video clip to review the difference between exothermic and endothermic reactions [Google search string:] BBC Bitesize – GCSE chemistry – Endothermic and exothermic reactions. [O1, O2]

Standard and high demand

- Students divide the reactions tested into exothermic like those used in self-heating cans, and endothermic like those used in some sport injury pads. [O2]

Consolidate and apply

- Students use Graph plotters 3.10.1 and 3.10.2 to complete questions 3 and 4 in Section 3.10 of the Student Book. [O3]

- Challenge groups to **suggest** ways of getting a larger temperature change from one of the reactions they tested. They should **appreciate** that the concentration of the solution, the mass of solid added and the chemicals used could all affect the temperature change. [O3]

Extend

- Ask students able to progress further to **classify** the following chemical changes as exothermic or endothermic and give reasons: photosynthesis, using electrolysis to split water into hydrogen and oxygen and heating copper carbonate to make it decompose. [O1]

Plenary suggestion

Ask students to draw a large triangle with a smaller inverted triangle that just fits inside it (giving four triangles). In the outer three ask them to write something they've seen, something they've done and something they've **discussed**. Then add in the central triangle something they've **learned**.

Answers to Worksheet 3.10

Answers will depend on the reaction choosen and the independent variable selected to investigate

Lesson 11: Reaction profiles

Lesson overview

OCR Specification reference

OCR C3.2

Learning objectives

- Use diagrams to show the energy changes during reactions.
- Show the difference between exothermic and endothermic reactions using energy profiles.
- Find out why many reactions start only when energy or a catalyst is added.

Learning outcomes

- Draw simple reaction profiles (energy level diagrams). [O1]
- Use reaction profiles to identify reactions as exothermic or endothermic. [O2]
- Explain what activation energy is. [O3]

Skills development

- WS 1.2 Recognise and interpret diagrams.
- WS 3.2 Translate data from one form to another.
- WS 4.1 Use scientific vocabulary, terminology and definitions.

Resources needed Pair of strongly bonded Molymod atoms; equipment as listed in Technician's notes 3.11; Worksheets 3.11.1 and 3.11.2

Digital resources Presentation 3.11 'Reaction profiles'

Key vocabulary exothermic, endothermic, product, profile, reactant

Teaching and learning

Engage

- Display a beaker of ethanol. Explain that burning ethanol is an exothermic reaction – but you need to supply a little energy to start the reaction. [O3]

Challenge and develop

- Demonstrate the alcohol 'gun' following the instructions in Technician's notes 3.11 and relate the apparatus to the spark plugs used in petrol engines. [O3]
- Show a video of hydrogen exploding after energy is used to start the reaction [Google search string:] Hydrogen – Periodic Table of videos. [O3]

Explain

- Introduce the term 'activation energy' to describe the energy needed to start a reaction and explain that for some reactions the activation energy is so low that sparks or heat is not needed to start them off. [O3]
- All students can be asked to read 'Colliding particles' in Section 3.11 of the Student Book and answer question 1. [O3]

Consolidate and apply

Low and standard demand

- Use Presentation 3.11 'Reaction profiles' and Worksheet 3.11.1 to guide students through the process of **drawing** reaction profiles. [O1, O2]

Standard and high demand

- Demonstrate that energy needs to be put in to pull the atoms in a Molymod molecule apart. It is an endothermic process and the model reflects that. However, the model is not completely realistic. When real atoms bond together energy is released. Students could complete Worksheet 3.11.2 to consolidate these ideas.

High demand

- Students can be asked to read 'Endothermic and exothermic reactions' in Section 3.11 of the Student Book and answer questions 2 and 3. [O1, O2]

Extend

Ask students able to progress further to **design** a quiz to test understanding of energy profiles.

Plenary suggestion

Ask students to list three things they know about activation energies and then join into groups of four to come up with an **agreed** list of points.

Answers to Worksheet 3.11.1

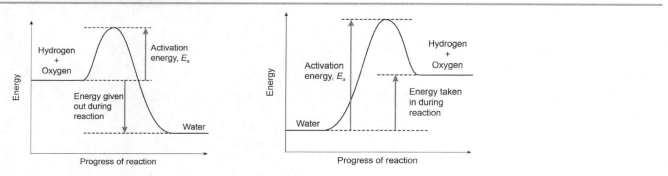

Answers to Worksheet 3.11.2

1. Endothermic; energy is required to break the bonds in the copper carbonate;
2. and 3.

Bonds break   Bonds form

4. The energy needed to break bonds must be less than the energy released when new bonds form.

Lesson 12: Energy change of reactions

Lesson overview

OCR Specification reference

OCR C3.2

Learning objectives

- Identify the bonds broken and formed during a chemical reaction.
- Consider why some reactions are exothermic and others are endothermic.
- Use bond energies to calculate overall energy changes.

Learning outcomes

- Describe the energy changes during bond breaking and bond making. [O1]
- Explain how a reaction is endothermic or exothermic overall. [O2]
- Calculate the energy transferred in chemical reactions using bond energies. [O3]

Skills development

- WS 1.2 Calculate quantities based on a model.
- WS 3.5 Use data to make predictions.
- WS 4.2 Recognise the importance of scientific quantities.

Maths focus Substitute numerical values into equations

Resources needed Equipment as listed in Technician's notes 3.12; Worksheets 3.12.1 and 3.12.2

Digital resources Presentation 3.12 'Bond energies'

Key vocabulary bond breaking, bond making, endothermic, exothermic

Teaching and learning

Engage

- Remind students that bond breaking is endothermic and vice versa. Then display slide 1 of Presentation 3.12 'Bond energies' and elicit students' ideas about whether this energy change is the same for every bond. [O1]

Challenge and develop

- Show a video to introduce the idea of bond energies [Google search string:] Introduction to Bond Energies (enthalpies). [O1, O2, O3]

- Ask pairs to make molecular models – one for methane and two for oxygen – and draw their displayed formulae. Then simulate the combustion of methane and draw displayed formulas of the products – Worksheet 3.12.1 structures the task. Students can then use bond energies to **calculate** the overall energy change for the reaction. [O3]

Explain

- Ask students to read 'Bond energies' in Section 3.12 in the Student Book and to answer question 1. [O1]

Standard demand

- **Pair talk** Ask students to use ideas about bond energies to **explain** why burning methane is exothermic. [O2]

High demand

- Ask students to use ideas about bond energies to **explain** why some reactions are exothermic and others are endothermic. They could **compare** explanations with another student and then join up into groups of four to **compare** ideas. [O2]

Consolidate and apply

Standard demand

- Pairs **practice** simplified bond energy calculations using Worksheet 3.12.2. [O3]

High demand

- Ask students to read 'Calculations' in Section 3.12 in the Student Book and to answer questions 2 and 3. [O3]

Extend

- Ask students able to progress further to answer question 4 in Section 3.12 in the Student Book. [O3]

Plenary suggestion

- Pairs write instructions for **calculating** the energy transferred in a chemical reaction. [O3]

Answers to Worksheet 3.12.1

1. and 2.

$$H-\underset{\underset{H}{|}}{\overset{\overset{H}{|}}{C}}-H \quad + \quad 3O{=}O \longrightarrow 2H{-}O{-}H \quad + \quad O{=}C{=}O$$

3. Table as below; 4. −280 kJ/mol

Bonds broken	How many?	Bond energy (kJ/mol)	Energy put in (kJ/mol)	Bonds formed	How many?	Bond energy (kJ/mol)	Energy released (kJ/mol)
C–H	4	412	1648	C=O	2	532	1064
O=O	2	498	996	H–O	4	465	1860
		Total =	2644			Total =	2924

Answers to Worksheet 3.12.2

1. −200 kJ/mol, exothermic; 2. −200 kJ/mol, exothermic

Lesson 13: Maths skills: Recognise and use expressions in decimal form

Lesson overview

OCR Specification reference

OCR C3.2

Learning objectives

- Read scales in integers and using decimals.
- Calculate the energy change during a reaction.
- Calculate energy transferred for comparison.

Learning outcomes

- Measure temperature changes accurately. [O1]
- Use temperature changes to compare the energy released by different fuels. [O2]
- Calculate the energy released per gram of fuel. [O3]

Skills development

- WS 2.2 Apply understanding of apparatus and techniques to suggest a procedure.
- WS 2.6 Read measurements off a scale in a practical context.
- WS 4.2 Recognise the importance of scientific quantities.

Maths focus Read scales in integers and using decimals and substitute numerical values into equations

Resources needed Equipment as listed in the Technician's notes; Practical sheet 3.13; Worksheet 3.13; Technician's notes 3.13

Key vocabulary decimals, integers, scale, fraction

Teaching and learning

Engage

- Display a selection of fuel burners containing different alcohols and the appearance of the flames produced when they burn. Students should **recognise** that the fuels can be used to heat water. The more energy a fuel releases, the hotter the water gets. [O2]

Challenge and develop

- Stress the need to make precise measurements in order to compare fuels fairly. Temperature changes are difficult to measure accurately because the water needs to be stirred to spread the heat and the temperatures at the start and at the end may not be whole numbers. Heat will also escape into the environment. To minimise heat losses the water temperature should only be raised by about 10 °C. [O1]

Low demand

- Students read 'Temperature changes and energy changes' in Section 3.13 of the Student Book and answer questions 1 and 2. [O1]

Standard and high demand

- Students use thermometers to measure room temperature and compare the values obtained. [O1]

Low and standard demand

- Demonstrate how to use a digital balance and stress that the mass of fuel used may be less than a gram so it needs to be measured to as many decimal places as possible. Students read 'Decimal form' in Section 3.13 of the Student Book and answer questions 3–5. [O3]

Explain

Low and standard demand

- Students complete Worksheet 3.13. Check their results and explain that neither of the methods described gives very accurate results. Taj found that a lot of the fuel evaporated from the open dish before he started to burn it. Sara found that it was very difficult to stop the fuel burning when the temperature rise was exactly 30 °C. Better results can be obtained if their methods are combined so that both the mass of fuel burnt and the temperature change are measured accurately. [O3]

Standard and high demand

- Students read 'Comparing fuels' in Section 3.13 of the Student Book and answer questions 6 and 7. [O3]

Consolidate and apply

- Check that students understand what data they need to collect in order to **compare** fuels fairly.

- Pairs select one of the available fuels, measure the mass of fuel needed to raise the temperature of 100 g of water by about 10 °C and **calculate** the energy released per gram of fuel burned. [O2, O3]

Extend

Ask students able to progress further to **suggest** why the values obtained in the lab are usually (much) lower than the true values. [O2]

Plenary suggestion

Ask students to list all the values that need to be measured in order to **compare** the energy released when two fuels burn, and to list the control variables that need to be kept the same for each fuel.

Answers to Worksheet 3.13

Both tables show that butanol releases most energy per gram; and methanol least

Temperature

Fuel	Start temperature (°C)	Final temperature (°C)	Temperature change (°C)
Methanol	24.5	56.0	31.5
Ethanol	18.0	53.5	35.5
Propanol	18.5	54.5	36.0
Butanol	24.0	64.0	40.0

Mass

Fuel	Start mass (g)	Final mass (g)	Mass of fuel burned (g)
Methanol	154.3	150.7	3.6
Ethanol	213.4	210.6	2.8
Propanol	185.8	183.4	2.4
Butanol	198.5	196.3	2.2

Lesson 14: Oxidation and reduction in terms of electrons

Lesson overview

OCR Specification reference

OCR C3.3

Learning objectives

- Observe some reactions between metal atoms and metal ions.
- Learn to write ionic equations and half equations.
- Classify half equations as oxidation or reduction.

Learning outcomes

- Use experimental results of displacement reactions to confirm the reactivity series. [O1]
- Write ionic equations for displacement reactions. [O2]
- Identify in a half equation which species are oxidised and which are reduced. [O3]

Skills development

- WS 2.6 Make and record observations.
- WS 3.5 Interpret observations.
- WS 4.1 Use scientific vocabulary, terminology and definitions.

Resources needed Equipment as listed in Technician's notes 3.14; Practical sheet 3.14; Worksheet 3.14

Digital resources Presentation 3.14.1 'Displacement'; Presentation 3.14.2 'One of these is not like the others'

Key vocabulary displacement, half equations, ionic equations, oxidation

Teaching and learning

Engage

- Demonstrate what happens when iron nails are added to a solution containing copper, as detailed in Technician's notes 3.14, and show Presentation 3.14.1 'Displacement'. Challenge groups to **explain** the colour changes that take place.

Challenge and develop

- Check that students **recognise** that the more reactive metal – iron – is pushing copper out of the solution and taking its place. [O1]
- Pairs use displacement reactions to **deduce** the order of reactivity of zinc, magnesium, copper and lead following the method on Practical sheet 3.14. [O1]

Explain

Low demand

- Students write word equations for any displacement reactions that take place. [O1]

Standard and higher demand

- Students write symbol equations for any reactions that take place. [O2]

Consolidate and apply

Use a video clip to demonstrate how beautiful a displacement reaction can look [Google search string:] Beautiful Reactions: Metal Displacement.

Low demand

- Students read 'Experimental results' and 'Displacement equations' in Section 3.14 in the Student Book and answer questions 1–3. Then they **practice** writing equations for displacement reactions using Worksheet 3.14. [O1]

Standard and higher demand

- Use slides 4–6 of Presentation 3.14.1 'Displacement' to introduce the ionic equation for the displacement of copper by iron. Students read 'Ionic and half equations' in Section 3.14 in the Student Book and answer questions 4–6. [O3]

Extend

- Ask students able to progress further to write ionic equations for the displacement reactions they witnessed. [O3]

Plenary suggestion

Use Presentation 3.14.2 'One of these is not like the others' to display groups of three formulae and ask them to say why one of them is not like the others. Encourage students to find as many possible answers as they can, e.g. in slide 1 Cu^{2+} is the only ion, the only one that is part of a compound, the only one that dissolves and the only one that can be reduced or displaced; in slide 2, Cu^{2+} is the only one that can be reduced by carbon and the only one that can be displaced by zinc, iron or magnesium; in slide 3, $MgSO_4$ is the only one that zinc or iron cannot displace from its solution.

Answers to Worksheet 3.14

1. Magnesium + silver nitrate → magnesium nitrate + silver; 2. Copper nitrate + zinc → zinc nitrate + copper; 3. Silver nitrate + iron → iron nitrate + silver; 4. Copper + magnesium sulfate → no reaction; 5. $ZnSO_4$ + Mg → $MgSO_4$ + Zn; 6. Cu + $MgSO_4$ → no reaction; 7. Mg + $FeSO_4$ → $MgSO_4$ + Fe; 8. $CuSO_4$ + Fe → $FeSO_4$ + Cu; 9. Zn + $PbSO_4$ → Pb + $ZnSO_4$; 10. $PbSO_4$ + Fe → Fe SO_4 + Pb

Answers to Practical sheet 3.14

A more reactive metal displaces a less reactive metal from its compound: so changes occur in the magnesium well in row 1, the magnesium, zinc and lead wells in row 2 and the magnesium and zinc wells in row 3 of the spotting tile; Mg + $ZnSO_4$ → $MgSO_4$ + Zn, Mg + $CuSO_4$ → $MgSO_4$ + Cu, Zn + $CuSO_4$ → $ZnSO_4$ + Cu, Pb + $CuSO_4$ → $PbSO_4$ + Cu, Mg + $PbSO_4$ → $MgSO_4$ + Pb, Zn + $PbSO_4$ → $ZnSO_4$ + Pb

© HarperCollins*Publishers* Limited 2016

Lesson 15: Key concept: Electron transfer, oxidation and reduction

Lesson overview

OCR Specification reference

OCR C3.3

Learning objectives

- Review ion formation.
- Classify half equations as oxidation or reduction.
- Review patterns in reactivity.

Learning outcomes

- Explain why atoms lose or gain electrons. [O1]
- Explain oxidation and reduction by electron transfer. [O2]
- Relate the ease of losing electrons to reactivity. [O3]

Skills development

- WS 1.2 Use models in explanations.
- WS 1.2 Make predictions based on a model.
- WS 4.1 Use scientific vocabulary, terminology and definitions.

Resources needed Periodic tables; Worksheet 3.15

Digital resources Presentation 3.15 'Oxidation and reduction'

Key vocabulary oxidation, reduction, electron loss, electron gain

Teaching and learning

Engage

Display slide 1 of Presentation 3.15 'Oxidation and reduction' and ask groups to **decide** how many of the statements are true. [O1, O2]

Challenge and develop

- Check that students **recognise** that only the first and last statements are true. Slides 2–5 of Presentation 3.15 'Oxidation and reduction' show the correct responses. [O1, O2]
- Students read 'Losing and gaining electrons' in Section 3.15 of the Student Book and complete questions 1 and 2. [O1]

Explain

Low and standard demand

- Students complete Worksheet 3.15. [O1, O2]

High demand

- Students draw dot-and-cross diagrams to show how ions form when magnesium reacts with oxygen. [O1, O2]

Consolidate and apply

- Students read 'Oxidation and reduction' and 'Redox reactions and ease of transfer' in Section 3.15 of the Student Book and complete questions 3–5. [O2, O3]

- If appropriate, use a video clip to summarise the reasons for the trends in the reactivity of Group 1 and Group 7 elements [Google search string:] Bitesize GCSE Chemistry Reactivity of Group 1 and Group 7 elements. [O3]

Extend

- Ask students able to progress further to write half equations for the reactions in question 4. [O2]

Plenary suggestion

Ask students to write one thing about oxidation and reduction they are sure of, one thing they are unsure of and one thing they need to know more about. Then let groups of five or six students **formulate** group lists to **feed back** to the rest of the class.

Answers to Worksheet 3.15

Electrons should be added to each electron shell to match the electron configurations shown

118   © HarperCollins*Publishers* Limited 2016

Lesson 16: Neutralisation of acids and salt production

Lesson overview

OCR Specification reference

OCR C3.3

Learning objectives

- React an acid and an alkali to make a salt.
- Predict the formulas of salts.
- Write balanced symbol equations.

Learning outcomes

- Describe one way that salts can be made. [O1]
- Predict products from given reactants. [O2]
- Deduce the formulae of salts from the formulae of common ions. [O3]

Skills development

- WS 1.2 Use models in explanations.
- WS 1.4 Describe and explain how salts are formed.
- WS 2.2 Describe how a soluble salt can be made from an acid.

Resources needed Equipment as listed in Technician's notes 3.16; Practical sheet 3.16; Worksheets 3.16.1 and 3.16.2

Key vocabulary alkali, base, carbonate, salt

Teaching and learning

Engage

- Ask each group to collect a burette and stand and elicit ideas about what it used for. Students should **recognise** that the scale is designed to measure the volume of a liquid released from it. [O1]

Challenge and develop

- Demonstrate how to prepare a burette safely by fixing it below eye level. Use a small funnel to pour in a few cubic centimetres of 0.4 mol/dm^3 hydrochloric acid with the tap open and a beaker under it. Then, when the tip of the burette is full of solution, close the tap and add more solution up to the zero mark. [O1]

- Show the colours of methyl orange in acidic and alkaline solutions and demonstrate how to swirl a flask as acid is added to alkali to monitor the colour continuously. [O1]

- Pairs prepare sodium chloride crystals **following** the method in Practical sheet 3.16. [O1]

Explain

Low demand

- Students complete Worksheet 3.16.1 to **review** the method they used. [O1]

Standard and high demand

- Students **describe** how they obtained pure crystals of sodium chloride. [O1]

Consolidate and apply

Low and standard demand

- Students complete Worksheet 3.16.2 to **practice** writing equations for salt formation. [O2]

 © HarperCollins*Publishers* Limited 2016

Standard and high demand

- Students read 'Formulae of salts from ions' from Section 3.16 of the Student Book and answer questions 3–6. [O3]

Extend

- Ask students able to progress further to write the formulae of the salts aluminium forms with each acid. [O3]

Plenary suggestion

Students could **peer mark** each other's descriptions of making sodium chloride for accuracy and clarity.

Answers to Worksheet 3.16.1

1. Correct labels; 2. Correct labels; 3. The indicator changes colour; 4. Neutralising sodium hydroxide with indicator present shows the volume of hydrochloric acid needed; adding the same volume of acid with no indicator present allows pure sodium chloride to be made

Answers to Worksheet 3.16.2

1. $KOH + HCl \rightarrow KCl + H_2O$; 2. $2NaOH + H_2SO_4 \rightarrow Na_2SO_4 + 2H_2O$; 3. $LiOH + HNO_3 \rightarrow LiNO_3 + H_2O$; 4. $NaOH + HCl \rightarrow NaCl + H_2O$; 5. $2KOH + H_2SO_4 \rightarrow K_2SO_4 + 2H_2O$; 6. $NaOH + HNO_3 \rightarrow NaNO_3 + H_2O$; 7. $2LiOH + H_2SO_4 \rightarrow Li_2SO_4 + 2H_2O$; 8. $KOH + HNO_3 \rightarrow KNO_3 + H_2O$; 9. $LiOH + HCl \rightarrow LiCl + H_2O$

Lesson 17: Soluble salts

Lesson overview

OCR Specification reference

OCR C3.3

Learning objectives

- React an acid and a metal to make a salt.
- Predict the formulae of salts.
- Write balanced symbol equations and half equations.

Learning outcomes

- Describe how to make pure, dry samples of soluble salts. [O1]
- Explain how to name a salt. [O2]
- Derive a formula for a salt from its ions. [O3]

Skills development

- WS 1.2 Use models in explanations.
- WS 1.4 Describe and explain how salts are formed.
- WS 2.2 Describe how a soluble salt can be made from an acid.

Resources needed Equipment as listed in Technician's notes 3.17; Practical sheet 3.17; Worksheets 3.17.1 and 3.17.2

Digital resources Presentation 3.17.1 'Copper sulfate'; Presentation 3.17.2 'Name those reactants'

Key vocabulary concentrate, crystallise, filter, filtrate

Teaching and learning

Engage

- Display Presentation 3.17.1 'Copper sulfate' and elicit ideas about how copper sulfate could be made. Students should **recognise** that copper will not react with sulfuric acid because it is less reactive than hydrogen. [O1]

Challenge and develop

- Explain that copper(II) oxide is a base like sodium hydroxide, but insoluble. Then give students the choice of making copper(II) sulfate from either copper(II) oxide or copper(II) carbonate. [O1, O2]
- Pairs **prepare** crystals of copper(II) sulfate from their chosen starting materials using Practical sheet 3.17 as a guide. [O1, O2]

Explain

Low demand

- Students complete Worksheet 3.17.1 to **review** the methods that can be used to make copper sulfate. [O1]

Standard and high demand

    Students **describe** how they obtained pure crystals of copper sulfate. [O1]

Consolidate and apply

Low demand

- Students **describe** how they made their salt crystals. [O1] They then complete Worksheet 3.17.2 to **practice** writing equations for salt formation. [O2]

Standard and high demand

- Students read Section 3.17 from the Student Book and answer questions 3–6. [O3]

Extend

- Ask students able to progress further to write symbol equations for the reactions of zinc oxide and magnesium carbonate with sulfuric, hydrochloric and nitric acid. [O2]

Plenary suggestion

Challenge students to list the reactants needed to make the salts shown in Presentation 3.17.2 'Name those reactants'.

Answers to Worksheet 3.17.1

1. An insoluble base; 2. Once the acid has been used up; no more black copper oxide can react; 3. Copper sulfate and water; 4. *Similarities:* both produce the same salt and water; and excess of each powder can be filtered out; *Differences:* the carbonate releases carbon dioxide when it reacts with acid; but the oxide doesn't; so only the carbonate bubbles as it reacts

Answers to Worksheet 3.17.2

1. $MgO + 2HCl \rightarrow MgCl_2 + H_2O$; 2. $ZnO + H_2SO_4 \rightarrow ZnSO_4 + H_2O$; 3. $MgO + H_2SO_4 \rightarrow MgSO_4 + H_2O$; 4. $CaCO_3 + H_2SO_4 \rightarrow CaSO_4 + H_2O + CO_2$; 5. $ZnCO_3 + 2HCl \rightarrow ZnCl_2 + H_2O + CO_2$; 6. $MgCO_3 + H_2SO_4 \rightarrow MgSO_4 + H_2O + CO_2$; 7. $MgCO_3 + 2HCl \rightarrow MgCl_2 + H_2O + CO_2$

 © HarperCollins*Publishers* Limited 2016

Lesson 18: Reactions of metals with acids

Lesson overview

OCR Specification reference

OCR C3.3

Learning objectives

- React an acid and a metal to make a salt.
- Predict the formulas of salts.
- Write balanced symbol equations and half equations.

Learning outcomes

- Describe how to make salts from metals and acids. [O1]
- Write full balanced symbol equations for making salts. [O2]
- Use half equations to describe oxidation and reduction. [O3]

Skills development

- WS 1.2 Use models in explanations.
- WS 1.4 Describe and explain how salts are formed.
- WS 2.2 Describe how a soluble salt can be made from an acid.

Resources needed Equipment as listed in Technician's notes 3.18; Practical sheet 3.18; Worksheets 3.18.1 and 3.18.2

Digital resources Presentation 3.18.1 'Acids'; Presentation 3.18.2 'Half equations'

Key vocabulary acid, half equation, metal, salt

Teaching and learning

Engage

- Display Presentation 3.18.1 'Acids' and elicit ideas about why hydrochloric acid, sulfuric acid and nitric acid all have the same effect on universal indicator even though their chemical formulae are very different.

Challenge and develop

- Explain that each 'H' in an acid's formula represents a hydrogen ion (H^+) and that these give acids their properties. Emphasise that the positive charge is not shown in the formula of an acid because the negative charge on the rest of the acid cancels it out. [O1, O2]

- Use slide 2 of Presentation 3.18.1 'Acids' to show that salts form when the hydrogen in an acid is replaced by a metal. [O1, O2]

- Pairs prepare crystals of zinc sulfate by reacting zinc with sulfuric acid following the guidance in Practical sheet 3.18. [O1, O2]

Explain

Low demand

- Students complete Worksheet 3.18.1 to **review** the method they used to make zinc sulfate. [O1]

Standard and high demand

- Students use diagrams to **explain** how they obtained pure crystals of zinc sulfate by reacting zinc with sulfuric acid. [O1]

Consolidate and apply

Low demand

- Students complete Worksheet 3.18.2 to **practice** writing symbol equations for reactions between acids and metals. [O2]

Standard and high demand

- Use Presentation 3.18.2 'Half equations' to point out that magnesium atoms are oxidised to ions as the metal reacts with acid, and the hydrogen ions in the acid are reduced. Students read 'Forming ions' in Section 3.18 of the Student Book and answer questions 3–7. [O3]

Extend

- Ask students able to progress further to write symbol and ionic equations for the reaction of aluminium with hydrochloric acid and sulfuric acid. [O2]

Plenary suggestion

Challenge groups to **discuss** three things they didn't know an hour ago.

Answers to Worksheet 3.18.1

1.They all contain hydrogen; 2. Hydrogen; 3. To make sure all the acid reacts; 4. The excess metal is filtered out; 5. The solution is evaporated to make it more concentrated; and then left to crystallise; 6. a) Zinc + hydrochloric acid → zinc chloride + hydrogen; b) magnesium + sulfuric acid → magnesium sulfate + hydrogen; c) hydrochloric acid + iron → iron chloride + hydrogen

Answers to Worksheet 3.18.2

1. $Zn + H_2SO_4 \rightarrow \mathbf{ZnSO_4} + H_2$; 2. $Fe + 2HCl \rightarrow FeCl_2 + H_2$; 3. $Ca + H_2SO_4 \rightarrow \mathbf{CaSO_4 + H_2}$; 4. $\mathbf{Ni + H_2SO_4} \rightarrow NiSO_4 + H_2$; 5. $\mathbf{Co + 2HCl} \rightarrow CoCl_2 + H_2$; 6. $Fe + \mathbf{H_2SO_4} \rightarrow FeSO_4 + H_2$; 7. $Mn + H_2SO_4 \rightarrow \mathbf{MnSO_4 + H_2}$; 8. $Ca + 2HCl \rightarrow \mathbf{CaCl_2 + H_2}$; 9. $Zn + \mathbf{2HCl} \rightarrow ZnCl_2 + H_2$; 10. $Ni + 2HCl \rightarrow \mathbf{NiCl_2 + H_2}$; 11. $Be + H_2SO_4 \rightarrow \mathbf{BeSO_4 + H_2}$

© HarperCollins*Publishers* Limited 2016

Lesson 19: Practical: Preparing a pure, dry sample of a soluble salt from a soluble oxide or carbonate

OCR Specification reference

OCR C3.3

Lesson overview

Learning objectives

- React a carbonate with an acid to make a salt.
- Describe each step in the procedure.
- Determine the purity of the product.

Learning outcomes

- Describe a practical procedure for producing a salt using a solid and an acid. [O1]
- Explain the apparatus, materials and techniques used for making the salt. [O2]
- Describe how to manipulate apparatus safely and accurately a measure melting point. [O3]

Skills development

- WS 2.3 Select the apparatus needed to prepare a soluble salt.
- WS 2.4 Carry out experiments appropriately.
- WS 3.8 Communicate the scientific rationale for the method used to make a soluble salt.

Resources needed Equipment as listed in Technician's notes 3.19; Practical sheet 3.19

Digital resources Presentation 3.19.1 'Magnesium sulfate'; Presentation 3.19.2 'Why?'

Key vocabulary insoluble, oxide, carbonate, filter, crystallisation

Teaching and learning

Engage

- Display Presentation 3.19.1 'Magnesium sulfate' and elicit ideas about how magnesium sulfate could be made. Students should **recognise** that the reaction could be done using magnesium ribbon, magnesium oxide or magnesium carbonate. [O1]

Challenge and develop

- Explain that magnesium carbonate will be used to prepare the salt and ask groups to **select** the apparatus they need. [O1]

Low demand

- Pairs prepare crystals of magnesium(II) sulfate using Practical sheet 3.19 as a guide. [O1]

Standard and high demand

- Pairs prepare an outline plan and check it with you before beginning to prepare their magnesium sulfate. [O1]

Explain

- Students read 'Carrying out procedures' and 'Using apparatus and techniques' in Section 3.19 of the Student Book and answer questions 1–9. [O1, O2]

Consolidate and apply

- Students read 'Measuring accurately' in Section 3.19 of the Student Book and answer questions 10–15. [O3]

Extend

- Ask students able to progress further to answer question 16 in the Student Book. [O3]

Plenary suggestion

Challenge groups to **explain** why the steps listed in Presentation 3.19.2 'Why?' are needed.

Lesson 20: pH and neutralisation

Lesson overview

OCR Specification reference

OCR C3.3

Learning objectives

- Estimate the pH of solutions.
- Identify weak and strong acids and alkalis.
- Investigate pH changes when a strong acid neutralises a strong alkali.

Learning outcomes

- Describe the use of universal indicator to measure pH. [O1]
- Use the pH scale to identify acidic or alkaline solutions. [O2]
- Recognise how the pH changes when a strong acid neutralises a strong alkali. [O3]

Skills development

- WS 1.2 Use models in explanations.
- WS 3.5 Interpret observations.
- WS 4.1 Use scientific vocabulary, terminology and definitions.

Resources needed Equipment as listed in Technician's notes 3.20.1, 3.20.2 and 3.20.3; Practical sheet 3.20; Worksheet 3.20

Digital resources Presentation 3.20 'Using indicators'

Key vocabulary hydroxyl ions, neutralisation, pH, universal indicator

Teaching and learning

Engage

- Remind pupils of their previous experiences with universal indicator by adding some to flasks of water, acids and alkalis to demonstrate the colours produced. A video clip could also be used [Google search string:] Identify which solution is more acidic or alkaline using pH values.

Challenge and develop

- Pairs use universal indicator to **classify** solutions as weak or strong acids or alkalis. Section 3.20 in the Student Book structures the task. [O1, O2]

- Introduce the idea that acidic and alkaline solutions can neutralise each other. Add some indicator to 5 cm^3 of alkali and demonstrate how to use a calibrated dropping pipette to add 1 cm^3 aliquots of an acid to it. Pairs can then **observe** how the pH changes when a strong acid is added to a strong alkali, and vice versa, by completing Practical sheet 3.20. [O3]

Explain

- Use Presentation 3.20 'Using indicators' to **show** that all alkalis contain hydroxide ions (OH^-). Give each group a model hydrogen ion from an acid and a model hydroxide ion from an alkali and elicit students' ideas about how they are able to neutralise each other. They should **recognise** that the ions combine to make water molecules. They can **model** neutralisation by sticking the ions together with Blu-Tack and turning them over to show the uncharged H_2O molecules formed.

- An animation could be used to help students to **visualise** the changes taking place during a neutralisation reaction [Google search string:] neutralisation.swf.

Consolidate and apply

- Explain that a pH probe can also be used to measure pH changes. Run sodium hydroxide solution into hydrochloric acid as detailed in Technician's notes 3.20.1 to demonstrate the pH changes during the reaction. As an alternative, a virtual titration could be used to confirm the pH changes that occur when a strong acid is neutralised by a strong alkali [Google search string:] Acid–base titration (New).swf.

- Use Presentation 3.20 'Using indicators' to emphasise that the ionic equation is the same when any acid is neutralised by any alkali.

Low and standard demand

- Students complete Worksheet 3.20. [O3]

High demand

- Ask students to **explain** why the same ionic equation can be used when any acid is neutralised by any alkali. [O3]

Extend

- Ask students able to progress further to find the pH values at which phenolphthalein and methyl orange indicators change colour and **explain** why they are more useful than universal indicator for determining when a strong acid has been neutralised by a strong alkali. [O3]

Plenary suggestion

Show a video clip of solid carbon dioxide being added to an alkaline solution and ask them to **create** a 'voiceover' for the video [Google search string:] Universal indicator with dry ice.

Answers to Worksheet 3.20

1. H^+; 2. OH^-; 3. H^+ and OH^- ions combine; 4. a) 1; b) 25 cm^3; c) $H^+(aq) + OH^-(aq) \rightarrow H_2O(l)$

Lesson 21: Strong and weak acids

Lesson overview

OCR Specification reference

OCR C3.3

Learning objectives

- Explore the factors that affect the pH of an acid.
- Find out how the pH changes when an acid is diluted.
- Find out how the concentrations of solutions are measured.

Learning outcomes

- Explain weak and strong acids in terms of their degree of ionisation. [O1]
- Explain neutrality and acidity in terms of hydrogen ion concentrations and pH. [O2]
- Explain dilute and concentrated in terms of the amount of substance per dm^3. [O3]

Skills development

- WS 1.2 Use models in explanations.
- WS 1.4 Describe and explain how the strength and concentration of an acid affects its pH.
- WS 4.1 Use scientific vocabulary, terminology and definitions.

Maths focus Make order of magnitude calculations

Resources needed Technician's notes 3.21; Worksheet 3.21

Digital resources Presentation 3.21 'Weak or strong'

Key vocabulary concentration, dilution, ionisation, strength

Teaching and learning

Engage

- Display slide 1 of Presentation 3.21 'Weak or strong' and elicit ideas about how hydrochloric acid and ethanoic acid can have the same concentration but different pH values. [O1]

Challenge and develop

- Demonstrate that magnesium ribbon reacts much faster with hydrochloric acid than with ethanoic acid. [O1]

- Compare the formulae of the two acids using slide 2 of Presentation 3.21 'Weak or strong' and/or space-filling models and remind students that hydrochloric acid is completely ionised and that the hydrogen ions are responsible for its acidity. Show students how ethanoic acid ionises and explain that only about 1% of the molecules in the ethanoic acid solution are ionised. [O1]

- A video clip could also be used to reinforce this idea [Google search string:] Distinguish strong and weak acids. [O1]

Explain

Low and standard demand

- Students read 'Ionisation' in Section 3.21 of the Student Book and answer questions 1 and 2. They then **describe** the differences between dilute and concentrated acids and between strong and weak acids. [O1]

High demand

- Students **explain** why it is possible for a concentrated weak acid to have a similar pH to a dilute strong acid. [O1]

Consolidate and apply

Low demand

* Students use the 'pH, neutralisation and titration curves' part of Section 3.21 in the Student Book to complete Worksheet 3.21. [O2]

Standard and high demand

* Students use the 'pH, neutralisation and titration curves' part of Section 3.21 in the Student Book to **remind** themselves that the concentrations of H^+ and OH^- ions are equal at pH 7. [O2]

* Students read 'Concentration' in Section 3.21 and answer question 3. [O3]

Extend

* Ask students able to progress further to **investigate** indicator pH ranges [Google search string:] Indicators – Royal Society of Chemistry and **explain** why different indicators are needed for different acid–base titrations. [O2]

Plenary suggestion

Use slide 3 of Presentation 3.21 'Weak or strong' to check that students can distinguish weak and strong acids.

Answers to Worksheet 3.21

1. Hydroxide, OH^-; 2. Neutralisation; H^+ and OH^- ions combine to form water, $H^+ + OH^- \rightarrow H_2O$; 3. The labelled diagram should resemble Figure 3.67 in the Student Book; 4. 24 cm^3; 5. Slightly more concentrated; 6. The curve would end at a higher pH (e.g. pH 4 or 5); 7.The curve would start at a lower pH (e.g. pH 10 or 11)

Lesson 22: Maths skills: Make order of magnitude calculations

Lesson overview

OCR Specification reference

OCR C3.3

Learning objectives

- Explore the factors that affect the acidity of rain.
- Find out how acid concentrations are compared.
- Explore the link between hydrogen ion concentration and pH.

Learning outcomes

- Use graphs and diagrams to apply the pH scale to acid rain distribution. [O1]
- Calculate the concentration of acids. [O2]
- Calculate the effect of hydrogen ion concentration on the numerical value of pH. [O3]

Skills development

- WS 1.2 Use models in explanations.
- WS 1.4 Describe and explain how the strength and concentration of an acid affects its pH.
- WS 4.1 Use scientific vocabulary, terminology and definitions.

Maths focus Make order of magnitude calculations

Resources needed Equipment as listed in Technician's notes 3.22; Practical sheet 3.22

Digital resources Presentation 3.22 'Acid rain'

Key vocabulary magnitude, logarithmic scale, derived, relative

Teaching and learning

Engage

- Display slide 1 of Presentation 3.22 'Acid rain' and check that students **remember** what causes acid rain. Then display slide 2 to show that the contribution of sulfur dioxide emissions to the acidity of rain has fallen in recent years. [O1]

Challenge and develop

- Explain that the amount of acid in samples of rain water can be compared by measuring their hydrogen ion concentrations or their pH values. [O2]
- Students follow the guidance on Practical sheet 3.22 to **explore** the pH changes that take place when a strong acid and a strong alkali are diluted by factors of 10. [O2]

Explain

- Students should **conclude** that the pH increases by one unit as the hydrogen ion concentration decreases by a factor of 10. They may also spot that the hydrogen ion and hydroxide ion concentrations are equal at pH 7. [O2]

Consolidate and apply

Low and standard demand

- Students compare the specimen results in Section 3.22 of the Student Book with their own and answer questions 5 and 6. [O3]

Standard and high demand

- Students read 'Concentrations of acid' in Section 3.22 of the Student Book and answer questions 3 and 4. [O2]

Extend

- Ask students able to progress further to watch a video exploring the exact relationship between pH and hydrogen ion concentration [Google search string:] 8.4.3 Each change of one pH unit represents a 10-fold change in [H+]. [O2]

Plenary suggestion

Students write one thing about pH or acid concentrations they are sure about, one thing they are unsure of and one thing they need to know more about. They then get into groups to **formulate** a group list to **feed back** to the rest of the class.

Lesson 23: Practical: Investigate the variables that affect temperature changes in reacting solutions such as acid plus metals, acid plus carbonates, neutralisations, displacement of metals

Lesson overview

OCR Specification reference

OCR C3.3

Learning objectives

- Devise a hypothesis.
- Devise an investigation to test your hypothesis.
- Decide whether the evidence supports your hypothesis.

Learning outcomes

- Use scientific theories and explanations to develop hypotheses. [O1]
- Plan experiments to make observations and test hypotheses. [O2]
- Evaluate methods to suggest possible improvements and further investigations. [O3]

Skills development

- WS 2.1 Suggest a hypothesis to explain given observations.
- WS 2.2 Apply understanding of apparatus and techniques to suggest a procedure.
- WS 2.6 Read measurements off a scale in a practical context.

Maths focus Translate information between numerical and graphical form

Resources needed Equipment as listed in Technician's notes 3.23; Practical sheet 3.23

Digital resources Presentation 3.23 'Comparing reactions'

Key vocabulary temperature, carbonate, thermometer, neutralisation

Teaching and learning

Engage

- Display Presentation 3.23 'Comparing reactions' and challenge groups to **suggest** as many variables as possible that could affect the temperature change achieved in a heat pack. [O1]

Challenge and develop

- Check that students **appreciate** that the masses of reactants, concentrations of solutions and the chemicals used could all affect the temperature change. Then show the range of chemicals available for them to test.

Low and standard demand

- Students read 'Developing a hypothesis' and 'Planning an investigation' in Section 3.23 in the Student Book and answer questions 1–5. Then pairs **plan** their own investigation using Practical sheet 3.23 as a guide. [O1, O2]

Higher demand

- Each student pair could **devise** a hypothesis and write a plan for an investigation. [O1, O2]
- Check each pair's plan before allowing them to collect any apparatus.

Explain

- Students **explain** how the evidence they have collected supports or refutes the hypothesis they tested. [O2]

Consolidate and apply

- Challenge groups to **suggest** ways of collecting stronger evidence to support their hypothesis. [O3]

Extend

- Ask students able to progress further to **suggest** further investigations. [O3]

Plenary suggestion

Ask pairs to **present** the hypothesis they tested and the conclusions they made to the rest of the class.

Answers to Practical sheet 3.23

Answers depend on the reaction chosen and the independent variable selected to investigate

Lesson 24: The process of electrolysis

Lesson overview

OCR Specification reference

OCR C3.4

Learning objectives

- Explore what happens when a current passes through a solution of ions.
- Find out what an electrolyte is and what happens when it conducts electricity.
- Find out how electricity decomposes compounds.

Learning outcomes

- Describe the electrolyte, anode and cathode in electrolysis. [O1]
- Identify reactions at electrodes during electrolysis. [O2]
- Write and balance half equations for the electrode reactions. [O3]

Skills development

- WS 1.2 Use models in explanations.
- WS 1.4 Explain how electrolysis works and why it is useful.
- WS 4.1 Use scientific vocabulary, terminology and definitions.

Resources needed Equipment as listed in Technician's notes 3.24; Practical sheet 3.24; Worksheet 3.24

Digital resources Presentation 3.24 'Moving ions'

Key vocabulary electrode, electrolysis, electrolyte, ion migration

Teaching and learning

Engage

- Display slide 1 of Presentation 3.24 'Moving ions' and explain that any conductor attached to a power supply becomes an electrode. The power supply pulls electrons off one electrode and pushes them onto the other. Introduce the terms 'anode' and 'cathode' and ask groups to **discuss** what would happen if an ionic compound was placed between the anode and cathode (solid sodium chloride appears between the electrodes in slide 2.) [O1]

Challenge and develop

- Check that students **recognise** that opposite charges attract, and the ions will tend to move because they are so much smaller than the electrodes. However, the attractive force would not be enough to pull the ions in a solid apart. [O1, O2]

- Students **explore** the effect of passing an electric current through a solution of potassium manganate(VII) following the guidance in Practical sheet 3.24. Set up one set of apparatus with dry filter paper as a control to show that the ions cannot move unless they have water to dissolve in. [O1]

Explain

- Display the results of the control experiment with dry potassium manganate(VII) for comparison with the students' observations and use slide 3 of Presentation 3.24 'Moving ions' to introduce the terms 'cation' and 'anion' and confirm that both ions in their experiment moved, even though only the manganate(VII) ions were visible. [O1]

- A video clip could be used to show two ions moving in opposite directions [Google search string:] Migration of ions in copper(II) chromate(VI) time lapse. [O1]

Consolidate and apply

- Use slides 4 and 5 of Presentation 3.24 'Moving ions' to show how ions are converted into atoms at each electrode. [O2]

Low demand

- Students complete Worksheet 3.24 to review the lesson. [O2]

Standard and high demand

- Students read Section 3.24 in the Student Book and answer questions 1–4. [O1, O2, O3]

Extend

Ask students able to progress further to write half equations for the reactions that occur at the electrodes when other molten compounds are electrolysed. [O3]

Plenary suggestion

Challenge pairs to **create** sentences using each of the words 'cation', 'anode' and 'electrolyte', and then join up into larger groups and choose the best sentences to feed back to the class.

Answers to Worksheet 3.24

1. The ions in solid sodium chloride are not free to move; 2. Appropriate labelling; 3. It receives an electron; and becomes an atom; 4. They lose one electron each; to become atoms and these pair up to make molecules; 5. Because electricity is used to decompose electrolytes; into their elements; 6. Lead would form at the negative cathode; and bromine at the positive anode

© HarperCollins*Publishers* Limited 2016

Lesson 25: Electrolysis of molten ionic compounds

Lesson overview

OCR Specification reference

OCR C3.4

Learning objectives

- Look in detail at the electrolysis of lead bromide.
- Communicate the science behind the extraction of elements from molten salts.
- Write balanced half equations for electrolysis reactions.

Learning outcomes

- Identify which ions migrate to the cathode and which to the anode. [O1]
- Explain how the ions of a molten electrolyte are discharged. [O2]
- Predict the products of electrolysis of molten binary compounds. [O3]

Skills development

- WS 1.2 Use models in explanations.
- WS 1.4 Describe and explain how elements can be extracted from molten compounds.
- WS 4.1 Use scientific vocabulary, terminology and definitions.

Resources needed Worksheet 3.25

Digital resources Presentation 3.25 'Electrolysis of molten compounds'

Key vocabulary anode, cathode, discharged, molten

Teaching and learning

Engage

Display Presentation 3.25 'Electrolysis of molten compounds' and ask groups to **decide** how many of the statements are true.

Challenge and develop

- Check that students **recognise** that electrolysis decomposes binary ionic compounds into their elements and that the process is expensive because most ionic compounds melt at only very high temperatures. Despite the expense, electrolysis is widely used to extract metals that are too reactive to be reduced by carbon. [O2]
- Stress that lead bromide is unusual because it can be melted in a Bunsen burner flame and show a video clip of this demonstration experiment [Google search string:] GCSE Science Revision Electrolysis of molten lead bromide. [O1, O2, O3]

Explain

- Students **create** posters, animations or videos to **explain** how molten ionic compounds are decomposed by electrolysis using Section 3.25 of the Student Book as a reference. [O1, O2, O3]

Consolidate and apply

- Use a video clip to review the reactions taking place at the electrodes when molten lead bromide is electrolysed [Google search string:] Electrolysis of Molten Compounds | Chemistry for All | The Fuse School. [O1, O2, O3]

Low demand

- Students complete Worksheet 3.25 to **review** the electrolysis of lead bromide. [O1, O2, O3]

Standard and high demand

- Students read 'Electrode half equations' in Section 3.25 of the Student Book and answer question 4. [O2, O3]

Extend

Ask students able to progress further to write half equations for the reactions that occur at the electrodes when these molten compounds are electrolysed: NaCl, $MgCl_2$, $AlCl_3$.[O2, O3]

Plenary suggestion

Groups take turns to **present** their posters, animations or videos and **offer** constructive comments to other groups.

Answers to Worksheet 3.25

1. Arrows should point from Br^- ions to the positive anode; and from Pb^{2+} ions to the cathode; 2. Positive anode; negative cathode; 3.Molten lead bromide; 4. Lead ions; attracted to the negative cathode where they gain electrons and are discharged to form a layer of lead over the electrode; 5. Bromide ions are discharged at the anode; and form bromine, a toxic liquid that is very volatile; 6. Brown vapour at the anode; a silver bead of molten lead under the cathode

Lesson 26: Electrolysis of aqueous solutions

Lesson overview

OCR Specification reference

OCR C3.4

Learning objectives

- Investigate the products formed when copper sulfate is electrolysed.
- Predict what products other solutions will give.
- Write half equations for reactions at electrodes.

Learning outcomes

- Explain the electrolysis of copper(II) sulfate using inert electrodes. [O1]
- Predict the products of the electrolysis of aqueous solutions. [O2]
- Represent reactions at electrodes by half equations. [O3]

Skills development

- WS 1.2 Use models in explanations.
- WS 1.4 Describe and explain how useful substances are obtained by electrolysis .
- WS 4.1 Use scientific vocabulary, terminology and definitions .

Resources needed Equipment as listed in Technician's notes 3.26; Worksheet 3.26; Practical sheet 3.26

Digital resources Presentation 3.26 'Copper(II) sulfate'

Key vocabulary electrode reaction, half equation, preferential discharge, relative reactivity

Teaching and learning

Engage

- Display slide 1 of Presentation 3.26 'Copper(II) sulfate' and ask students to **predict** what products will form at the electrodes when a solution of copper(II) sulfate is electrolysed. [O2]

Challenge and develop

- Students carry out the electrolysis of aqueous copper(II) sulfate following the guidance in Practical sheet 3.26. [O1]

Explain

- A video clip could be used to review students' observations [Google search string:] Electrolysis of copper(II) sulfate solution with graphite electrodes. [O1]
- Use slides 2 and 3 of Presentation 3.26 'Copper(II) sulfate' to confirm their observations and show half equations for the reactions at each electrode. [O2, O3]

Consolidate and apply

Low demand

- Students complete Worksheet 3.26 to consolidate their ideas about electrolysis. [O2]

Standard and high demand

- Use slides 4 and 5 of Presentation 3.26 'Copper(II) sulfate' to stress that, because electrons flow from the anode to the cathode, the numbers of electrons in each half equation have to balance. Oxidation at the anode supplies electrons for the reduction reaction at the cathode. [O3]
- Read Section 3.26 in the Student Book and answer questions 1–4. [O2, O3]

Extend

Ask students able to progress further to **describe** and **explain** what happens when copper(II) sulfate solution is electrolysed using pre-weighed copper electrodes.

Plenary suggestion

Students write one thing about electrolysis they are sure about, one thing they are unsure of and one thing they need to know more about. They then get into groups to **formulate** a group list to **feed back** to the rest of the class.

Answers to Worksheet 3.26

1. Cu^{2+}; H^+; 2. Cu^{2+}; 3. OH^- ions lose electrons; and release oxygen gas (which can react with the carbon electrode to make carbon dioxide); 4. Oxygen relights a glowing splint. 5. H^+ and SO_4^{2-} ions are left in solution; the H^+ ions make the solution acidic

Lesson 27: Practical: Investigating what happens when aqueous solutions are electrolysed using inert electrodes

Lesson overview

OCR Specification reference

OCR 3.4

Learning objectives

- Devise a hypothesis.
- Devise an investigation to test your hypothesis.
- Decide whether the evidence supports your hypothesis.

Learning outcomes

- Use scientific theories and explanations to develop hypotheses. [O1]
- Plan experiments to make observations and test hypotheses. [O2]
- Apply a knowledge of the apparatus needed for electrolysis. [O3]

Skills development

- WS 1.2 Use models in explanations.
- WS 2.1 Suggest a hypothesis to explain given observations.
- WS 2.2 Apply understanding of apparatus and techniques to suggest a procedure.

Resources needed Equipment as listed in Technician's notes 3.27; Practical sheet 3.27

Digital resources Presentation 3.27.1 'Developing hypotheses'; Presentation 3.27.2 'Competing ions'

Key vocabulary electrolysis, electrode, inert, electrolyte, cathode, anode

Teaching and learning

Engage

- Display slide 1 of Presentation 3.27.1 'Developing hypotheses' to remind students what a hypothesis is. Then display slide 2 and ask groups to **devise** a hypothesis to **explain** the observations presented. [O1]

Challenge and develop

- Display slide 3 of Presentation 3.27.1 'Developing hypotheses' to prompt groups to **plan** investigations to **test** their hypotheses. [O2, O3]

Low demand

- Practical sheet 3.27 could be used to support students who find **planning** their investigation difficult. [O2, O3]

Explain

Low demand

- Ask students to **compare** their own results with those presented in Section 3.27 of the Student Book and answer questions 1–5. [O2]

Standard and high demand

- Students **explain** whether their results support or refute their group's hypothesis. [O2]

Consolidate and apply

- Use Presentation 3.27.2 'Competing ions' to confirm students' observations and show what happens at each electrode. [O2]

 © HarperCollins*Publishers* Limited 2016

Extend

- Ask students able to progress further to write half equations to show what happened at the cathode for each electrolyte they tested.

Plenary suggestion

- Students **compile** a guide to help someone with no knowledge of chemistry predict the elements formed at the anode and cathode when an aqueous solution is electrolysed.

When and how to use these pages: Check your progress, Worked example and End of chapter test

Check your progress

Check your progress is a summary of what students should know and be able to do when they have completed the chapter. Check your progress is organised in three columns to show how ideas and skills progress in sophistication. Students aiming for top grades need to have mastered all the skills and ideas articulated in the final column (shaded pink in the Student book).

Check your progress can be used for individual or class revision using any combination of the suggestions below:

- Ask students to construct a mind map linking the points in Check your progress
- Work through Check your progress as a class and note the points that need further discussion
- Ask the students to tick the boxes on the Check your progress worksheet (Teacher Pack CD). Any points they have not been confident to tick they should revisit in the Student Book.
- Ask students to do further research on the different points listed in Check your progress
- Students work in pairs and ask each other what points they think they can do and why they think they can do those, and not others

Worked example

The worked example talks students through a series of exam-style questions. Sample student answers are provided, which are annotated to show how they could be improved.

- Give students the Worked example worksheet (Teacher Pack CD). The annotation boxes on this are blank. Ask students to discuss and write their own improvements before reviewing the annotated Worked example in the Student Book. This can be done as an individual, group or class activity.

End of chapter test

The End of chapter test gives students the opportunity to practice answering the different types of questions that they will encounter in their final exams. You can use the Marking grid provided in this Teacher Pack or on the CD Rom to analyse results. This shows the Assessment Objective for each question, so you can review trends and see individual student and class performance in answering questions for the different Assessment Objectives and to highlight areas for improvement.

- Questions could be used as a test once you have completed the chapter
- Questions could be worked through as part of a revision lesson
- Ask Students to mark each other's work and then talk through the mark scheme provided
- As a class, make a list of questions that most students did not get right. Work through these as a class.

© HarperCollins*Publishers* Limited 2016

Marking Grid for Single Chemistry End of Chapter 3 Test 3816

Student Name	Q. 1 (AO1) Getting Started [Foundation Tier] 1 mark	Q. 2 (AO1) 1 mark	Q. 3 (AO1) 2 marks	Q. 4 (AO1) 1 mark	Q. 5 (AO2) 1 mark	Q. 6 (AO2) 2 marks	Q. 7 (AO1) 2 marks	Q. 8 (AO1) Going Further [Foundation and Higher Tiers] 1 mark	Q. 9 (AO1) 1 mark	Q. 10 (AO2) 2 marks	Q. 11 (AO1) 4 marks	Q. 12 (AO2) 2 marks	Q. 13 (AO1) More Challenging [Higher Tier] 1 mark	Q. 14 (AO2) 1 mark	Q. 15 (AO2) 2 marks	Q. 16 (AO2) 2 marks	Q. 17 (AO3) 4 marks	Q. 18 (AO3) Most demanding [Higher Tier] 2 marks	Q. 19 (AO3) 4 marks	Q. 20 (AO3) 4 marks	Total marks	Percentage

Check your progress

You should be able to:

state the law of the conservation of mass →	explain how to balance equations in terms of numbers of atoms on both sides of the equation →	explain the meaning of subscripts within a formula and multipliers before a formula in a balanced equation
explain that when there is a mass change in a reaction it may be because a gas is being given off →	explain why there appears to be a mass change when metal carbonates are heated or metals are heated in oxygen →	explain observed changes in mass in non-enclosed systems and explain the changes in terms of the particle model
describe the measurement of amounts of substance in moles →	calculate the number of moles in a given mass →	calculate the mass of a given number of moles
calculate the masses of substances in a balanced symbol equation →	calculate the masses of reactants and products from balanced symbol equations →	calculate the mass of a given reactant or product
identify exothermic and endothermic reactions from temperature changes →	identify exothermic reactions as causing a temperature rise →	identify endothermic reactions as causing a temperature decrease
identify examples of exothermic reactions →	identify examples of endothermic reactions →	explain and evaluate the uses of some exothermic and endothermic reactions
investigate changes in temperature of different reactions →	investigate the variables that affect temperature changes in reacting solutions →	explain how the variables investigated affect temperature changes
recognise that energy transfer during a reaction is due to bonds being broken and then new bonds being made →	describe the energy changes in bond breaking as endothermic and bond making as exothermic and explain how the energy of a reaction is calculated overall →	calculate the energy transferred in chemical reactions using bond energies
describe how to make pure, dry samples of soluble salts →	explain how to name a salt →	derive a formula for a salt from its ions
describe the use of universal indicator to measure pH →	use the pH scale to identify acidic or alkaline solutions →	investigate pH changes when a strong acid neutralises a strong alkali.
explain weak and strong acids by the degree of ionisation →	describe neutralisation through the effect on hydrogen ions and pH →	explain the terms dilute and concentrated as the amounts of substances dissolved
explain why some metals need to be extracted by electrolysis →	explain the process of the electrolysis of aluminium oxide →	explain which non-metals are formed at the anode in preference
use apparatus to electrolyse aqueous solutions in the laboratory →	explain which metals (or hydrogen) are formed at the cathode in preference →	predict the products of the electrolysis of aqueous solutions containing a single ionic compound
explain the electrolysis of copper sulfate using inert electrodes →	predict the products of the electrolysis of aqueous solutions →	represent reactions at electrodes by half equations

Worked example

Kim and Jo are electrolysing dilute sulfuric acid.

1 **Identify the substance seen at the anode.**

    a nitrogen   **b** hydrogen   **c** sulfur   (**d** oxygen)

> The answer oxygen is correct.

2 **Describe how they will test for hydrogen gas.**

    It pops with a lighted splint.

> This test is correct.

3 **Construct the half equation for the discharge of hydrogen at the electrode.**

    $H^+ + e^- \rightarrow H_2$

> The charge on the ion and the gain of an electron are correct. The molecule H_2 is correct. The equation needs to be balanced,
> $2H^+ + 2e^- \rightarrow H_2$

4 **Next Kim and Jo want to electrolyse copper sulfate. Jo says that they cannot use solid copper sulfate. Explain why.**

    The ions need to be free to move.

> This answer is partly correct. The ions need be free to move to conduct electricity, so need to be molten or in solution.

5 **They choose to use a solution. They pass a current through the solution of copper sulfate, using carbon electrodes.**

    a **Describe what will happen at the cathode.**

        It will get a coat of pink/brown copper.

> The answer is correct.

    b **Describe what they will see at the anode.**

        It will disintegrate.

> The student needs to be clear that this happens with copper electrodes, not carbon, and is used in the purification process. The answer should be that bubbles of oxygen will appear.

    c **What will they see happening to the copper sulfate solution?**

        It will stay blue.

> The answer is again confused with the purification process. The correct answer is that the blue colour disappears. This is due to the copper depositing from the solution on to the electrode.

6 **Construct the half equation for the discharge of copper at an electrode and explain whether this reaction is oxidation or reduction. Explain why copper is deposited and hydrogen is not evolved.**

    $Cu^{2+} + 2e^- \rightarrow Cu$ This is reduction as electrons are gained (RIG). Copper is deposited because it is more reactive than hydrogen.

> The half equation is correct. Reduction is correct – no need for the memory aid. Copper is less reactive than hydrogen, (which is why it is deposited).

© HarperCollins*Publishers* Limited 2016

Predicting & identifying reactions & products: Introduction

When and how to use these pages

This unit builds students understanding of patterns in the Periodic Table.

Overview of the unit

The students will explore the trends in the properties of elements in groups 1, 7, 0 and the transition metals and learn how to identify common gases, anions and cations. They will investigate methods used to identify substances including flame tests and the precipitation of metal hydroxides, halides, sulfates and carbonates. Then find out why instrumental methods are preferred to chemical tests.

Students will carry out experiments, make and record observations, describe patterns in data and use data to make predictions. They will interpret observations from chemical tests and identify unknown substances.

The unit offers a number of opportunities for students to use mathematics. They use ratios to write simple formulae, substitute numerical values into chemical equations to balance them, translate data between graphical and numerical form, plot variables against each other to describe trends and relationships; and interpolate and extrapolate graphs to make predictions.

Obstacles to learning

Students need good laboratory skills to complete the required practicals. They will need to work in an organised, tidy and clean manner, and take care not to contaminate the test substances. They will be using techniques and skills acquired over several lessons so they need to record their work carefully.

Practicals in this unit

In this unit students will do the following practical work:

- React magnesium, zinc, iron and copper with hydrochloric acid and with sulfuric acid.
- Carry out tests and observations to identify gases
- Carry out precipitation reactions to identify metal ions
- Carry out a microscale investigation to find out which carbonates are in water.
- Use test tube reactions to identify carbonates, chlorides, bromides and iodides
- Carry out flame tests to identify metal ions
- Use chemical tests to identify the ions in an unknown ionic compound

Students will also observe:

- Trends in the reactivity of the Group 1 elements with water
- The appearance of the halogens
- The reaction between sodium and chlorine
- Displacement reactions which show the trend in the reactivity of Group 7 elements
- The physical properties of transition metals
- The catalytic activity of a transition metal

	Lesson title	Overarching objectives
1	Exploring Group 0	Explain their lack of reactivity and trends in their physical properties.
2	Exploring Group 1	Describe their high reactivity and trends in their chemical properties.
3	Exploring Group 7	Describe their high reactivity and trends in their physical and chemical properties.
4	Transition metals	Describe similarities and differences between transition metals and contrast them with those of Group 1 elements.
5	Reaction trends and prediction reactions	Explain trends in reactivity in Groups 1 and 7 and changes across a period.
6	Reactivity series	Deduce an order of reactivity of metals based on their reactions, if any, with water or dilute acids.
7	Test for gases	Recall the tests for four common gases and use them to identify unknown samples.
8	Metal hydroxides	Know that the different colours of insoluble metal hydroxides can be used in precipitation reactions to identify some metal ions.
9	Tests for anions	Carry out tests to identify halide ions, sulfate ions and carbonate ions.
10	Flame tests	Know how flame tests can be used to identify some metal ions.
11	Instrumental methods	State the advantages of using instrumental methods compared with chemical tests to identify unknown substances.
12	Practical: Use chemical tests to identify the ions in unknown single ionic compounds	Use chemical tests to identify the ions in unknown single ionic compounds.

Lesson 1: Exploring Group 0

Lesson overview

OCR Specification reference

OCR C4.1

Learning objectives

- Explore the properties of noble gases.
- Find out how the mass of their atoms affects their boiling points.
- Relate their chemical properties to their electronic structures.

Learning outcomes

- Describe the properties of noble gases. [O1]
- Predict and explain the trends of the boiling points of the noble gases (going down the group). [O2]
- Explain how properties of the elements in Group 0 depend on their electron configurations. [O3]

Skills development

- WS 1.2 Match features of a model to data from observations.
- WS 3.5 Recognise and describe patterns in data.
- WS 4.1 Use scientific vocabulary, terminology and definitions.

Maths focus Plot two variables from data using negative numbers

Resources needed Internet access; graph paper; Worksheet 4.1

Digital resources Graph plotter 4.1; Presentation 4.1.1 'Argon'; Presentation 4.1.2 'Group 0'

Key vocabulary elements, helium, neon, argon, density, unreactive

Teaching and learning

Engage

- Display Presentation 4.1.2 'Group 0' to prompt small-group **discussion** about the elements in Group 0. [O2]

Challenge and develop

- Use a video clip to give an overview of the physical properties of these elements. [Google search string:] Noble gases Open University video.

- Challenge students to **find** an important use for each element and **describe** the properties that use depends on. Divide the class into groups of up to six and allocate a different Group 0 element to each student. They should regroup into 'expert' groups to **research** their element and then **share** their findings with their home group. [O1]

Explain

- Explain that although all Group 0 elements are unreactive, they show trends in their properties that make them useful for different jobs. Draw attention to the table of boiling points in Section 4.1 of the Student Book and the atomic masses listed in the periodic table. [O2]

Low demand

- Pairs use Graph Plotter 4.1 to **display** the relationship between atomic mass and the boiling points of Group 0 elements. Then they **describe** the pattern in the graph. [O2]

Standard and high demand

- Students **display** the relationship between atomic mass and the boiling points of Group 0 elements using graph paper. Then they **describe** the pattern in the graph. [O2]

- Students read 'Patterns in Group 0' in the Student Book and answer questions 1 and 2. [O2]

Consolidate and apply

- Display Presentation 4.1.1 'Argon' and ask what is the same about the electronic structures of the elements in Group 0 and whether or not this could explain their lack of reactivity. [O3]

Low demand

- Students complete Worksheet 4.1. [O3]

Standard and high demand

- Students read 'Why do elements in Group 0 exist as single atoms?' from Section 4.1 in the Student Book and answer questions 3 and 4. [O3]

Extend

Ask students able to progress further to **research** the relationship between the density of Group 0 elements and their atomic masses.

Plenary suggestion

Show Presentation 4.1.2 'Group 0' again and prompt pairs to **discuss** their conclusion. Then ask pairs to join up to make groups of six and **agree** a response to **feed back** to the rest of the class.

Answers to Worksheet 4.1

1. They all have full outer shells; 2. Helium's outer shell can only hold two electrons; 3. They have no 'spare' outer electrons to give away and there are no gaps in their outer shells to fill

Lesson 2: Exploring Group 1

Lesson overview

OCR Specification reference

OCR C4.1

Learning objectives

- Explore the properties of Group 1 metals.
- Compare their reactivity.
- Relate their reactivity to their electronic structures.

Learning outcomes

- Explain why Group 1 metals are known as the alkali metals. [O1]
- Use the trends down the group to make predictions. [O2]
- Relate the properties of alkali metals to their electron configurations. [O3]

Skills development

- WS 2.4 Identify the main hazards associated with Group 1 metals.
- WS 2.6 Make and record observations.
- WS 4.1 Use scientific vocabulary, terminology and definitions.

Maths focus Interpolate and extrapolate graphs

Resources needed Equipment as listed in the Technician's notes; Worksheets 4.2.1, 4.2.2 and 4.2.3; Technician's notes 4.2

Digital resources Presentation 4.2 'Ions'

Key vocabulary alkali, density, indicator, ion, reactivity, stable electronic structure

Teaching and learning

Engage

- Show the bottles of Group 1 metals (Technician's notes 4.2) and elicit students' ideas about why they are stored under oil and why the pieces being used are so small. [O1]

Challenge and develop

- Demonstrate the reactions of lithium, sodium and potassium with water following the method detailed in Technician's notes 4.2. Emphasise the similarities between them, the way they become softer and less dense going down the group and the way their reactivity increases going down the group. [O1]
- Follow up the demonstration with a video clip so students can see close-ups of the metals and their reactions with water. [Google search string:] Alkali metals in water [O1]

Explain

Low demand

- Students read 'Properties of Group 1 elements' and 'Reaction trends of alkali metals' in Section 4.2 in the Student Book and answer questions 1–5. [O1]

Standard and high demand

- Students **record** the similarities and differences between the reactions of lithium, sodium and potassium with water. [O1]

Consolidate and apply

Low and standard demand

- Students use the trends down the group to **make predictions** using Worksheets 4.2.1 and 4.2.2. [O2]
- Students complete Worksheet 4.2.3. [O3]

High demand

- Students use the trends in melting point down the group to **make predictions** using Worksheet 4.2.2. [O2]
- Point out francium at the bottom of Group 1. Explain that it is very rare and that all its isotopes are radioactive, so it is nearly impossible to study. Ask students to **predict** how it reacts with water and to **justify** their predictions using evidence from reactions they have seen. [O2]
- Show Presentation 4.2 'Ions' and discuss the similarity between the electronic structures of lithium and sodium. [O3]
- Ask pairs or small groups to **explain** how the electronic structures of these elements make them have similar chemical properties. [O3]

Extend

Ask students able to progress further to **research** the reactions of beryllium, magnesium and calcium with water, **compare** them with Group 1 elements and **describe** any similarities and differences they notice.

Plenary suggestion

Ask students to draw a large triangle with a smaller inverted triangle that just fits inside it (so they have four triangles). In the outer three ask them to write something they've seen; something they've done; and something they've discussed. Then add: something they've **learned** in the central triangle.

Answers to Worksheet 4.2.1

1. They will react more violently with water than potassium; because the reactivity of Group 1 metals increases going down the group; 2. $2Li + 2H_2O \rightarrow 2LiOH + H_2$; $2K + 2H_2O \rightarrow 2KOH + H_2$; $2Rb + 2H_2O \rightarrow 2RbOH + H_2$; 3. Caesium hydroxide; hydrogen; 4. The hydroxides dissolve in water to make alkaline solutions

Answers to Worksheet 4.2.2

Using graphs to make predictions: potassium, melting point 55–65 °C; francium, melting point 26–30 °C

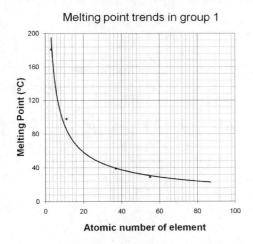

Melting point trends in group 1

Answers to Worksheet 4.2.3

1. They both have one electron in their outer shell; 2. 2,8,8,1; 3. One of the positive charges in the nucleus is no longer cancelled out; 4. Na^+; K^+

 © HarperCollins*Publishers* Limited 2016

Lesson 3: Exploring Group 7

Lesson overview

OCR Specification reference

OCR C4.1

Learning objectives

- Explain why Group 7 non-metals are known as 'halogens'.
- Compare their reactivity.
- Relate their reactivity to their electronic structures.

Learning outcomes

- Recall that fluorine, chlorine, bromine and iodine are non-metal elements called halogens. [O1]
- Use the trends down the group to make predictions. [O2]
- Construct balanced symbol equations for the reactions. [O3]

Skills development

- WS 2.4 Carry out experiments appropriately.
- WS 2.6 Make and record observations.
- WS 4.1 Use scientific vocabulary, terminology and definitions.

Resources needed Equipment as listed in the Technician's notes; Worksheets 4.3.1, 4.3.2 and 4.3.3; Technician's notes 4.3.1 and 4.3.2

Digital resources Presentation 4.3 'Halogens'

Key vocabulary bromine, chlorine, halogen, iodine

Teaching and learning

Engage

- Read lines 9–14 of the Wilfred Owen poem *Dulce et Decorum Est* and ask students which gas the author is referring to. Stress that the gas is toxic at high concentrations, but only kills bacteria at the concentrations found in swimming pools. [Google search string:] Wilfred Owen Dulce et Decorum Est [O1]

Challenge and develop

- Demonstrate the appearance of chlorine, bromine and iodine following the guidance in Technician's notes 4.3.1 or show images of the elements. Stress that at room temperature bromine is a very volatile liquid rather than a gas, and iodine is a grey solid that sublimes to form a purple vapour when warmed. [O1]

- Demonstrate the reaction between sodium and chlorine, or show a video of it. Stress that every Group 1 element reacts with every Group 7 element and all the compounds formed in this way are salts. That's why the group are called halogens – which means 'salt-producing'. [O1]

Explain

Low and standard demand

- Using Worksheet 4.3.1, students **practice** writing equations for similar reactions to the one they have witnessed. [O3]

High demand

- Ask students to read 'The halogens' from Section 4.3 in the Student Book and answer question 1. [O3]
- Students could choose pairs of elements from Groups 1 and 7 that will react in a similar way to sodium and bromine and write symbol equations to **describe** their reactions. [O3]

Consolidate and apply

- Display Presentation 4.3 'Halogens' and discuss the trends in their colours and states as their atoms get bigger going down the group. [O2]

Low and standard demand

- Students **practice** using a graph with negative axes to make predictions by completing Worksheet 4.3.2. [O2]

High demand

- Ask students to read 'Group 7 trends' from Section 4.3 in the Student Book and answer questions 2–4. [O2]

- Demonstrate the reaction between chlorine water and potassium bromide and prompt groups to come up with a hypothesis to **explain** the colour change. Use the term 'displacement' to describe this sort of reaction. [O2]

- Show students a solution of iodine and ask them which halogen is dissolved in the water. **Explain** that iodine molecules are purple but they form brown solutions. Ask them to **predict** what will happen if chlorine water is added to potassium iodide. Demonstrate this reaction using Technician's notes 4.3.2 and add hexane to the products to support the conclusion that iodine has been displaced by revealing its purple colour. [O2]

- Check students' understanding by asking them to **predict** what will happen when bromine is added to potassium chloride. Students could complete Worksheet 4.3.3 to **practice** writing equations for this sort of reaction. [O3]

Extend

Ask students able to progress further by considering that the compounds the halogens make with Group 1 metals are all white crystalline solids. They can **research** some of the molecular compounds the halogens make with other non-metals and **describe** their properties.

Plenary suggestion

Name a halogen and challenge groups to list five things they know about it.

Answers to Worksheet 4.3.1

1. a) → lithium iodide; b) → potassium bromide; 2. a) $2Na + Cl_2 → 2NaCl$; 3. a) $2Na + Br_2 → 2NaBr$; b) $2K + Cl_2 → 2KCl$; c) $2K + I_2 → 2KI$; d) $2Na + At_2 → 2KAt$; e) $2Cs + I_2 → 2CsI$; f) $2Li + Br_2 → 2LiBr$; 4. White crystals; that will dissolve in water

Answers to Worksheet 4.3.2

290–300 °C.

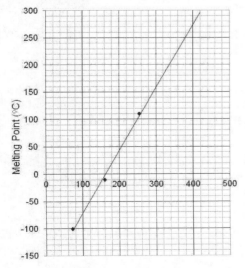

Molecular mass of element

Answers to Worksheet 4.3.3

1. Pale green; colourless; brown; colourless; 2. Displacement; 3. C; D; 4. a) $Br_2 + 2KI → I_2 + 2KBr$; b) $I_2 + 2KAt → At_2 + 2KI$; c) $Cl_2 + 2NaI → I_2 + 2NaCl$; d) $F_2 + 2NaBr → Br_2 + 2NaF$; e) $F_2 + 2KCl → Cl_2 + 2KF$

Lesson 4 Transition metals

Lesson overview

OCR Specification reference

OCR C4.1

Learning objectives

- Compare the properties of transition metals with those of Group 1 metals.
- Explore the uses of transition metals.
- Find out why they can form compounds with different colours.

Learning outcomes

- Describe the properties of transition metals. [O1]
- Relate these properties to their uses. [O2]
- Describe the ions that transition metals form. [O3]

Skills development

- WS 1.2 Use models in explanations.
- WS 2.6 Make and record observations.
- WS 3.5 Recognise and describe patterns.

Resources needed Equipment as listed in the Technician's notes; Worksheet 4.4; Technician's notes 4.4

Digital resources Presentation 4.4 'Transition metals'

Key vocabulary catalyst, chromium, cobalt, manganese, nickel, transition element

Teaching and learning

Engage

- Display a range of unlabelled coloured transition metal compounds and elicit students' ideas about what they have in common. [O1]
- Contrast coloured transition metal compounds with the white compounds formed by Group 1 elements. [O1]
- Use the cannon-fire audio experiment to demonstrate the catalytic activity of a manganese compound, following the instructions in Technician's notes 4.4. [O1]

Challenge and develop

- Display slide 1 of Presentation 4.4 'Transition metals' and give groups a minute to **suggest** as many differences as possible between the physical properties of Group 1 metals and those of transition metals. [O1]
- Use the Royal Society of Chemistry Periodic Table [Google search string:] rsc periodic-table trends, to compare visually the densities and melting points of Group 1 and transition metals. [O1]

Explain

- Explain that, like any group of elements, the transition metals share properties but are not identical. Students complete Worksheet 4.4 to **link** the properties of transition metals to their common uses. [O1, O2]

Low demand

- Students choose one example of a catalyst from Section 4.4 in the Student Book and record the name of the transition metal and the process it speeds up. [O2]

Standard and high demand

- Ask students to **research** the Haber process and the conversion of vegetable oils into margarine. For each process they should find the name of the catalyst used, the reaction it speeds up and why the reaction is important.[O2]

Consolidate and apply

- Display slide 2 of Presentation 4.4 'Transition metals' and explain that the roman numerals in a compound's name show which ion is present. [O3]

Low and standard demand

- Students **practice** writing formulas for ions using slide 3 of Presentation 4.4 'Transition metals'. [O3]

High demand

- Students read 'Ions and coloured compounds' from Section 4.4 in the Student Book and answer questions 2–4. [O3]

Extend

Ask students able to progress further to **find out** what ions elements 21–30 can form. Should zinc and scandium be called transition metals? [O3]

Plenary suggestion

Show students a video of the cobalt-catalysed reaction between a tartrate and hydrogen peroxide. [Google search string:] cobalt magic video. Ask groups to deduce what is happening using what they know about the compounds of transition metals like cobalt. Cobalt(II) ions are pink in solution. They catalyse the reaction and in the process are converted into green cobalt(III) ions and then back to pink cobalt(II) ions.

Answers to Worksheet 4.4

Chromium, F; manganese, G; iron, A; cobalt, D; nickel, B; copper, E; osmium, C

Lesson 5: Reaction trends and predicting reactions

Lesson overview

OCR Specification reference

OCR C4.1

Learning objectives

- Review the patterns in the periodic table.
- Compare the trends in Group 1 and Group 7.
- Relate these trends to the way atoms form ions.

Learning outcomes

- Predict the reactions of elements with water, dilute acid and oxygen from their position in the periodic table. [O1]
- Explain why the trends down the group in Group 1 and in Group 7 are different. [O2]
- Explain the changes across a period. [O3]

Skills development

- WS 1.2 Use models in explanations.
- WS 1.2 Make predictions based on a model.
- WS 3.5 Recognise and describe patterns or trends.

Resources needed Large wall-mounted periodic table; individual periodic tables; Worksheet 4.5

Digital resources Presentation 4.5 'Explaining reactivity'

Key vocabulary electron arrangement, reactivity, trend

Teaching and learning

Engage

- Ask students to **predict** which Group 1 metal and Group 7 non-metal would react together most violently. If francium is suggested, explain that because all its isotopes are radioactive, and decay very rapidly, it is impossible to get enough of the element to do the experiment with. [O1]
- Show a video clip to demonstrate the reaction between fluorine with caesium. [Google search string:] Reacting fluorine with caesium – First Time on Camera. Remind students that we can predict how an element will react by looking at its position in the periodic table because the element's position reflects its electronic structure. [O1]

Challenge and develop

- Display slide 1 of Presentation 4.5 'Explaining reactivity' and discuss the differences between the electronic structures of lithium and sodium and how that affects how easily they lose their outer electrons. Then compare fluorine and chlorine using slide 2 to explain why smaller non-metal atoms are more reactive. [O2]
- Use a video clip to summarise the trends in the reactivity of the elements in Group 1 and Group 7. [Google search string:] Bitesize – GCSE Chemistry – Reactivity of Group 1 and 7 elements. [O2]

Explain

Low demand

- Students complete the questions on Worksheet 4.5. [O2]

Standard and high demand

- Ask students to read 'Opposite trends' and 'Trends across the Periodic table' from Section 4.5 in the Student Book and answer questions 2–4. [O2, O3]

High demand

- Students use diagrams to **explain** how the size of their atoms affects the reactivity of the elements in Group 1 and Group 7. [O2]

Consolidate and apply

- Use a video clip to review the trend in properties across a period from Group 1 to Group 3 and from Group 6 to Group 8. [Google search string:] Science in Focus Periodic Table Blind Date. [O3]

High demand

- Use slide 3 of Presentation 4.5 'Explaining reactivity' to summarise the trends in atomic radius across Periods 1–4 of the periodic table. Group 2 metals are generally less reactive than Group 1 elements. Ask small groups to **suggest** two possible explanations for this. They should spot that the outer electrons are closer to the nucleus in Group 2 elements and that two electrons need to be lost to form an ion, rather than one. [O3]

Extend

Ask students able to progress further to **create** a costume design and script for an element from Group 1 or 7 that will be appearing on the next episode of 'Periodic Table Blind Date'. [O1]

Plenary suggestion

Display the names of elements from Groups 1, 7 and 0, and challenge small groups to list as many things about each element as they can in 30 seconds.

Answers to Worksheet 4.5

1. They lose their outer electron; 2. Sodium; it is further from the nucleus; 3. Potassium's outer electron is further from the nucleus; 4. They attract an electron into their outer shell; 5. Chlorine; 6. Its outer shell is closer to the nucleus; so it attracts electrons more strongly

© HarperCollins*Publishers* Limited 2016

Lesson 6: Reactivity series

Lesson overview

OCR Specification reference

OCR C4.1

Learning objectives

- Compare the reactivity of metals.
- Observe some reactions between metal atoms and metal ions.
- Consider why some metals are more reactive than others.

Learning outcomes

- Describe the reactions, if any, of metals with water or dilute acids. [O1]
- Deduce an order of reactivity of metals based on experimental results. [O2]
- Explain how the reactivity is related to the tendency of the metal to form its positive ion. [O3]

Skills development

- WS 2.2 Identify variables.
- WS 3.3 Find the mean of a set of data.
- WS 3.5 Interpret observations.

Maths focus Calculate mean values

Resources needed Equipment as listed in Technician's notes 4.6; Practical sheet 4.6; Worksheet 4.6

Digital resources Presentation 4.6.1 'Comparing metals'; Presentation 4.6.2 'Forming ions'

Key vocabulary displacement, positive ion, reactivity, tendency

Teaching and learning

Engage

- Display Presentation 4.6.1 'Comparing metals' and **challenge** groups to list ways in which sodium, magnesium and zinc react in a similar way, and also the differences between their reactions.

Challenge and develop

- Use video clips to remind students how sodium reacts with water [Google search string:] Sodium in water; and show that it reacts even faster with acids [Google search string:] Sodium metal exploding in dilute hydrochloric acid. [O1]
- Demonstrate that zinc, magnesium, copper and iron have no observable reaction with water. Then ask pairs to use their reactions with hydrochloric acid and sulfuric acid to **arrange** them in order of reactivity. Practical sheet 4.6 structures the investigation. [O1, O2].

Explain

- Use Section 4.6 of the Student Book to explain that it is difficult to compare all metals using a single reaction because their reactivities are so different. But we can observe how quickly they displace hydrogen from water and from acids. [O1]

Low demand

- Students write word equations for any reactions they observed. [O1]

Standard and high demand

- Students write symbol equations for the reactions they observed using Presentation 4.6.1 'Comparing metals'. [O2]

Consolidate and apply

Low demand

- Students complete Worksheet 4.6 to practice **interpreting** results. [O1, O2]

Standard and high demand

- Use Presentation 4.6.2 'Forming ions' to point out that magnesium atoms are oxidised to ions when magnesium reacts with acids, and that more reactive metals have a higher tendency to form positive ions. [O3]

- Read 'Positive ions' in Section 4.6 of the Student Book and answer questions 2 and 3. [O3]

Extend

- Ask students able to progress further to write symbol equations for the reaction of sodium with hydrochloric acid and with sulfuric acid. [O1]

Plenary suggestion

Predictions Manganese lies between magnesium and zinc in the reactivity series. Ask groups to **deduce** as many facts as they can about its reactions with acid.

Answers to Worksheet 4.6

1. *Dependent variable* is the volume of hydrogen released in 1 minute; *independent variable* is the metal used; *control variables* are the mass (and surface area) of metal, volume and concentration of acid, and type of acid; 2. From the most reactive to the least reactive: Mg (31 cm^3), Zn (11 cm^3), Fe (4 cm^3), Cu (0 cm^3); 3. The second result for zinc (15 cm^3); a higher mass of metal may have been added; 4. Describes a recognisable way of comparing the reactions of metals with an acid

Lesson 7: Test for gases

Lesson overview

OCR Specification reference

OCR C4.2

Learning objectives

- Recall the tests for four common gases.
- Identify the four common gases using these tests.
- Explain why limewater can be used to detect carbon dioxide.

Learning outcomes

- Distinguish gases using chemical tests. [O1]
- Describe how each test works. [O2]
- Recognise that carbon dioxide reacts with limewater to form an insoluble carbonate. [O3]

Skills development

- WS 2.3 Select the apparatus that should be used for a particular purpose and explain why.
- WS 2.4 Identify the main hazards in specified practical contexts and work safely.
- WS 2.6 Make and record observations.

Resources needed Equipment as listed in the Technician's notes; Practical sheet 4.7; Technician's notes 4.7

Digital resources Presentation 4.7 'Identifying gases'

Key vocabulary carbon dioxide, chlorine, hydrogen, oxygen

Teaching and learning

Engage

- Display slide 1 of Presentation 4.7 'Identifying gases' and prompt small groups to recall the tests for each of the gases shown. Section 4.7 in the Student Book can be used for reference if necessary. [O1]

Challenge and develop

- A video clip could be used to remind students what a positive result for each gas looks like [Google search string:] Gas Tests. [O1]

- Present the challenge outlined in Practical sheet 4.7 and stress that groups need to think carefully before they begin the tests. There are four different gases but each group is allowed a maximum of two tubes of each gas. To succeed they either need to **collaborate** with other groups or **decide** which tests to carry out first in order to reduce the number of possible gases in the remaining tubes. They should **record** each test they do and the results they obtain before **planning** their next step.

  Students may notice that the introduction to the task allows them to **eliminate** one possibility for each tube. They may be able to detect a faint colour in one tube, indicating chlorine, and a piece of litmus paper held near the bottom of the test-tube should turn red then turn colourless due to chlorine's bleaching effect. The rubber bung could be used to stopper one tube to allow it to be shaken with limewater. Students should appreciate that hydrogen and oxygen are less dense than air so any tubes removed from the water should be kept inverted or stoppered while they prepare to test the gases in them. [O1]

Explain

- Use slide 2 of Presentation 4.7 'Identifying gases' to explain the chemistry involved in the test for carbon dioxide, and why the cloudy precipitate observed to form in the limewater can disappear if excess carbon dioxide is present.

- A video clip could be used to highlight the chemical reactions each test depends on [Google search string:] Testing for hydrogen, oxygen, carbon dioxide and chlorine. [O1]

Consolidate and apply

- Low and standard demand

  Students read Section 4.7 of the Student Book and answer questions 1–7. [O2]

- High demand

  Students **describe** how each test works and give equations for the chemical reactions involved. [O2, O3]

Extend

Ask students able to progress further to find out what reaction takes place when calcium carbonate reacts with a solution of carbon dioxide (carbonic acid).

Plenary suggestion

Use slides 3–7 of Presentation 4.7.1 'Identifying gases' as a whole class quiz. Students should be encouraged to say whether the result allows them to be definite about the identity of a gas, or only **suggest** what it might be.

Answers to plenary quiz

1. Definitely oxygen; 2. Definitely chlorine; 3. Possibly carbon dioxide; other gases will also put out a lighted splint; 4. Definitely carbon dioxide; 5. Possibly hydrogen; other gases (e.g. methane) react explosively with oxygen, but only hydrogen burns with a squeaky pop

 © HarperCollins*Publishers* Limited 2016

Lesson 8: Metal hydroxides

Lesson overview

OCR Specification reference

OCR C4.2

Learning objectives

- Recognise the precipitate colour of metal hydroxides.
- Explain how to use sodium hydroxide to test for metal ions.
- Write balanced equations for producing insoluble metal hydroxides.

Learning outcomes

- Know that copper(II) compounds form a blue precipitate, iron(II) forms a light green precipitate, iron(III) forms a brown precipitate and solutions of aluminium, calcium and magnesium form white precipitates when sodium hydroxide solution is added. [O1]
- Describe the practical procedure to use sodium hydroxide solution to precipitate metal hydroxides. [O2]
- Write balanced chemical equations, including state symbols, for precipitating copper(II) hydroxide, iron(II) hydroxide, iron(III) hydroxide, aluminium hydroxide, calcium hydroxide and magnesium hydroxide in chemical reactions. [O3]

Skills development

- WS 2.2 Plan experiments or devise procedures to make observations, produce or characterise a substance, test hypotheses, check data or explore phenomena.
- WS 2.4 Carry out experiments appropriately having due regard for the correct manipulation of apparatus, the accuracy of measurements and health and safety concerns.
- WS 4.1 Use scientific vocabulary, terminology and definitions.

Resources needed Practical sheet 4.8; Worksheet 4.8; Technician's notes 4.8

Key vocabulary hydroxide, precipitate, gelatinous, ion

Teaching and learning

Engage

- Remind students that in this chapter they have found out how chromatography and flame tests are used as identification tools. Ask students what they identify. Establish that chromatography can be used to help identify coloured pigments and that flame tests can be used to identify some cations/metal ions. Explain that we can also use chemical reactions to identify some metal ions/cations. [O1]

Challenge and develop

Low demand

- Demonstrate adding dilute sodium hydroxide solution to a solution of copper(II) sulfate. Explain that sodium sulfate and copper(II) hydroxide are produced. Copper(II) hydroxide is insoluble in water, so it forms as a solid, sometimes gelatinous. Tell students this is called a *precipitate*. Explain that many metal hydroxides are insoluble and that we can identify the metal ion present from the colour of the hydroxide. [O1]

- Students can answer question 1 in Worksheet 4.8. [O1]

Standard demand

- Introduce Practical sheet 4.8. Students follow instructions to **determine** the colours of the precipitates formed between dilute sodium hydroxide and copper(II) sulfate, iron(II) nitrate, iron(III) chloride, aluminium chloride, calcium chloride and magnesium chloride. Emphasise the importance of not contaminating the reactants and of labelling the test tubes. You may need to remind them that 'a few drops' does not mean a dropper full. Students can work in pairs and carry out the practical, recording their observations in the table provided. [O2]

- Discuss their results and conclude how some metal ions can be identified by their reactions with sodium hydroxide solution. [O2]

Explain

High demand

- Ask students to write the word equation and a balanced chemical equation for the reaction between sodium hydroxide solution and copper(II) sulfate solution. Students can use Section 4.8 of the Student Book if help is needed. Explain how an ionic equation can be used to show the precipitate formation. Emphasise that only the ions needed to make the precipitate are included in an ionic equation. [O3]

- Students can answer question 2 on Worksheet 4.8. They are writing word equations, balanced chemical equations and ionic equations for the reactions in Practical sheet 4.8. [O3]

Consolidate and apply

- Students can answer question 3 on Worksheet 4.8. They are **suggesting** procedures they could carry out to identify unknown compounds. This should include flame tests as well as precipitating hydroxides. [O2]

Extend

Ask students able to progress further to:

- **Discuss** why all the reactants in Practical 4.8 have to be soluble in water. [O2]

- **Find out** how state symbols can be used in equations to indicate that a precipitate is produced. [O3]

- **Find out** why Group 1 metal ions do not produce precipitates with sodium hydroxide solution. [O3]

Plenary suggestions

Ask students to draw a large triangle on a piece of paper with a smaller inverted triangle inside it. In the three outer triangles, they can write something they have seen, something they have done and something they have **discussed**. In the central triangle, they write something they have learned.

Ask students to write down the most important thing they have learned in this lesson. They can join up in pairs, double up and **share** their ideas. A class list could be **compiled**.

Answers to Worksheet 4.8

1. a) A blue gelatinous precipitate; b) It is insoluble in water; c) Precipitate; d) Blue; e) In solution; 2. a)(i) sodium hydroxide + copper sulfate → sodium sulfate + copper hydroxide, $2NaOH + CuSO_4 → Na_2SO_4 + Cu(OH)_2$, $Cu^{2+} + 2OH^- → Cu(OH)_2$; b) Sodium hydroxide + iron(II) sulfate → sodium sulfate + iron(III) hydroxide, $2NaOH + FeSO_4 → Na_2SO_4 + Fe(OH)_2$, $Fe^{2+} + 2OH^- → Fe(OH)_2$; c) Sodium hydroxide + iron(III) chloride → sodium sulfate + iron(III) chloride, $3NaOH + FeCl_3 → 3NaCl + Fe(OH)_3$, $Fe^{3+} + 3OH^- → Fe(OH)_3$; d) Sodium hydroxide + aluminium chloride → sodium sulfate + aluminium hydroxide, $3NaOH + AlCl_3 → 3NaCl + Al(OH)_3$, $Al^{3+} + 3OH^- → Al(OH)_3$; e) Sodium hydroxide + calcium chloride → sodium chloride + calcium hydroxide, $2NaOH + CaCl_2 → 2NaCl + Ca(OH)_2$, $Ca^{2+} + 2OH^- → Ca(OH)_2$; f) Sodium hydroxide + magnesium chloride → sodium chloride + magnesium hydroxide, $2NaOH + MgCl_2 → 2NaCl + Mg(OH)_2$, $Mg^{2+} + 2OH^- → Mg(OH)_2$; 3. a) Make solutions, add dilute sodium hydroxide, calcium chloride produces a white precipitate, copper chloride produces a blue precipitate; b) Make solutions, add a few drops of dilute sodium hydroxide, all give a white precipitate, continue adding dilute sodium hydroxide, aluminium hydroxide dissolves, carry out flame tests to identify calcium ions, give a red flame; c)(i) Flame test; (ii) Flame test, precipitate hydroxide if magnesium not present; (iii) Precipitate hydroxide; (iv) Flame test or precipitate hydroxide; (v) Precipitate hydroxide, dissolves in excess sodium hydroxide; (vi) Flame test

Answers to Practical sheet 4.8

1. Copper(II) hydroxide, iron(II) hydroxide, iron(III) hydroxide; 2. Aluminium hydroxide, calcium hydroxide, magnesium hydroxide; 3. Aluminium hydroxide dissolves in excess sodium hydroxide solution; calcium hydroxide and magnesium hydroxide do not

Lesson 9: Tests for anions

Lesson overview

OCR Specification reference

OCR C4.2

Learning objectives

- Identify the tests for carbonates.
- Explain the tests for halides and sulfates.
- Identify anions and cations from the results of tests.

Learning outcomes

- Describe that carbonates react with dilute acids and carbon dioxide gas is given off, which turns limewater milky. [O1]
- Explain that acidified silver nitrate solution reacts with halides to produce a silver halide precipitate that can identify the halide present; explain that barium chloride solution reacts with sulfates to produce a white precipitate of barium sulfate. [O2]
- Identify carbonate, chloride, bromide and iodide anions from the results of tests. [O3]

Skills development

- WS 2.4 Carry out experiments appropriately having due regard for the correct manipulations of apparatus, the accuracy of measurements and health and safety considerations.
- WS 3.1 Present observations and other data using appropriate methods.
- WS 3.5 Interpret observations and other data (presented in verbal, diagrammatic, graphical, symbolic or numerical form), including identifying patterns and trends, making inferences and drawing conclusions.

Resources needed Practical sheets 4.9.1 and 4.9.2; Worksheet 4.9; Technician's notes 4.9.1 and 4.9.2

Key vocabulary carbonates, halides, sulfates, limewater

Teaching and learning

Engage

- Remind students of the previous two lessons in which they investigated some chemical tests used to identify metal ions/cations – namely flame tests and precipitating hydroxides. Ask them to work in pairs and list some common anions with their formulae and charges. Revise the formulae of carbonates, chlorides, bromides, iodides and sulfates. You may need to remind them that '-ate' at the end of a name means the anion contains oxygen, and '-ide' at the end means the anion contains only one type of atom, such as chloride, Cl^-. [O1]
- You could practice writing chemical formulae if needed. [O3]

Challenge and develop

Low demand

- Tell students that most carbonates are insoluble, but some are soluble. They can carry out the investigation on Practical sheet 4.9.1. This is a microscale investigation in which students add metal ions and carbonate ions on a clear plastic sheet to find which form a precipitate, and therefore which carbonates are insoluble. Practical 4.9.1 should only take 10–15 minutes if the solutions of ions are in dropper bottles (or dropping pipettes). Students can answer the questions in the Evaluation section. You may need to discuss these first. [O1]
- Students are now going use Practical sheet 4.9.2 to carry out some test tube reactions used to identify carbonates, halides and sulfates. You may wish to demonstrate the method used to test for carbon dioxide gas using a dropping pipette. Students use several different solutions. You may wish to reinforce good lab practice and remind them that they must not contaminate solutions by transferring droppers between solutions, labelling as necessary, keeping an organised work space and wiping up any spillages immediately. Students can carry out the carbonate test. [O1]

Standard demand

- Emphasise that all carbonates react with dilute acids and give off carbon dioxide gas. Students can read about carbonates in Section 4.9 of the Student Book.

- Students can carry out the tests for halides and sulfates using Practical sheet 4.9.2. Explain that chlorides, bromides and iodides are collectively called *halides*. The test with acidified silver nitrate solution distinguishes between the halides by producing different coloured precipitates. Encourage students to use the terms 'white', 'cream' and 'yellow' to **describe** the colours of the precipitates. [O2]

- Students can **identify** the unknown anions in two samples X and Y. [O1, O2]

- Students can answer the Evaluation section of Practical sheet 4.9.2 and question 1 in Worksheet 4.9. [O2]

Explain

- Work through 'Balanced equations for precipitation' in Section 4.9 of the Student Book. Emphasise the use of state symbols for showing that a precipitate has been formed. Students can answer question 2 in Worksheet 4.9. [O2]

Consolidate and apply

High demand

- Students can answer question 3 in Worksheet 4.9. They are required to identify the anions and cations present from the results of identification tests. This includes flame tests and precipitation of hydroxides. [O3]

Extend

Ask students able to progress further to:

- **Find out** why silver nitrate solution is acidified with dilute nitric acid in the halide test, and why barium chloride solution is acidified with dilute hydrochloric acid in the sulfate test. [O2]

- Write ionic equations for the reactions in question 1 in Worksheet 4.9. [O2]

Plenary suggestions

Write the word 'precipitate' on the board. Students have 1 minute to work in pairs to name three chemical reactions that produce a precipitate. They can **peer assess** their reactions with another pair. This could be extended using the term 'no precipitate forms'.

Students can work in pairs and write 'Identifying anions and cations' in the centre of a sheet of paper. They can add arrows and labels to include all the methods they have learned about in the last few lessons.

Answers to Worksheet 4.9

1. a) sodium chloride + silver nitrate → sodium nitrate + silver chloride, $NaCl(aq) + AgNO_3(aq) → NaNO_3(aq) + AgCl(s)$; b) sodium bromide + silver nitrate → sodium nitrate + silver bromide, $NaBr(aq) + AgNO_3(aq) → NaNO_3(aq) + AgBr(s)$; c) sodium iodide + silver nitrate → sodium nitrate + silver iodide, $NaI(aq) + AgNO_3(aq) → NaNO_3(aq) + AgI(s)$; d) potassium bromide + silver nitrate → potassium nitrate + silver bromide, $KBr(aq) + AgNO_3(aq) → KNO_3(aq) + AgBr(s)$; 2. a) sodium sulfate + barium chloride → sodium chloride + barium sulfate, $Na_2SO_4(aq) + BaCl_2(aq) → 2NaCl(aq) + BaSO_4(s)$; b) potassium sulfate + barium chloride → potassium chloride + barium sulfate, $K_2SO_4(aq) + BaCl_2(aq) → 2KCl(aq) + BaSO_4(s)$; c) copper(II) sulfate + barium chloride → copper(II) chloride + barium sulfate, $CuSO_4(aq) + BaCl_2(aq) → CuCl_2(aq) + BaSO_4(s)$; 3. a) A is copper(II) chloride, B is aluminium carbonate, C is potassium sulfate, D is sodium iodide, E is iron(II) bromide; b)(i) $CuCl_2(aq) + 2AgNO_3(aq) → Cu(NO_3)_2(aq) + 2AgCl(s)$; (ii) $CuCl_2(aq) + 2NaOH(aq) → Cu(OH)_2(s) + 2NaCl(aq)$

Answers to Practical sheet 4.9.1

1. Magnesium ions, calcium ions, iron(II) ions, copper(II) ions; 2. Sodium and potassium carbonates are soluble; magnesium, calcium, iron(II) and copper carbonates are insoluble; 3. Soluble carbonates contain Group 1 metal ions

Answers to Practical sheet 4.9.2

1. a) Add dilute hydrochloric acid, test any gas given off with limewater; carbonates give off carbon dioxide which turns limewater milky; b) Add acidified silver nitrate solution; chlorides give a white precipitate, bromides a cream precipitate and iodides a yellow precipitate; c) Add acidified barium chloride solution; sulfates produce a white precipitate; 2. As supplied by technician

Lesson 10: Flame tests

Lesson overview

OCR Specification reference

OCR C4.2

Learning objectives

- Carry out flame-test procedures.
- Identify the colours of flames of ions.
- Identify substances from the results of the tests.

Learning outcomes

- Carry out flame tests using a flame-test wire to produce a recognisable coloured flame. [O1]
- Identify the colours of flames produced by lithium ions, sodium ions, potassium ions, calcium ions and copper ions. [O2]
- Interpret results from flame tests to identify the cations present. [O3]

Skills development

- WS 2.2 Plan or devise procedures to make observations, produce or characterise a substance, test hypotheses, check data or explore phenomena.
- WS 2.4 Carry out experiments appropriately having due regard for the correct manipulation of apparatus, the accuracy of measurements and health and safety considerations.
- WS 2.7 Evaluate methods and suggest possible improvements and further investigations.

Resources needed Practical sheet 4.10; Worksheet 4.10; Technician's notes 4.10

Digital resources Video of fireworks (YouTube)

Key vocabulary flame test, lilac, crimson, yellow

Teaching and learning

Engage

- Show students a video of some fireworks exploding with different colours. There are several on YouTube. Ask them to **suggest** what they think a firework contains. Establish that fireworks are formulated and some of the ingredients produce different coloured flames. Tell them that the colours are produced by different metal ions in compounds and that these colours can be reproduced in the lab using flame tests. [O1]

Challenge and develop

- Introduce students to Practical sheet 4.10 and explain the procedure. They can work in pairs and carry out the procedure in the Method, steps 1–7. They initially use flame-test wires for the tests and record the results. Advise students to do the sodium chloride test last. If they do not clean the flame-test wire thoroughly between each test, their flame colours may be masked. A partially darkened room will make the colours more apparent. Students need to record their results in a table. They can **compare** colours with those of other pairs of students. [O1]

- Discuss the colours obtained and correct any words used to describe the coloured flames. These should be 'crimson' for the lithium ions, 'yellow' for sodium ions, 'lilac' for potassium ions, 'red' for calcium ions and 'green' for copper ions. Students who are colour blind may need some assistance. You may need to demonstrate flame tests for calcium ions and lithium ions side-by-side so that they can appreciate the difference between 'red' and 'crimson'. Discuss the results if the flame-test wires are not properly cleaned between tests. [O2]

- Students can answer question 1 in the Evaluation in Practical sheet 4.10. [O2]

- Allow students to carry out step 8 in the Method on Practical sheet 4.10. This is a repeat of the tests using soaked wooden splints instead of flame-test wires. The splints must be completely soaked along their whole length for at least 12 hours. A new splint is needed for each test, but the used parts can be cut off and the splint used again. Soaked wooden splints give the same results as flame-test wires if the splint is still wet. When it dries out, the wood starts to burn. Students are **comparing** and **evaluating** the two different methods of doing flame tests. They can answer question 2 in the Evaluation in Practical sheet 4.10.

Explain

- Explain that flame tests can be used to identify metal ions. This requires just one type of metal ion to be present so that colours are not masked. Students can **identify** the compounds in the table in Section 4.10 of the Student Book. [O3]

Consolidate and apply

- Students can complete Worksheet 4.10 to consolidate the procedure and interpret results. [O1, O2, O3]

- You could use the spectacular demonstration of flame-test colours on the RSC website. A blackout room will increase the drama. [O2]

- Ask students to list the differences between chromatography and flame tests. [O1, O2, O3]

Extend

Ask students able to progress further to:

- **Find out** what 'excited electrons' have to do with flame tests. They will need access to the internet. [O3]

- **Find out** what a spectroscope is and how it can be used to confirm the results of flame tests.

Plenary suggestions

Each student requires two strips of paper. Ask them to write a question on one strip about something they have learned in the lesson, and the answer on the other strip. They can work in groups of six to eight. Shuffle the strips and redistribute so that each student has one question and one answer. One student can ask their question, and the student holding the correct answer should **respond** and then ask their question.

Students can write a question and mark scheme that includes two or more points. They can try their questions out on each other.

Answers to Worksheet 4.10

1. a)(i) The colour of the ions will be visible/a yellow flame will mask the colour; (ii) To make sure the colour is not masked by another; (iii) To make the compound stick to the wire; (iv) Sodium ions produce a very strong colour which makes it difficult to clean the wire completely; b)(i) One mark; (ii) Marks lost are 'wrong type of wire is used', 'wire is not cleaned by dipping in hydrochloric acid and holding it in the Bunsen flame', 'wire is not cleaned between tests', 'wrong colour for potassium ions'; 2. ii, iii, v, vi

Answers to Practical sheet 4.10

1. a) Crimson; b) Yellow; c) Lilac; d) Brick-red; e) Blue-green; 2. Include flame-test wires need cleaning, wooden splints do not; wooden splints eventually burn, flame test wires do not; wooden splints are cheaper

Lesson 11: Instrumental methods

OCR Specification reference

OCR 4.2

Lesson overview

Learning objectives

- Identify advantages of instrumental methods compared with the chemical tests.
- Describe some instrumental techniques.
- Explain the data provided by instrumental techniques.

Learning outcomes

- Know that instrumental methods are quicker, more accurate and more sensitive than chemical tests. [O1]
- Describe some instrumental techniques such as spectroscopy, mass spectroscopy and separation techniques. [O2]
- Simply explain some data provided by instrumental techniques, such as mass number from the mass spectrum of an element. [O3]

Skills development

- WS 1.4 Explain everyday and technological applications of science; evaluate associated personal, social, economic and environmental implications; and make decisions based on the evaluation of evidence and arguments.
- WS 3.2 Translate data from one form to another.
- WS 3.8 Communicate the scientific rationale for investigations, methods used, findings and reasoned conclusions through written and electronic reports and presentations using verbal, diagrammatic, graphical, numerical and symbolic forms.
- WS 4.1 Use scientific vocabulary, terminology and definitions.

Maths focus Translate information between graphical and numerical forms

Resources needed Worksheets 4.11.1 and 4.11.2

Digital resources Class internet access

Key vocabulary radiation, spectroscopy, separation techniques, gas–liquid chromatography

Teaching and learning

Engage

- Allow students to read the introduction in Worksheet 4.11.1 about Mars Curiosity Rover and the onboard instrumentation. There are several images on Google images you could also use. NASA's website is also a useful source of information for more able students. [O1]

- Ask students to **recall** the different chemical tests they have used to identify anions and cations (chromatography, flame tests, precipitating hydroxides, test tube reactions for carbonates, halides and sulfates) and **discuss** whether or not these tests are suitable to use on Mars in the Curiosity Rover. Students might find it useful to list the problems associated with performing the chemical tests they have used in the previous few lessons. Establish that the Curiosity Rover needs to be able to identify samples quickly, accurately and to be able to detect very small amounts. [O1]

Challenge and develop

Low demand

- Students can answer question 1 in Worksheet 4.11.1. Check that students **understand** that instruments such as the mass spectrometer are quick, accurate and far more sensitive than the tests they have carried out in the lab. They will need to use Section 4.11 of the Student Book. [O1]

Standard demand

- Students can work in pairs and **carry out** the research in question 5 in Worksheet 4.11.1. They will need internet access. Each student in a pair **researches** how one type of instrumentation works and what it is used for. The methods are gas chromatography and mass spectrometry. They will find Section 4.11 in the Student Book useful. They then **report** back to each other, **explaining** their method. They may need some guidance on **interpreting** some websites, especially those **researching** mass spectrometry because these tend to be pitched at AS level. They can answer the question 2 in Worksheet 4.11.1. [O2]

- Students can use the Student Book to **find out** about other types of instrumentation. [O2]

Explain

High demand

- Tell students that the results from analyses using instrumental methods are usually shown in the form of a graph, a series of lines or of blocks. Explain that mass spectrometry produces a mass spectrum and that gas chromatography produces a gas chromatogram. Use Section 4.11 in the Student Book to explain what they show. [O3]

- Students can answer Worksheet 4.11.2. They are **interpreting** simple mass spectra and a gas chromatogram. Note that isotope peaks have not been included in the mass spectra. [O3]

Consolidate and apply

- Remind students that Buckminsterfullerene is a form of carbon discovered relatively recently. Tell them that its discovery was confirmed using mass spectrometry. When a sample containing C_{60} was fed into a mass spectrometer, a peak was obtained that corresponded to a molecule of C_{60}. They can **work out** the molecular mass shown by this peak and sketch a mass spectrum of C_{60}. [O3]

Extend

Ask students able to progress further to **find out**:

- what CG-MS is [O2]
- how blood and urine samples from athletes are screened for prohibited substances
- how infrared spectroscopy is used to measure the amount of alcohol in blood samples
- about flame emission spectroscopy.

Plenary suggestions

Ask students to write down three big ideas they have learned this lesson. They can **share** their ideas in groups and **rank** the combined ideas in order of importance. They can **share** these with the class.

Ask students to work in pairs to **compile** a list of differences between their school lab and a modern drugs testing lab used by organisers of the Olympic games. They can **share** their lists with the rest of the class.

Answers to Worksheet 4.11.1

1. Chromatography, flame tests, precipitating hydroxides, test tube reactions; 2. They take too long, are not very accurate, need human presence, etc.; 3. Quicker, more accurate, more sensitive; 4. Spectroscopy, mass spectroscopy, electrochemical analysis, thermal analysis, separation techniques, microscopy; 5. (Answers will vary) a) Should include *gas chromatograph*: mixture injected into carrier gas, passes along column and separates, detected; *mass spectrometer*: sample ionised, separated by magnet or ion drift, detected; b) *Gas chromatograph*: complicated mixtures of compounds; *mass spectrometer*: mass numbers of elements, molecular mass of molecules; c) Mass numbers of elements (isotopes), molecular mass of compounds, fragments; d) E.g. *gas chromatography* : to analyse urine samples for drugs; *mass spectrometry*: to analyse Martian soil samples

Answers to Worksheet 4.11.2

1. a)(i) 11, (ii) Boron; b)(i) 18, (ii) Water; 2. a) Six; b) F; c) A; d) B and C; e) 8 minutes

Lesson 12: Practical: Use chemical tests to identify the ions in unknown single ionic compounds

Lesson overview

OCR Specification reference

OCR C4.2

Learning objectives

- Describe how to carry out experiments safely using the correct manipulation of apparatus for the qualitative analysis of ions.
- Make and record observations using flame tests and precipitation methods.
- Identify unknown ions in chemical compounds.

Learning outcomes

- Describe how to carry out flame tests and precipitate hydroxides to identify cations and test tube reactions to identify carbonate, sulfate and halide anions. [O1]
- Carry out the above identification tests safely and effectively. [O2]
- Use the results of identification tests to identify the ions present in a single ionic compound. [O3]

Skills development

- WS 2.2 Plan experiments or devise procedures to make observations, produce or characterise a substance, test hypothesis, check data or explore phenomena.
- WS 2.3 Apply knowledge of a range of techniques, instruments, apparatus and materials to select those most appropriate to the experiment.
- WS 2.4 Carry out experiments appropriately having due regard for the correct manipulation of apparatus, the accuracy of measurements and health and safety concerns.
- WS 2.6 Make and record observations and measurements using a range of apparatus and methods.
- WS 2.7 Evaluate methods and suggest possible improvements and further investigations.
- WS 3.5 Interpret observations and other data (presented in verbal, diagrammatic, graphical, symbolic or numerical form), including identifying patterns and trends, making inferences and drawing conclusions.

Resources needed Practical sheets 4.12.1 and 4.12.2, Technician's notes 4.12

Key vocabulary precipitate, flame test, hydrochloric acid, barium chloride, silver nitrate

Teaching and learning

Engage

- Allow students time to read 'Carrying out analysis' in Section 4.12 of the Student Book. Ask them to work in pairs and write down briefly:
  - o how they would test for carbon dioxide, hydrogen, oxygen and chlorine gases
  - o what flame tests and precipitation reactions with sodium hydroxide solution test for
  - o how they would test for carbonate ions, sulfate ions and halide ions.
  Discuss their answers and refer students to the relevant sections in the Student Book. [O1]

- Students can answer questions 1, 2 and 3 in Section 4.12 in the Student Book. [O1]

Challenge and develop

- Introduce students to the practical. They are going to carry out tests to **identify** the ions present in an unknown ionic compound.

Low demand

- Students can use Practical sheet 4.12.1 to identify compound X. This sheet gives full instructions, step-by-step. Remind students about reducing the risk of contamination of the sample and reagents, and of working safely. You may want them to **produce** a risk assessment, or to **discuss** one. [O1, O2]

Standard and high demand

- Students can use Practical sheet 4.12.2 to **identify** compound X. This sheet requires them to **plan** their procedure before analysing compound X, and to **construct** a results table. They will find 'Making and recording observations' in Section 4.12 of the Student Book useful to plan the order of their tests. You will need to check their plans. [O1, O2]

Explain

- Students can **complete** the Evaluation sections of the relevant practical sheets to identify the unknown compound. They can **suggest** any improvements to the procedure they used and any changes to the apparatus used. [O3]

Consolidate and apply

- Students can **agree** on the identity of X. [O3]

- They can **identify** compounds A to E in 'Analysing unknown compounds' in Section 4.12 in the Student Book. [O3]

Extend

Ask students able to progress further to:

- Repeat the procedure to **identify** another compound, or a mixture of two compounds. [O2, O3]

- Given the concentrations of the reagents and volumes used (e.g. 2 cm^3), students can **calculate** the maximum possible mass of precipitate that can be made in precipitation reactions used. They can **suggest** how the actual yield could be measured and the percentage yield calculated. [O3]

Plenary suggestions

Students can work in pairs to **agree** on the most important three safety tips when doing practical work. They can double up and **share** their lists, and eventually **agree** as a class on the most important items. This procedure can be repeated, but this time **agreeing** on the three most important practical tips to ensure a valid result.

Answers to Evaluations on Practical sheets 4.12.1 and 4.12.2

Depends on the identity of compound X used. Suggestions on exprimental improvements depend on student's experiences, but may include the use of well plates, spotting tiles, spectroscopy, emission spectroscopy

When and how to use these pages: Check your progress, Worked example and End of chapter test

Check your progress

Check your progress is a summary of what students should know and be able to do when they have completed the chapter. Check your progress is organised in three columns to show how ideas and skills progress in sophistication. Students aiming for top grades need to have mastered all the skills and ideas articulated in the final column (shaded pink in the Student book).

Check your progress can be used for individual or class revision using any combination of the suggestions below:

- Ask students to construct a mind map linking the points in Check your progress
- Work through Check your progress as a class and note the points that need further discussion
- Ask the students to tick the boxes on the Check your progress worksheet (Teacher Pack CD). Any points they have not been confident to tick they should revisit in the Student Book.
- Ask students to do further research on the different points listed in Check your progress
- Students work in pairs and ask each other what points they think they can do and why they think they can do those, and not others

Worked example

The worked example talks students through a series of exam-style questions. Sample student answers are provided, which are annotated to show how they could be improved.

- Give students the Worked example worksheet (Teacher Pack CD). The annotation boxes on this are blank. Ask students to discuss and write their own improvements before reviewing the annotated Worked example in the Student Book. This can be done as an individual, group or class activity.

End of chapter test

The End of chapter test gives students the opportunity to practice answering the different types of questions that they will encounter in their final exams. You can use the Marking grid provided in this Teacher Pack or on the CD Rom to analyse results. This shows the Assessment Objective for each question, so you can review trends and see individual student and class performance in answering questions for the different Assessment Objectives and to highlight areas for improvement.

- Questions could be used as a test once you have completed the chapter
- Questions could be worked through as part of a revision lesson
- Ask Students to mark each other's work and then talk through the mark scheme provided
- As a class, make a list of questions that most students did not get right. Work through these as a class.

Marking Grid for End of Single Chemistry of Chapter 4 Test 3816

Student Name																			
Getting Started [Foundation Tier]																			
Q. 1 (AO1) 1 mark																			
Q. 2 (AO1) 1 mark																			
Q. 3 (AO1) 2 marks																			
Q. 4 (AO1) 1 mark																			
Q. 5 (AO2) 2 marks																			
Q. 6 (AO1) 1 mark																			
Q. 7 (AO1) 1 mark																			
Q. 8 (AO1) 2 marks																			
Going Further [Foundation and Higher Tiers]																			
Q. 9 (AO1) 1 mark																			
Q. 10 (AO1) 1 mark																			
Q. 11 (AO1) 2 marks																			
Q. 12 (AO3) 2 marks																			
Q. 13 (AO1) 1 mark																			
Q. 14 (AO2) 2 marks																			
Q. 15 (AO1) 2 marks																			
More Challenging [Higher Tier]																			
Q. 16 (AO2) 2 marks																			
Q. 17 (AO3) 6 mark																			
Q. 18 (AO2) 2 marks																			
Q. 19 (AO2) 1 mark																			
Q. 20 (AO2) 2 marks																			
Most demanding [Higher Tier]																			
Q. 21 (AO3) 4 marks																			
Q. 22 (AO1) 2 marks																			
Q. 23 (AO3) 4 marks																			
Total marks																			
Percentage																			

Check your progress

You should be able to:

describe the unreactivity of the noble gases →	explain the trend down Group 0 of increasing boiling point →	explain the trend down Group 0 of increasing boiling point in terms of atomic mass
predict the reactions with water of Group 1 elements lower than potassium →	predict and explain the relative reactivity down the groups →	explain the trend down the group of increasing reactivity by electron structure
recall the colours of the halogens and the order of reactivity of chlorine, bromine and iodine →	describe the order of reactivity and explain the displacement of halogens →	predict displacement reaction outcomes of halogens other than chlorine, bromine and iodine.
explain that a stable outer shell of electrons makes noble gases unreactive →	predict the properties of 'unknown' elements from their position in the group →	explain the trend of increasing reactivity in terms of electron structure
explain that transition metals have higher melting points and are stronger and harder than Group 1 metals →	explain that transition metals are less reactive than Group 1 metals and form coloured solutions →	explain that transition metals form ions of different charges and are useful as catalysts
describe the reactions, if any, of metals with water or dilute acids to place these metals in order of reactivity →	explain how the reactivity is related to the tendency of the metal to form its positive ion →	deduce an order of reactivity of metals based on experimental results
use experimental results of displacement reactions to confirm the reactivity series →	use the reactivity series to predict displacement reactions →	write ionic equations for displacement reactions
carry out flame test procedures →	identify the colours of flames of ions →	identify substances from the results of the tests
recognise the precipitate colour of metal hydroxides →	explain how to use sodium hydroxide to test for metal ions →	write balanced equations of producing insoluble hydroxides
identify the tests for carbonates →	explain the tests for halides and sulfates →	identify anions and cations from the results of tests
identify advantages of instrumental methods compared with the chemical tests →	describe some instrumental techniques →	explain the data provided by instrumental techniques
describe flame emission spectroscopy →	identify the advantages of instrumental methods compared with the chemical tests →	interpret an instrumental result given appropriate data in chart or tabular form, using a reference set

© HarperCollinsPublishers Limited 2016

Worked example

Sam and Alex are researching some properties of Group 1 metals.

1 **Shade the section of the periodic table where the Group 1 metals are found.**

1 **H** hydrogen 1.0			
3 **Li** lithium 6.9	4 **Be** beryllium 9.0		
11 **Na** sodium 23.0	12 **Mg** magnesium 24.3		
19 **K** potassium 39.1	20 **Ca** calcium 40.1	21 **Sc** scandium 45.0	22 **Ti** titanium 47.9
37 **Rb** rubidium 85.5	38 **Sr** strontium 87.6	39 **Y** yttrium 88.9	40 **Zr** zirconium 91.2
55 **Cs** caesium 132.9	56 **Ba** barium 137.3	57-71 lanthanides	72 **Hf** hafnium 178.5
87 **Fr** francium	88 **Ra** radium	89-103 actinides	104 **Rf** rutherfordium

> This is incorrect. The first column needs to be shaded.

The two metals they are researching are sodium and potassium.

2 **Write down two properties that these metals have.**

They are shiny when cut

They have a very high density

> The first property is correct. However, sodium and potassium float on water so have a density less than water. The student may be confusing Group 1 metals with transition metals, which have high density.

Sam and Alex find out that sodium and potassium react with water. They find that sodium reacts with water to make sodium hydroxide and that hydrogen is given off.

3 **Write a word equation for the reaction**

sodium + water → sodium hydroxide

> The reactants are correct but hydrogen needs to be written as a product on the right hand side.

Sam says that potassium reacts more vigorously than sodium but Alex says that they are in the same group so they react the same.

4 **a Explain why Sam is correct about the trend.**

The lower down the group the better they react

> This answer could be expressed more clearly by substituting the word 'better' with 'more vigorously'.

b Explain why Sam is correct using ideas about the structure of atoms.

The bigger the atom the quicker the reaction

> This answer needs more detail. The further away the outer electron is from the nucleus the more easily it is 'lost', as the pull by the positive nucleus on the negative electron is less.

 © HarperCollins*Publishers* Limited 2016

Monitoring & controlling chemical reactions: Introduction

When and how to use these pages

This unit:builds on ideas introduced earlier about bonding and develops students' abilities to analyse data and use symbol equations to describe reactions.

Overview of the unit

In this unit, students will use reacting masses to balance equations, learn about gas volumes, calculate theoretical and percentage yields and use titration to find the concentration of an acid. They use a range of methods to measure reaction rates; identify ways of speeding up reactions, and use collision theory and ideas about activation energy to make predictions. They also explore reversible reactions and use Le Chatelier's principle to predict the effects of changing temperatures, pressures and concentrations on equilibrium systems.

As part of the practical work for this unit students will identify independent, dependent and control variables; identify the main hazards in practical contexts; plan experiments; carry out experiments appropriately; describe techniques; read measurements off scales; make and record observations; present data appropriately; recognise and describe patterns and trends; use models in explanations; use data to make predictions, and communicate findings and reasoned conclusions.

There are a number of opportunities for the students to use mathematics. They will find the mean of sets of experimental values; translate information between graphical and numeric forms; use ratios to write simple formulae; substitute numerical values into chemical equations to balance them; plot variables from experimental data; draw a tangent to a curve and use its gradient as a measure of the reaction rate.

Obstacles to learning

Students may struggle to apply collision theory to explain changes in reaction rates. In particular they may not appreciate that breaking a solid into smaller particles exposes a larger overall surface area to collisions. They may need to be reminded that it is the collision rate, rather than the total number of collisions, that determines the reaction rate. They may need help to distinguish between the effect of a catalyst, which provides a lower activation energy; the effect of increasing the concentration, which increases the collision frequency; and the effect of raising the temperature, which increases both the collision frequency and the fraction of the collisions that supply the activation energy.

Practicals in this unit

In this unit students will do the following practical work:

- Use a titration to find the concentration of hydrochloric acid.

- React magnesium with hydrochloric acid and determine the total volume of hydrogen produced every 10 seconds in order to measure the reaction rate.

- Measure the total volume of carbon dioxide produced when varying amounts of acid are reacted with excess marble chips.

- React marble chips with hydrochloric acid and determine the total mass loss every 10 seconds in order to measure the reaction rate.

- Measure the effect of temperature on the reaction between sodium thiosulfate and hydrochloric acid.

- Investigate the effect of concentration on the rate of the reaction between magnesium with hydrochloric acid.

- React different amounts of magnesium ribbon with hydrochloric acid and measure the volume of hydrogen produced.

- Test potential catalysts for the reaction between aluminium foil and hydrochloric acid.

They will observe the following demonstrations:

- The preparation of known concentrations of potassium manganate(VII) solution.

- Finding the volume of one mole of hydrogen gas.

- The reversible dehydration of blue copper(II) sulfate(VI)-5-water.

- How a reversible reaction at equilibrium can be disturbed by changing a concentration.

- How a reversible reaction at equilibrium can be disturbed by changing the temperature

 © HarperCollins*Publishers* Limited 2016

	Lesson title	Overarching objectives
1	Concentration of solutions	Know that the concentration of a solution can be measured in g/dm^3 and in mol/dm^3.
2	Using concentrations of solutions	Use a titration to find the concentration of one of the reactants, given the concentration of the other reactant.
3	Practical: Finding the reacting volumes of solutions of acid and alkali by titration	Find the reacting volumes of solutions of acid and alkali by titration
4	Amounts of substance in volumes of gases	Calculate the volumes of gaseous reactants and products from balanced symbol equations, given that one mole of gas occupies 24 dm^3 at room temperature and pressure.
5	Key concept: Percentage yield	Calculate percentage yields from actual yields and from theoretical yields calculated according to balanced equations and a given amount of reactant.
6	Atom economy	Calculate atom economy and be able to select reaction routes to give high atom economies.
7	Maths skills: Change the subject of an equation	Use balanced symbol equations to determine the masses of reactants needed or the masses of products expected.
8	Measuring rates	Explain how to measure the rate of a reaction and interpret graphs showing the stages of a reaction.
9	Calculating rates	Plot and interpret graphs showing product formed, or reactant used up, against time, calculate average reaction rates and use the gradients of tangents to measure of the reaction rates at specific times.
10	Factors affecting rates	Identify factors which affect the rates of reactions and predict their affects.
11	Collision theory	Use collision theory to explain activation energy and rate changes.
12	Catalysts	Explain how catalysts work.
13	Factors increasing the rate	Predict the effects of changing conditions on reaction rates and recognise proportional relationships.
14	Practical: Investigate how changes in concentration affect the rate of reactions by a method involving the production of a gas and a method involving a colour change	Investigate how changes in concentration affect the rate of reactions by a method involving the production of a gas and a method involving a colour change
15	Reversible reactions and energy changes	Explain what a reversible reaction is and describe the energy changes involved.
16	Equilibrium	Describe how a dynamic equilibrium is reached and predict the effects of adding reactant or product.
17	Changing concentration and equilibrium	Apply Le Chatelier's principle to predict how changing reactant or product concentrations shifts the position of equilibrium.
18	Changing temperature and equilibrium	Predict the effect of temperature changes on the position of equilibrium for exothermic and endothermic reactions.
19	Changing pressure and equilibrium	Use Le Chatelier's principle to make predictions about changing pressures.
21	Maths skills: Use the slope of a tangent as a measure of rate of change	Plot variables from experimental data; draw a tangent to a curve and use its gradient as a measure of the reaction rate.

179 © HarperCollins*Publishers* Limited 2016

Lesson 1: Concentration of solutions

Lesson overview

OCR Specification reference

OCR C5.1

Learning objectives

- Relate mass, volume and concentration.
- Calculate the mass of solute in solution.
- Relate concentration in mol/dm^3 to mass and volume.

Learning outcomes

- Understand that the concentration of a solution depends on the mass of solute present and the volume of the solution. [O1]
- Calculate the mass of solute in a given volume of solution of known concentration in terms of mass per given volume of solution. [O2]
- Explain how the concentration of a solution in mol/dm^3 is related to the mass of solute and the volume of solution and carry out calculations based on concentration in mol/dm^3. [O3]

Skills development

- WS 3.2 Translate data from one form to another.
- WS 4.1 Use scientific vocabulary, terminology and definitions.
- WS 4.3 Use SI units (e.g. kg, g, mg, km, m, mm, kJ, J) and IUPAC nomenclature unless inappropriate.
- WS 4.5 Interconvert units.

Maths focus Recognise and use expressions in decimal form; use ratios, fractions and decimal form; change the subject of an equation; substitute numerical values into algebraic using appropriate units for physical quantities

Resources needed Practical sheet 5.1, Worksheet 5.1, Technician's notes 5.1

Key vocabulary solute, solvent, solution, concentration

Teaching and learning

Engage

- Remind students that many of the chemicals they use are solutions – examples are dilute and concentrated acids and alkalis, solutions of salts, etc. Students can list the solutions they use outside the lab (e.g. bleach, some soft drinks, vinegar, etc.). Remind students about formulated products and explain that maintaining the same concentration is essential. [O1]

Challenge and develop

High demand

- Ask students what they would need to measure to ensure that the concentration of all bottles of lemonade were the same. Establish that the mass of solute (sugar, flavouring) and the volume of the solution need to be measured and kept constant. [O1]

- Introduce Practical sheet 5.1. This is a demonstration of making up two solutions of potassium manganate(VII) of different concentrations. The sheet includes details for making up the solutions and student questions at appropriate stages. Make up solution A. Emphasise that it is the volume of solution that is measured, *not* the volume of solvent used; hence, the solution has a volume of 100 cm^3, not 100 cm^3 of water used. [O1]

Explain

High demand

- Explain that the concentration of solution A can be expressed in g/100 cm^3 solution. Tell students that concentrations are also frequently measured in g/dm^3. (Remind students that 1 dm^3 = 1000 cm^3.) Either work through 'Checking the correct units' in Section 5.1 in the Student Book, or explain how g/100 cm^3 is converted into g/dm^3. [O2]

- Students can answer questions 2(a) and 2(b) in Practical sheet 5.1 and question 1 in Worksheet 5.1.

- **Explain** that concentrations are also measured in mol/dm^3. Introduce the formula triangle in Section 5.1 of the Student Book and show students how it is used to calculate the amount in moles, the concentration in mol/dm^3 or the volume in dm^3. Emphasise that you must use the correct units in the triangle and may need to convert these first. You can work through the example in the Student Book and students can answer question 2 on Worksheet 5.1. Check their answers. [O3]

Consolidate and apply

- Using Practical sheet 5.1 make up solution B. This is a more dilute solution of potassium manganate(VII). Students can answer the questions in step 4. Check their answers. [O2, O3]

- Tell students that the dilute sodium hydroxide used in the lab is 2 mol/dm^3. The technician needs to make up 1 dm^3 of dilute sodium hydroxide solution. Ask what mass of sodium hydroxide must be weighed out. [O3]

Extend

Ask students able to progress further to:

- **Convert** the concentrations in question 1(b) and 1(c) in Worksheet 5.1 to concentrations in mol/dm^3. [O3]

- **Find out** how volumetric flasks are used to make up solutions and **explain** why they are more accurate than measuring cylinders. [O1]

- **Suggest** how solution B in Practical sheet 5.1 can be used to make a solution of potassium manganate(VII) ten times more dilute (0.0002 mol/dm^3) than solution B.

Plenary suggestions

Students work in pairs. One is 'g/dm^3' and the other is 'mol/dm^3'. You can allow them a few minutes to read over the lesson's work. They can then **explain** to each other in turn how concentration is measured using their units.

One student can sit in the hot seat. The remainder **compile** a question on the lesson's content. Students can ask their questions in turn. They take the 'hot seat' when an answer is wrong.

Answers to Worksheet 5.1

1 a)(i) 4 g/100 cm^3; (ii) 15 g/100 cm^3; b)(i) 68 g/dm^3; (ii)1.0 g/dm^3; (iii) 19 g/dm^3; (iv) 0.1 g/dm^3; (v) 27 g/dm^3; c)(i) 5 g/dm^3; (ii) 10 g/dm^3; (iii) 9.2 g/dm^3; (iv) 0.4 g/dm^3;(v) 20 g/dm^3; 2. a) 0.2 mol; b) 4 mol/dm^3; c) 200 cm^3; d) 0.02 mol/dm^3; e) 0.1 mol; f) 1.0 mol/dm^3

Answers to Practical sheet 5.1

1. a) 3.16 g/cm^3; b) 31.6 g/dm^3; c) 0. 2 mol.dm^3; 2, a) 0.31 g/100 cm^3; b) 3.1 g/dm^3; c) 0.02 mol/dm^3; d) Solution B is paler than solution A

Lesson 2: Using concentrations of solutions

Lesson overview

OCR Specification reference

OCR C5.1

Learning objectives

- Describe how to carry out titrations.
- Calculate the concentrations in titrations in mol/dm^3 and in g/dm^3.
- Explain how the concentration of a solution in mol/dm^3 is related to the mass of the mass of the solute and the volume of the solution.

Learning outcomes

- Describe how a burette and pipette are used to carry out a titration with a strong acid and a strong alkali using an indicator. [O1]
- Use data from the results of a titration to calculate the unknown concentration of a reactant in mol/dm^3 and in g/dm^3. [O2]
- Explain how the concentration of a solution in mol/dm^3 can be used to calculate the amount of solute in a given volume. [O3]

Skills development

- WS 4.2 Recognise the importance of scientific quantities and understand how they are determined.
- WS 4.3 Use SI units (e.g. kg, g, mg, km, m, mm, kJ, J) and IUPAC nomenclature unless inappropriate.
- WS 4.6 Use an appropriate number of significant figures in calculations.

Maths focus Recognise and use expressions in decimal form; use ratios, fractions and percentages; change the subject of an equation; substitute numerical values into algebraic equations using appropriate units for physical quantities

Resources needed Practical sheet 5.2; Worksheet 5.2; Technician's notes 5.2

Digital resources YouTube videos – titrations

Key vocabulary titration, burette, pipette, indicator

Teaching and learning

Engage

- Remind students how concentrations can be expressed in mol/dm^3. Set a puzzle – if the volumes of two solutions that react completely are known, and the concentration of one of them, how do you know the concentration of the other solution? Tell students the answer is do a *titration*. [O1]

Challenge and develop

High demand

- Introduce Practical sheet 5.2 and set the problem above. Students are provided with a 0.1 mol/dm^3 solution of sodium hydroxide. They are also given a dilute solution of hydrochloric acid. They have to carry out a titration to determine the concentration of the acid. Explain that this involves adding a precise volume of acid from the burette that reacts completely with a known volume of sodium hydroxide solution. [O1]

- Ask students to write a balanced chemical equation for the reaction. Discuss what will be present when exact (equivalent) volumes of acid and alkali are reacted completely. Establish that only sodium chloride solution should be present. Ask students how pH values will change and inform them that we can tell when exact volumes have been added from pH values. [O1]

- Show students the apparatus they will be using – a burette, a pipette and a pipette filler. Explain/demonstrate how to set up and fill a burette (it is assumed the burettes are clean and dry). Demonstrate how to use the tap on the burette and run a small volume of acid through the burette to fill the tip. Demonstrate how to use the pipette filler to measure 25 cm^3 of solution. You may wish students to **practise** this procedure first using water. Explain the use of phenolphthalein indicator (methyl orange can be used as an alternative). Describe how to carry out the titration, taking the initial and final readings from the bottom of the meniscus and to one decimal place. Worksheet 5.2 asks for a trial run and three titres within 0.1 cm^3. You may wish to amend this depending on the time available and the ability of the students. There are several useful videos on YouTube you could use to demonstrate the procedure. [O1]

- Students can answer question 1 in Worksheet 5.2. [O1]

- Allow students to carry out their titrations and **record** their burette readings in the table in Worksheet 5.2. [O1]

Explain

High demand

- Explain how to calculate an average titre. Students can use 'Calculating with concentrations' in Section 5.2 of the Student Book. They can complete the Evaluation section in Practical sheet 5.2 to **calculate** the concentration of their hydrochloric acid. You could explain this step-by-step, allowing students to use the Student Book or just Practical sheet 5.2. Compare the different values obtained for the concentration of hydrochloric acid. [O2, O3]

- Students can answer question 2 in Worksheet 5.2 to **practise** titration calculations. [O2, O3]

Consolidate and apply

- Ask students to close all books, including worksheets and practical sheets. Ask them to work in pairs to **compile** a list of instructions that could be used by another group to carry out a titration. [O1]

- Explain or remind students that vinegar contains an acid called ethanoic acid. Ask them to suggest how they could find the concentration of ethanoic acid in vinegar. [O1, O2, O3]

Extend

Ask students able to progress further to:

- **Suggest** how a sample of dry sodium chloride could be obtained from the products of the titration. The sodium chloride must not be contaminated with indicator. This could be a practical exercise if time allows. [O1]

- **Consider** why different groups obtained different answers for the concentration of the hydrochloric acid. Students could list different sources of error. [O2]

- **Find out** what a 'standard solution' is and how one is used. [O1]

Plenary suggestions

Organise students into groups and assign each group a word relating to the lesson, such as 'burette', 'titre', 'pipette', 'concentration', etc. Ask some questions to which the answers are these words. Students can 'claim' their words.

Ask students to draw a large triangle on a sheet of paper with a smaller inverted triangle inside. In the three outer triangles, they write something they have done, something they have discussed and something they have seen. In the outer triangle, they write something they have **learned**.

Answers to Worksheet 5.2

1. a)(i) To measure a variable exact volume of a reactant; (ii) To measure a fixed exact volume of a reactant; (iii) To draw a solution into a pipette; (iv) Marks the end point of the reaction; (v) To see the colour of the indicator clearly; b)(i) From the bottom of the meniscus; (ii) When a permanent colour change is obtained; (iii) Support the burette on the bench/floor and use a small funnel; 2. a)(i) 0.00198 mol; (ii) 0.00198 mol; (iii) 0.079 mol/dm^3; (iv) 3.17 g/dm^3; b) 0.20 mol/dm^3

Answers to Practical sheet 5.2

Depends on student's experimental results

Lesson 3: Practical: Finding the reacting volumes of solutions of acid and alkali by titration

Lesson overview

OCR Specification reference

OCR C5.1

Learning objectives

- Use an acid to neutralise a known volume of alkali.
- Use a burette to determine the volume of an acid needed.
- Use the results to determine the concentration of an alkali.

Learning outcomes

- Describe how safety, the correct manipulation of apparatus and the accuracy of measurements are managed when titrations are carried out. [O1]
- Make and record observations and accurate measurements using burettes. [O2]
- Calculate the concentration of a solution from the concentration and volume of another. [O3]

Skills development

- WS 2.2 Describe how to carry out a titration.
- WS 2.4 Carry out a titration accurately.
- WS 2.6 Make and record measurements.

Resources needed Equipment as listed in Technician's notes 5.3; Practical sheet 5.3

Digital resources Presentation 5.3.1 'Titration'; Presentation 5.3.2 'Improving accuracy'

Key vocabulary titration, concentration, acid, alkali

Teaching and learning

Engage

- Display slide 1 of Presentation 5.3.1 'Titration' to remind students how they neutralised sodium hydroxide with hydrochloric acid in Practical sheet 5.3. Ask how the sodium hydroxide concentration affects the end-point. Students should **recognise** that the more concentrated the alkali, the more acid needs to be added to reach the end-point. [O1]

Challenge and develop

- Demonstrate how to prepare a burette safely by clamping it below eye level. Use a small funnel to pour in a few cubic centimetres of 0.4 mol/dm^3 hydrochloric acid with the tap open and a beaker under it. Then, once the tip of the burette is full of solution, close the tap and add more solution up to the zero mark. [O1]

- Demonstrate how to dispense exactly 20 cm^3 of sodium hydroxide solution into a conical flask using a volumetric pipette and add a few drops of indicator. [O1]

- Show the colours of methyl orange in acidic and alkaline solutions and demonstrate how to swirl the flask as acid is added to alkali to monitor the colour continuously. [O1]

- Pairs **determine** the volume of acid need to neutralise 20 cm^3 of sodium hydroxide solution following the method on Practical sheet 5.3. [O2]

Explain

- Students read 'Managing titrations safely' and 'Obtaining accurate measurements in a titration' in Section 5.3 of the Student Book. [O1]

© HarperCollins*Publishers* Limited 2016

Low demand

- Pairs answer questions 1–7 from Section 5.3 of the Student Book. [O1]

Standard and high demand

- Pairs describe how to carry out a titration safely and list ways of making the results as accurate as possible. [O1]

Consolidate and apply

- Use slide 2 of Presentation 5.3.1 'Titration' to show students how to use their results to **calculate** the concentration of the sodium hydroxide they used. Then collect the results of their calculations to display the range of values obtained. [O3]

Extend

- Ask students able to progress further to read the final part of Section 5.3 of the Student Book and answer questions 8–11. [O3]

Plenary suggestion

Display Presentation 5.3.2 'Improving accuracy' and ask groups to **suggest** how each of the procedures listed improves the accuracy of a titration.

 © HarperCollins*Publishers* Limited 2016

Lesson 4: Amounts of substance in volumes of gases

Lesson overview

OCR Specification reference

OCR C5.1

Learning objectives

- Explain that the same amount of any gas occupies the same volume at room temperature and pressure (rtp).
- Calculate the volume of a gas at rtp from its mass and relative formula mass.
- Calculate the volumes of gases from a balanced equation and a given volume of a reactant or product.

Learning outcomes

- Reason that the same amount of moles of any gas occupies the same volume at rtp, regardless of the mass of the gas molecules. [O1]
- Convert the mass of a gas into the amount in moles and calculate the volume occupied by that gas at rtp, given that 1 mole occupies 24 dm^3. [O2]
- Calculate the volumes of gaseous reactants and products from a balanced symbol equation given the volume of a gaseous reactant or product. [O3]

Skills development

- WS 1.2 Use a variety of models, such as representational, spatial, descriptive, computational and mathematical, to solve problems, make predictions and develop scientific explanations and understanding of familiar and unfamiliar facts.
- WS 4.1 Use scientific vocabulary, terminology and definitions.
- WS 4.2 Recognise the importance of scientific quantities and understand how they are determined.
- WS 4.3 Use SI units (e.g. kg, g, mg, km, m, mm, kJ, J) and IUPAC nomenclature unless inappropriate.
- WS 4.6 Use an appropriate number of significant figures in calculation.

Maths focus Recognise and use expressions in decimal form; use ratios, fractions and percentages; change the subject of an equation; substitute numerical values into algebraic equations using appropriate units for physical quantities

Resources needed Practical sheet 5.4; Worksheet 5.4; Technician's notes 5.4

Key vocabulary gas volume, room temperature and pressure, decimetre cubed, atmospheric pressure

Teaching and learning

Engage

- Write the equation 'Mg + 2HCl → $MgCl_2$ + H_2' on the board and ask students to calculate the mass of hydrogen gas produced by reacting 2.4 g magnesium (A_r of Mg is 24, A_r of H is 1, A_r of Cl is 35.5). [O1]
- Ask them how easy it is to weigh hydrogen gas or any gas. Establish that we usually measure volumes of gases. [O1]

Challenge and develop

High demand

- Students can read 'Comparing volumes of gases' in Section 5.4 of the Student Book. Reinforce the idea that the same amount of all gases in moles occupies the same volume. You can also explain this by saying that the size/mass of gas molecules is negligible compared to the relatively vast distances between them. [O1]
- Use Practical sheet 5.4 to demonstrate an experiment to measure the volume of gas produced when a known mass of magnesium reacts with dilute hydrochloric acid. (You may need to practise the technique of inverting the burette before the lesson.) Students can answer the Evaluation in Practical sheet 5.4 and

calculate the number of moles of magnesium reacted and the volume of 1 mole of hydrogen gas. You may prefer to work through this with them. They should obtain an answer close to 24 dm^3. [O2]

Explain

High demand

- Explain that 1 mole of any gas occupies 24 dm^3 at rtp, and this also means the formula mass of any gas occupies 24 dm^3. Show students how to calculate the volume of a gas at rtp from the amount of moles. For example, what volume does two moles of oxygen gas occupy at rtp? [O2]

- Show them how to calculate the volume occupied by a given mass of gas. For example, what volume does 6 g of hydrogen gas occupy? You can also use 'Moles of gases' in Section 5.4 of the Student Book. They can answer question 1 on Worksheet 5.4. Check their answers. [O2]

- Ensure that students understand that 'equal amounts of gases in moles occupy the same volume under the same conditions of temperature and pressure' (from the Student Book) and that this is 24 dm^3 at rtp. [O2]

- Show students how volumes of gases can be calculated from balanced symbol equations. You can work through the example of $CH_4 + 2O_2 \rightarrow CO_2 + 2H_2O$, showing initially that (at rtp) 24 dm^3 of methane reacts with 48 dm^3 of oxygen to produce 24 dm^3 of carbon dioxide and 48 dm^3 of water vapour. Ask students what volume of carbon dioxide is produce when 1.2 dm^3 of methane burns. They can answer question 2 in Worksheet 5.4. Check their answers. [O3]

Consolidate and apply

- Ask students to explain why nitrogen and hydrogen are mixed in a 1:3 molar ratio in the Haber process when ammonia is manufactured. [O1, O3]

- Remind them that methane and propane are both used as cooking and heating fuels, and that carbon dioxide is a greenhouse gas. Ask them to **calculate** which produces the larger volume of carbon dioxide gas per mole of fuel burned. [O3]

Extend

Ask students able to progress further to:

- **Suggest** how changing the temperature and pressure will affect the volume of 1 mole of gas. [O2]

- **Calculate** the volume of carbon dioxide produced when 2.2 g of propane burns completely in oxygen. [O3]

Plenary suggestions

Ask students to **compile** a question and an answer on gas volumes from this lesson. They can pair up and test their question.

Ask them to work in pairs to **construct** a list of 'What we have learned today'. Encourage them to include the different types of calculations they have done involving gas volumes. They can **share** their lists with the rest of the class.

Answers to Worksheet 5.4

1. a)(i) 2.4 dm^3; (ii) 48 dm^3; (iii) 2.4 dm^3; (iv) 12 dm^3; (v) 12 dm^3; b)(i) 32 g; (ii) 14 g; (iii) 1.6 g; (iv) 40 g; (v) 0.4 g; c)(i) 48 dm^3; (ii) 2.4 dm^3; (iii) 240 dm^3; (iv) 6 dm^3; (v) 12 dm^3; 2. a)(i) 4.8 dm^3; (ii) 4.8 dm^3; b)(i) 48 dm^3; (ii) 18 dm^3

Answers to Practical sheet 5.4

Depend on student's results

Lesson 5: Key concept: Percentage yield

Lesson overview

OCR Specification reference

OCR C5.1

Learning objectives

- Calculate the percentage yield from the actual yield.
- Identify the balanced equation needed for calculating yields.
- Calculate theoretical product amounts from reactant amounts.

Learning outcomes

- Use the formula

$$\text{percentage yield} = \frac{\text{mass of product actually made}}{\text{maximum theoretical mass of product}} \cdot 100$$

  to calculate percentage yields given actual and theoretical yields. [O1]
- Be able to write balanced symbol equations that can be used to calculate percentage yields. [O2]
- Use balanced symbol equations and formula masses to calculate theoretical amounts, given the amount of a reactant. [O3]

Skills development

- WS 3.3 Carry out and represent mathematical and statistical analysis.
- WS 4.2 Recognise the importance of scientific quantities and understand how they are determined.
- WS 4.6 Use an appropriate number of significant figures in calculations.

Maths focus Recognise and use expressions in decimal form; use ratios, fractions and percentages; use an appropriate number of significant figures; change the subject of an equation

Resources needed Two fly swats; Practical sheet 5.5; Worksheet 5.5; Technician's notes 5.5

Key vocabulary yield, percentage yield, theoretical yield, actual yield

Teaching and learning

Engage

- Remind students that reducing industrial waste in chemical reactions is a priority for environmental and economic reasons. Explain that it is important that chemists use exact amounts that react together. These can be calculated as well as the amount of product that can be produced. Explain that the amount of product made is called the *yield*. You can tell students that their experiment to produce ammonium phosphate should have produced 1 g of product. This is called the *theoretical yield*. [O1]

Challenge and develop

Low demand

- Students should have **measured** the mass of ammonium phosphate produced and hopefully realised it is less than 1 g. Explain that the amount they obtained is called the *actual yield*. Discuss reasons why actual yields are lower than calculated yields. Students can read 'Product yield' in Section 5.5 of the Student Book and answer questions 1 and 2. They can answer question 1 on Worksheet 5.5. [O1]

Explain

- Explain that we compare actual yields with theoretical yields, called the *percentage yield*, using the equation:

$$\text{percentage yield} = \frac{\text{mass of product actually made}}{\text{maximum theoretical mass of product}} \cdot 100$$

Students can **calculate** the percentage yield they obtained in the ammonium phosphate preparation. [O1]

Standard demand

- Students can read 'Calculating the percentage yield' in Section 5.5 of the Student Book and answer questions 3 and 4. They can answer question 2 on Worksheet 5.5. Check student's answers and that they have used the correct number of significant figures. This should be the same number used in the question. [O1]

High demand

- Work through examples to show students how to calculate theoretical yields. Tell them that copper can be obtained from copper sulfide by roasting it in oxygen. The equation is: $CuS + O_2 \rightarrow Cu + SO_2$. Show students how to calculate formula masses so that 96 g copper sulfide produces 64 g copper. Explain how, for example, the theoretical yield of 480 g copper sulfide can be calculated and then used to work out the percentage yield. Another suitable example involves the Haber process; you can calculate the maximum theoretical yield obtained from 60 g of hydrogen gas. [O2, O3]

- Students can answer question 3 on Worksheet 5.5. They can read 'Calculating theoretical yields' in Section 5.5 of the Student Book and answer questions 5 and 6. [O2, O3]

Consolidate and apply

Either

- Use Practical sheet 5.5 to demonstrate how to prepare a sample of magnesium oxide from magnesium. Ask students to itemise the measurements they would make and how they would use them to **calculate** the percentage yield for the reaction. [O2, O3]

Or

- Demonstrate the experiment on Practical sheet 5.5 and allow students to **calculate** the percentage yield for the reaction. [O2, O3]

Extend

Ask students able to progress further to:

- **Discuss** how the Haber process manipulates conditions to increase the percentage yield of ammonia obtained. [O1]

- **Explain** why the reactant used to calculate the theoretical yield should be the limiting factor. [O2, O3]

- **Explain** the difference between percentage yield and atom economy. [O1, O2, O3]

Plenary suggestions

Write the terms 'actual yield', 'theoretical yield' and 'percentage yield' on the board. Two students are in the limelight, each equipped with a fly swat. The others can ask questions in turn to which the answers are one of the terms on the board. Students **compete** to see who can hit the correct answer first.

Each student needs two strips of paper. They write a question on one strip and the answer on the other. Shuffle and redistribute the strips so that each student has one question and one answer. Students can ask their questions in turn. The student with the correct answer responds and asks the next question.

Answers to Worksheet 5.5

1. a)(i) The amount obtained practically; (ii) The calculated yield; (iii)

$$\text{percentage yield} = \frac{\text{mass of product actually made}}{\text{maximum theoretical mass of product}} \cdot 100$$

b) Not all reactants reacted; product was lost during the stages; unwanted reactions; c) The calculated maximum yield is obtained; d) 50% of the calculated yield is obtained; 2. a) 66.7%; b) 95.7%; c) 50%; d) 57%; e) 52.4%; 3. a)(i) 1700 g; (ii) 29.4%; b)(i) 6600 g; (ii) 87.88%; c)(i) 820 tonnes; (ii) 87.8%

Answers to Practical sheet 5.5

1. Depends on results; 2. $2Mg + O_2 \rightarrow 2MgO$; 3. Depends on results; 4. Depends on results

Lesson 6: Atom economy

Lesson overview

OCR Specification reference

OCR C5.1

Learning objectives

- Identify the balanced equation of a reaction.
- Calculate the atom economy of a reaction to form a product.
- Explain why a particular reaction pathway is chosen.

Learning outcomes

- Balance a symbol equation for a chemical reaction to produce a specified product. [O1]
- Use the formula:

$$\text{atom economy} = \frac{M_r \text{ desired product}}{\text{Sum of } M_r \text{ of all reactants}} \times 100$$

to calculate the atom economy of a reaction to produce a product from a balanced symbol equation. [O2]
- Given alternative reaction pathways to produce a product, calculate and use atom economy data to compare the sustainability of each pathway. [O3]

Skills development

- WS 1.4 Explain everyday and technological applications of science, evaluate associated personal, social, economic and environmental implications and make decisions based on the evaluation of evidence and argument.
- WS 3.3 Carry out and represent mathematical and statistical analysis.
- WS 3.8 Communicate the scientific rationale for investigations, methods used, findings and reasoned conclusions, through written and electronic reports and presentations using verbal, diagrammatic, graphical, numerical and symbolic forms.

Maths focus Recognise and use expressions in decimal form; use ratios, fractions and percentages; change the subject of an equation

Resources needed Worksheets 5.6.1 and 5.6.2

Digital resources RSC website: Green Chemistry, atom economy and sustainable development

Key vocabulary atom economy, sustainability, reaction pathway, by-product

Teaching and learning

Engage

- Ask students to work in pairs and write down what they think 'sustainable' means. Remind them that they have probably heard this word many times in the media. They can double up with another pair and **agree** on a joint definition. These can be **shared** with the class. [O1]

- Ask students to read the information in question 1 on Worksheet 5.6.1. They may need some help interpreting the second principle because the language is more complex. Discuss with the class what these two principles mean for the chemist making a product such as ibuprofen or rocket fuel. Students can answer question 1 on Worksheet 5.6.1. Discuss their answers to 1a, 1b and 1c. You may find selected parts of the [Google search string:] Green Chemistry, atom economy and sustainable development worksheet on the RSC website useful if students are finding the concepts difficult. (This also includes percentage yield, which is covered Lesson 5.5.). [O1]

Challenge and develop

- Low demand: Tell students that before chemists can choose chemical reactions that produce minimum waste, they must know the correctly balanced symbol equations for the reactions. Students can answer question 2 on Worksheet 5.6.1. [O1]

- Standard demand: Introduce the idea of atom economy as a way of measuring the amount of starting material that ends up as useful product, and hence the amount of waste produced. Students can read 'The atom economy' in Section 5.6 in their Student Book. You could refer back to the equations students balanced in question 2 on Worksheet 5.6.1 as additional examples of reactions with a 100% atom economy and reactions with a lower than 100% atom economy. Students can answer questions 1 and 2 in Section 5.6 of the Student Book, and question 1 on Worksheet 5.6.2. [O2]

Explain

- Explain that reactions with a 100% atom economy are desirable in the chemical industry and that the atom economy for a reaction can be calculated from its balanced symbol equation. Students can read 'Calculating atom economy' in Section 5.6 of the Student Book. You may need to revisit calculating relative formula mass. Students can answer questions 3 and 4 in the Student Book and question 2 on Worksheet 5.6.2. [O2]

- High demand: Reinforce the idea that in order to make the chemical industry sustainable, reactions with a high atom economy are desirable because there is less waste/potential pollution, etc. Explain that many products can be manufactured using different chemical reactions and chemists can choose the best reaction, or pathway. Discuss the use of by-products to increase the atom economy of a reaction. Students can read 'Choosing reaction pathways' in Section 5.6 in the Student Book and answer question 3 on Worksheet 5.6.2. Check their answers. [O3]

Consolidate and apply

- Tell students that when ibuprofen was first manufactured, it involved six steps, and so six chemical reactions. More waste than medicine was produced, most of which ended up on the scrap heap, and its atom economy was 40%. Today, ibuprofen is manufactured in three steps and the atom economy is 77%. Ask students to work in pairs and list three advantages of using the modern method to manufacture Ibuprofen. **Share** these as a class. [O2, O3]

- Students can calculate the atom economies for the reactions to produce calcium sulfate in question 2b on Worksheet 5.6.1. [O2]

Extend

Ask students able to progress further to:

- **Suggest** what the other principles of green chemistry might be, or **carry out** an online search to find them out. [O1]

- **Discuss** what other factors may lead to a reaction with the highest atom economy not being chosen as the manufacturing reaction. [O3]

Plenary suggestions

Ask students to work in groups of two or three and **produce** a short (maximum 30 seconds) mini presentation to explain to a chemical manufacturer why they need to use chemical reactions with high atom economies. They can give their mini presentations to the class.

Ask students work in pairs and list the three most important ideas they have learned this lesson. They can **share** their ideas with the class and possibly **compile** a class list.

Answers to Worksheet 5.6.1

1. a) Include pollution, treatment of toxic waste, disposal; b) Include cost, resources not wasted; c) Choose a reaction with few/no other products; 2. a)(i) 37.5% and 60%; (ii) $TiO_2 \rightarrow Ti + O_2$; b)(i) 46.7% and 47.0%; (ii) $2NH_3 + H_2O_2 \rightarrow N_2H_4 + 2H_2O$

Answers to Worksheet 5.6.2

1. a), b), e) are 100%; c), d) are 0–100%; 2. a) 56%; b) 49.3%; c) 70%; d) 45.9%; e) 51.1%; f) 17.6%; 3. a)(i) $2NH_3 + H_2SO_4 \rightarrow (NH_4)_2SO_4$; (ii) $NH_3 + HNO_3 \rightarrow NH_4NO_3$; b)(i) $CaO + H_2SO_4 \rightarrow CaSO_4 + H_2O$; (ii) $Ca(OH)_2 + H_2SO_4 \rightarrow CaSO_4 + 2H_2O$; (iii) $CaCO_3 + H_2SO_4 \rightarrow CaSO_4 + H_2O + CO_2$; (iv) $2SO_2 + 2CaCO_3 + O_2 \rightarrow 2CaSO_4 + 2CO_2$; c)(i) $NH_2Cl + NH_3 \rightarrow N_2H_4 + HCl$; (ii) $2NH_3 + H_2O_2 \rightarrow N_2H_4 + 2H_2O$

Lesson 7: Maths skills: Change the subject of an equation

Lesson overview

OCR Specification reference

OCR C5.1

Learning objectives

- Use equations to demonstrate conservation.
- Rearrange the subject of an equation.
- Carry out multi-step calculations.

Learning outcomes

- Use balanced chemical equations to calculate the mass of a reactant or product given the reacting masses of all the other reactants and products. [O1]
- Understand and be able to apply the mathematical steps involved in rearranging the subject of equations. [O2]
- Carry out multi-step calculations, such as calculating a concentration from titration results, or calculating a mass of product or reactant from a balanced chemical equation. [O3]

Skills development

- WS 3.3 Carry out and represent mathematical and statistical analysis.
- WS 4.6 Use an appropriate number of significant figures in calculations.

Maths focus Recognise and use expressions in decimal form; use ratios, ratios and percentages; change the subject of an equation

Resources needed Periodic table; Worksheet 5.7

Digital resources Computer and data projector, TES Powerpoint – Changing the subject of an equation, class internet access

Key vocabulary equation, rearrange, multi-step calculation, multiply

Teaching and learning

Engage

Ask students if they think we can totally destroy matter. Discuss their answers and explain that mass is always conserved. Write 'A = BC' on the board and ask them to write expressions for B and for C. Discuss their answers and assess their current understanding. Explain that this is a very important skill that is used frequently in chemistry calculations. Students could **recall** where and when they have had to **rearrange** equations in this topic. [O2]

Challenge and develop

Low demand

- Explain that if we start with 100 g of reactants in a chemical reaction, we will make 100 g of products. Write the equation '$2NaOH + CuSO_4 \rightarrow Na_2SO_4 + Cu(OH)_2$' on the board. Students can use A_r values to **calculate** the reacting masses of each substance in the equation (80 g + 160 g → 142 g + 98 g). Ask them to work out the sum of the reactants and the sum of the products. They can try this out with other examples. [O1]

- Students can read 'Conservation of mass' in Section 5.7 of the Student Book to find out how to calculate the mass of a missing reactant or product. They can answer question 1 in Worksheet 5.7. Check their answers. [O1]

192

© HarperCollins*Publishers* Limited 2016

Explain

Standard demand

- You could ask students to write the missing step that most of them have probably done mentally in question 1(b) in Worksheet 5.7. Explain that if the numbers are more complex, they will need to go through this step, so they need to understand it. Students can read 'Calculating the mass needed' in Section 5.7 of the Student Book. You could show them a simpler method whereby 40 g MgO → 120 g MgSO$_4$;

  and so, 2 g MgO → $\frac{120}{40}$ × 2 g MgSO$_4$ or 6 g. Some students will find this easier.

- If students need help in changing the subject of an equation, you could show them the TES Powerpoint on the subject. You will need to review its suitability first. [O2]

- They can answer questions 1–4 in Section 5.7 in the Student Book. They can answer question 2 in Worksheet 5.7.1. [O2]

High demand

- Students can read 'Finding the concentration of an acid' in Section 5.7 in the Student Book. They can answer questions 5 and 6. They can answer question 3 in Worksheet 5.7. This involves changing the subject of the

  equation in: concentration (mol/dm^3) = $\dfrac{\text{number of moles}}{\text{volume (dm}^3)}$ and carrying out calculations. [O3]

Consolidate and apply

- Ask students to work in pairs to **compile** a step-by-step guide to changing the subject of an equation without using the Student Book or notes. They can try their guides out on other pairs. [O2]

- Alternatively, students could **devise** their own Powerpoint presentation on 'How to change the subject of an equation'. [O2]

Extend

Ask students able to progress further to **devise** an equation to find a missing reacting mass in the reaction A + B → C + D when all other masses are known. Masses are based on 1 mol each of A, B, C and D reacting. They can **devise** four equations – one for A, one for B, etc. [O1]

Plenary suggestions

Students can make up an equation using A, B and C. They can **devise** a question on changing the subject of their equation. They can work in pairs to test their questions out on each other. This could be made more difficult by introducing D into the equation.

Ask students to work in pairs to list three big ideas about quantitative chemistry they have learned from this lesson. They can **share** their ideas with another pair and with the class to **produce** a class list of important ideas.

Answers to Worksheet 5.7

1. a)(i) 56 g + 36.5 g → 74.5 g + 18 g; (ii) Total mass reactants = 92.5 g, total mass products = 92.5 g; 2. a) 44 g; b) 2.2 g; 3. a) 11.1 g; b) 246 g; 4. a)(i) Number of moles = concentration (mol/dm^3) × volume (dm^3);

(ii) volume (dm^3) = $\dfrac{\text{number of moles}}{\text{concentration (mol/dm}^3)}$; b) 0.01 mol; c) 37.5 cm^3; d) 50 cm^3

© HarperCollins*Publishers* Limited 2016

Lesson 8: Measuring rates

Lesson overview

OCR Specification reference

OCR C5.2

Learning objectives

- Measure the volume of gas given off during a reaction.
- Use the results to measure the reaction rate.
- Explore how the rate changes during the reaction.

Learning outcomes

- Describe how to measure the volume of gas produced in a reaction. [O1]
- Explain how the reaction rate can be measured. [O2]
- Interpret graphical data showing the stages of a reaction. [O3]

Skills development

- WS 2.2 Describe a practical procedure for a specific purpose.
- WS 2.6 Read measurements off a scale in a practical context.
- WS 3.1 Present data using appropriate methods.

Maths focus Plot two variables from experimental data

Resources needed Equipment as listed in Technician's notes 5.8; Practical sheet 5.8; Worksheet 5.8

Digital resources Presentation 5.8.1 'A fizzy reaction'; Presentation 5.8.2 'Which method?'; Graph plotter 5.8

Key vocabulary balance, gas syringe, measuring cylinder, volume

Teaching and learning

Engage

- Let off a party popper and explain that a fast chemical reaction inside the popper produced enough gas to blow out the paper streamers. Then display some rusted iron or an image of something rusting and emphasise that some reactions are much slower. [O2]

- Display Presentation 5.8.1 'A fizzy reaction' and ask groups to **suggest** as many ways as possible of measuring how fast the reaction is. [O2]

Challenge and develop

- Demonstrate how to collect the hydrogen released in the reaction of magnesium with hydrochloric acid following the method on Practical sheet 5.8. Make students aware of the need to set the gas syringe to zero with the bung out of the conical flask. Insert the bung and start the stopclock promptly as the magnesium is dropped into the acid, and swirl gently throughout the reaction to prevent bubbles of hydrogen from adhering to the magnesium. [O1]

- Groups of 3 or 4 collect the hydrogen released by the reaction of magnesium with hydrochloric acid following the method on Practical sheet 5.8. [O1]

Explain

- Students could **plot** their graphs by hand or use Graph plotter 5.8 to plot their results as they get them. If the number of computers is limited, each group could plot their results on a different sheet. Warn groups to remember which number is theirs so that they can identify their results when they are printed out. [O3]

Low demand

- Compare the results obtained with Figure 5.8 in Section 5.8 of the Student Book and ask students to **identify** what is the same about their graphs and what is different. They should **label** each part of the graph to show what is happening. If there are problems with any groups' results, the first graph on Worksheet 5.8 could be used instead. [O3]

Standard and high demand

- Ask students to **describe** and **explain** the shape of their graph. [O3]

Consolidate and apply

- Show a video to summarise ways of following a reaction [Google search string:] 5.8.2 Describe experimental procedures for measuring rates of reactions. [O2]

Low demand

- Students complete Worksheet 5.8 to **practice** interpreting graphs. [O3]

Standard and high demand

- Ask students to read Section 5.8 in the Student Book and complete questions 1–5. [O1, O2]

Extend

- Ask students able to progress further to identify when their reaction stopped and calculate the average reaction rate in cm^3 of hydrogen per second. [O2]

Plenary suggestion

Display each slide of Presentation 5.8.2 'Which method?' in turn and challenge students to **think** of as many ways as possible to follow each reaction. [O1, O2]

Answers to Worksheet 5.8

1. 47.5 cm^3; 2. 44–46 s; 3. Faster; 4. The reaction rate is decreasing/changing; 5. The reaction has finished; 6. a) 24 cm^3; b) 7 cm^3

Lesson 9: Calculating rates

Lesson overview

OCR Specification reference

OCR C5.2

Learning objectives

- Find out how to calculate rates of reaction.
- Use graphs to compare reaction rates.
- Use tangents to measure rates that change.

Learning outcomes

- Calculate the mean rate of a reaction from the product formed, or reactant used up, and the time taken. [O1]
- Draw and interpret graphs that show the amount of product formed, or reactant used up, against time. [O2]
- Use the slope of a tangent to measure the rate of reaction. [O3]

Skills development

- WS 2.2 Describe a practical procedure for a specific purpose.
- WS 3.1 Present data using appropriate methods.
- WS 3.2 Translate data from one form to another.

Maths focus Plot two variables from experimental data; draw and use the slope of a tangent to a curve as a measure of the rate of reaction

Resources needed Equipment as listed in the Technician's notes 5.9; Practical sheet 5.9; Worksheet 5.9

Digital resources Presentation 5.9.1 'Comparing rates' and Presentation 5.9.2 'Which units?'

Key vocabulary axes, construction lines, gradient, rate of reaction

Teaching and learning

Engage

- Display slide 1 of Presentation 5.9.1 'Comparing rates' and challenge groups to use the numbers on the axes to **explain** the difference between the red and blue graphs. [O2]

Challenge and develop

- Introduce the idea of an average rate of reaction, using slides 2 and 3 of Presentation 5.9.1 'Comparing rates'. [O1]

Low and standard demand

- Students use Worksheet 5.9 to **calculate** the mean rates of reaction for the two graphs in Figure 5.21 in Section 5.9 of the Student Book. [O1]

Standard and high demand

- Use Figure 5.20 in Section 5.9 of the Student Book to demonstrate how to calculate a gradient. Students could **calculate** the initial gradients of the blue and red graphs on the copy of Figure 5.21 on Worksheet 5.9. [O2, O3]

Explain

Low demand

- Students write instructions to **explain** how to calculate the mean rate of a reaction. [O1]

- Use slide 4 of Presentation 5.9.1 'Comparing rates' to remind students that the rate usually changes during a reaction. [O3]

Consolidate and apply

- Students follow Practical sheet 5.9 to measure the mass loss as marble chips react with hydrochloric acid and then plot a graph of their results. If insufficient balances are available the reaction could be demonstrated, or gas volumes could be measured instead using the method detailed in Practical sheet 5.9. Spreadsheets could be used to plot the graphs to leave more time for their interpretation. [O3]

Low demand

- Calculate the mean rate of the reaction between marble chips and an acid. [O1]

Standard and high demand

- Calculate the initial rate of the reaction between marble chips and an acid. [O3]

Extend

- Ask students able to progress further to explain why reactions are usually faster at the start. [O2]

Plenary suggestion

Use Presentation 5.9.2 'Which units?' to review the units used to compare rates. [O2]

Answers to Worksheet 5.9

Mean rate of graph 1 = 1.75 cm^3/s; mean rate of graph 2 = 1.08 cm^3/s

Lesson 10: Factors affecting rates

Lesson overview

OCR Specification reference

OCR C5.2

Learning objectives

- Measure the time taken to produce a specific amount of product.
- See how a reactant's temperature or concentration can affect this time.
- Investigate the effect of breaking up a solid reactant into smaller pieces.

Learning outcomes

- Identify factors that affect the rates of reactions. [O1]
- Describe how these factors affect reaction rates. [O2]
- Explain how changes of surface area affect rates. [O3]

Skills development

- WS 2.2 Identify independent, dependent and control variables.
- WS 2.4 Identify the main hazards in a practical context.
- WS 2.6 Make and record observations and measurements.

Maths focus Understand the relationship between a reaction time and a reaction rate

Resources needed Equipment as listed in the Technician's notes 5.10; Practical sheet 5.10; Worksheets 5.10.1 and 5.10.2

Digital resources Graph plotter 5.10; Presentation 5.10 'Changing rates'

Key vocabulary catalyst, concentration, pressure, temperature

Teaching and learning

Engage

- Show a video clip of a fast reaction in an air bag [Google search string:] 'It's a Gas 28 - Airbag Sodium Azide', and elicit students' ideas about when it's useful for a reaction to be fast and when reaction speeds need to be controlled.

Challenge and develop

- Introduce the investigation outlined in Practical sheet 5.10 by demonstrating the reaction between warm sodium thiosulfate solution and hydrochloric acid. Depending on the students' prior experience, they may need support with the planning section of the investigation. Key points to bring out in the discussion are: they need results for five or more temperatures covering a good range; the main hazards are sulfur dioxide, the hydrochloric acid and the Bunsen burner; the risk of breathing difficulties can be minimised by good ventilation, keeping reactant temperatures below 65 °C and rinsing the chemicals away as soon as the measurements have been taken.

- Errors in measuring temperatures can be minimised by heating to just above the target temperature, stirring, and then cooling to the exact temperature required. Errors in timing can be minimised by using identical crosses.

- Students should appreciate the need to obtain multiple results at each temperature so that anomalous readings can be discarded and concordant results averaged to minimise the effects of random errors. However, groups are unlikely to have time to test more than one or two different temperatures, so a whole class table could be set up for each group to contribute results to. [O1]

Explain

Low demand

- Students **plot** a graph of the time needed to blot out the cross against temperature, and **describe** the results. Sheet 1 of Graph plotter 5.10 could be used for this. If students' results lie outside the range of the graph, right click on the vertical axis to change the maximum value displayed. [O2]

Standard and high demand

- Students **plot** a graph of the time against temperature using sheet 1 of Graph plotter 5.10. They then switch to sheet 2 of Graph plotter 5.10 to **compare** the reaction rates at different temperatures and **describe** the results. [O2]

Consolidate and apply

- Use slide 1 of Presentation 5.10 'Changing rates' to remind students about the reaction between magnesium and an acid and introduce the effect of increasing the concentration of a solution. [O1, O2]

Low demand

- Ask students to complete Worksheet 5.10.1 to practice **interpreting** experimental data. [O1, O2]

- Use slide 2 of Presentation 5.10 'Changing rates' to remind students about the reaction between marble chips and an acid and introduce the effect of reducing the size of solid particles. [O1, O2]

Standard and high demand

- Model the surface areas of large and small chips using centimetre cubes and students complete Worksheet 5.10.2. [O2]

Extend

- Ask students able to progress further to read 'Surface to volume ratio' in Section 5.10 in the Student Book and complete questions 4–6. [O2, O3]

Plenary suggestion

Give pairs of students a series of relationships (e.g. the effect of temperature on reaction rate, the volume of product against time at high and low temperatures) and ask them to draw the expected graphs on mini-whiteboards.

Answers to Worksheet 5.10.1

1. Accurately plotted graph; with a smooth line of best fit; 2. The higher the acid concentration, the faster the reaction rate

Answers to Worksheet 5.10.2

1. 24 cm^3; 2. 48 cm^3; 3. Twice as fast; the area in contact with the acid is twice as big

Lesson 11: Collision theory

Lesson overview

OCR Specification reference

OCR C5.2

Learning objectives

- Find out about the collision theory.
- Use collision theory to make predictions about reaction rates.
- Relate activation energies to collision theory.

Learning outcomes

- Describe the collision theory. [O1]
- Use collision theory to explain rate changes. [O2]
- Explain activation energy in terms of collision theory. [O3]

Skills development

- WS 1.2 Use models in explanations.
- WS 3.5 Use data to make predictions.
- WS 4.2 Recognise the importance of scientific quantities.

Resources needed Worksheets 5.11.1 and 5.11.2

Digital resources Presentation 5.11.1 'Colliding particles' and Presentation 5.11.2 'Changing temperatures'

Key vocabulary activation energy, collision, effective, frequency

Teaching and learning

Engage

- Use Molymod models of hydrogen and chlorine molecules to model the way they react to make hydrogen chloride. Remind students that new molecules can't form until the bonds in the original molecules have broken, and show that modelling the process lets us explain and predict how reaction rates can be changed. [O1]

Challenge and develop

- Use the 'Collisions' part in Section 5.11 in the Student Book to describe the collision theory, namely that reactions can only occur when reacting particles collide with sufficient energy to break bonds, and the minimum energy the particles must have is the activation energy. [O1]

- Use a video clip to illustrate this idea [Google search string:] How to speed up chemical reactions (and get a date). [O1, O2]

Explain

- Use slide 1 of Presentation 5.10 'Changing rates' to remind students about the effect of increasing the acid concentration when magnesium reacts with an acid. [O2]

Low demand

- Students complete the questions on Worksheet 5.11.1. Presentation 5.11.1 'Colliding particles' could be used to review their responses. [O2]

Standard and high demand

- Pairs of students use the collision theory to **explain** in words and pictures why magnesium reacts faster when the acid concentration is higher. Pairs then join up to **compare** ideas. [O2]

Consolidate and apply

- Use a video clip to illustrate the use of collision theory to explain how temperature, concentration and surface area affect reaction rates [Google search string:] Collision theory and how to speed up rates of reaction. [O1, O2]

Low demand

- Students complete Worksheet 5.11.2 using collision theory to explain why heating speeds up reactions. [O2]

Standard and high demand

- Display slide 1 of Presentation 5.11.2 'Changing temperatures' to emphasise that most collisions are unsuccessful. [O3]

- Pairs of students **collaborate** to produce a detailed answer to question 2 in the 'Activation energy' section of the Student Book. Slide 2 of Presentation 5.11.2 'Changing temperatures' could be used to emphasise the main points. [O3]

Extend

- Ask students able to progress further to **consider** how catalysts could increase the percentage of successful collisions without giving particles more energy. [O2]

Plenary suggestion

Display poor explanations for changing reaction rates and challenge pairs to **improve** them, e.g. particles are more crowded in concentrated solutions; hotter particles collide more often.

Answers to Worksheet 5.11.1

1. Every magnesium atom being hit by an acid particle should be coloured; 2. Extra acid particles should be added to the image; and also some extra collisions; 3. Three black gas particles and three white gas particles spread around the large box; and squashed closer together in the small box; 4. Increasing the concentration increases the rate of a reaction because there are more particles in the same volume; which causes more (frequent) collisions to occur; 5. Adding extra acid solution does not affect the concentration of acid particles around the magnesium; so it doesn't increase the frequency of collisions between acid and magnesium particles

Answers to Worksheet 5.11.2

1. Left-hand box – particles collide but do not react; right-hand box – particles collide and react to form products; 2. They move faster; 3. It increases; 4. Activation energy; 5. Faster particles collide more energetically/violently

Lesson 12: Catalysts

Lesson overview

OCR Specification reference

OCR C5.2

Learning objectives

- Investigate catalysts.
- Find out how catalysts work.
- Learn how they affect activation energy.

Learning outcomes

- Identify catalysts in reactions. [O1]
- Describe the properties of catalysts. [O2]
- Use ideas about activation energy to explain how catalysts work. [O3]

Skills development

- WS 2.2 Plan experiments to test hypotheses.
- WS 2.6 Make and record observations.
- WS 3.8 Communicate reasoned conclusions.

Resources needed Equipment as listed in Technician's notes 5.12; Practical sheet 5.12; Worksheet 5.12

Key vocabulary activation energy, catalysis, catalyst, catalytic action

Teaching and learning

Engage

- Add a spatula of manganese(IV) oxide to a flask of hydrogen peroxide solution so that students can witness the bubbles of gas escaping as it decomposes. Demonstrate that this is oxygen by holding a glowing splint inside the flask. It should relight time after time. Stress that the oxygen comes from the hydrogen peroxide – not the manganese(IV) oxide – which catalyses the reaction. [O1]

Challenge and develop

- Demonstrate that there is no visible reaction between aluminium and hydrochloric acid. Stress that aluminium is a reactive metal, but it appears to be unreactive because a layer of aluminium oxide protects its surface. A catalyst could make it react with hydrochloric acid at room temperature.

- Groups of three or four **investigate** potential catalysts for the reaction between aluminium and hydrochloric acid using the guidance on Practical sheet 5.12. [O1]

Explain

- All students can be asked to read 'Catalysts' in Section 5.12 of the Student Book and answer question 1. [O2]

Consolidate and apply

- Use a video clip to illustrate the use of catalysts in catalytic converters [Google search string:] Faces of chemistry: Catalysts (Johnson Matthey) – video 2.

Low demand

- Students could use Worksheet 5.12 to **review** their understanding of catalysts. [O1, O2]

 © HarperCollins*Publishers* Limited 2016

Standard and high demand

- Students read 'Activation energy pathways' in Section 5.12 in the Student Book and complete question 3. [O3]

Extend

- Challenge students able to progress further to **describe** the similarities and differences between catalysts and enzymes. [O2, O3]

Plenary suggestion

Ask students to draw a large triangle with a smaller inverted triangle that just fits inside it (so it has four triangles). In the outer three ask them to write something they've seen, something they've done and something they've **discussed**. Then add in the central triangle something they've learned.

Answers to Worksheet 5.12

1. The catalyst helps the hydrogen peroxide break down and release oxygen; 2. Catalysts are not used up during reactions; 3. Manganese(IV) oxide is best; the same volume of gas is produced more quickly; 4. The catalyst is made of a precious metal; which is expensive; 5. Provides a larger surface area; so it speeds up the reaction

 © HarperCollins*Publishers* Limited 2016

Lesson 13: Factors increasing the rate

Lesson overview

OCR Specification reference

OCR C5.2

Learning objectives

- Interpret graphs.
- Consider what determines the reaction rate.
- Explore the effect of changing the amounts of reactants used.

Learning outcomes

- Analyse experimental data on rates of reaction. [O1]
- Predict and explain the effects of changing reaction conditions. [O2]
- Recognise proportional relationships between variables. [O3]

Skills development

- WS 2.4 Make and record observations.
- WS 3.5 Recognise and describe patterns and trends.
- WS 3.8 Communicate findings.

Resources needed Equipment as listed in Technician's notes 5.13; Practical sheet 5.13; Worksheet 5.13

Digital resources Presentation 5.13 'Graphs'

Key vocabulary excess, analyse, graph, steeper

Teaching and learning

Engage

- Display Presentation 5.13 'Graphs' and ask groups to **suggest** why the red and blue graphs have different gradients but end at the same place. [O1, O2]

Challenge and develop

- Remind students about the reaction between magnesium and an acid and challenge groups to work collaboratively to **investigate** the effect of doubling the amount of magnesium. Practical sheet 5.13 provides some guidance but students could plan their own methods. These need checking to ensure that the capacity of the gas syringe is not exceeded and the acid remains in excess. (The largest amount of magnesium used should be 6 cm. If this is added to 50 cm^3 of 1 mol/dm^3 hydrochloric acid there is a three-fold excess of acid.) Students could be encouraged to work in large groups, divide the practical work between them and share results. [O1, O2]

Explain

- Students **analyse** their results and **explain** how well they support their hypothesis.

Consolidate and apply

Low and standard demand

- Students read 'Analysing data' and 'Increasing the rate of reaction' in Section 5.13 in the Student Book and complete questions 1–5 to practice **analysing** graphs. They will need to fill in Worksheet 5.13 to complete questions 3–5. [O1, O2]

Standard and high demand

- Students read 'Changing the mass' in Section 5.13 in the Student Book and complete questions 7 and 8. [O3]

Extend

- Ask students able to progress further to **explain** why increasing a reaction rate doesn't usually change the quantity of product formed. [O2]

Plenary suggestion

Present a graph of the volume of hydrogen released against time, suggest a change to the reaction conditions and ask students to **predict** what the new graph would look like and sketch it on a mini whiteboard. For example, 1 cm of magnesium ribbon in 50 cm^3 of 1 mol/dm^3 hydrochloric acid versus 2 cm in the same acid; 2 cm of magnesium ribbon in 50 cm^3 of 1 mol/dm^3 hydrochloric acid versus the same amount in 50 cm^3 of 2 mol/dm^3 acid.

Answers to Worksheet 5.13

Volume of gas made after 10 seconds (cm^3)	
Low acid concentration	High acid concentration
13	30

Volume of gas made after 10 seconds (cm^3)	
Low temperature	High temperature
13	30

Lesson 14: Practical: Investigate how changes in concentration affect the rates of reactions by a method involving the production of a gas and a method involving a colour change

Lesson overview

OCR Specification reference

OCR C5.2

Learning objectives

- Devise a hypothesis.
- Devise an investigation to test a hypothesis.
- Decide whether the evidence supports a hypothesis.

Learning outcomes

- Use scientific theories and explanations to develop a hypothesis. [O1]
- Plan experiments to test the hypothesis and check data. [O2]
- Make and record measurements using gas syringes. [O3]
- Evaluate methods and suggest improvements and further investigations. [O4]

Skills development

- WS 2.1 Suggest a hypothesis to explain given observations.
- WS 2.2 Apply understanding of apparatus and techniques to suggest a procedure.
- WS 2.6 Read measurements off a scale in a practical context.

Maths focus Plot two variables from experimental data

Resources needed Equipment as listed in Technician's notes 5.14; Practical sheets 5.14.1 and 5.14.2

Digital resources Presentation 5.14.1 'Comparing concentrations'; Presentation 5.14.2 'Evaluation'; Graph plotter 5.14

Key vocabulary concentration, volume, rate, tangent

Teaching and learning

Engage

- Display Presentation 5.14.1 'Comparing concentrations' and ask groups to develop hypotheses based on their existing knowledge of collision theory. [O1]

Challenge and develop

- Show students the equipment available and give them time to **plan** how to collect evidence to support or refute their hypotheses. If they have not used gas syringes before show them how to set the gas syringe to zero with the bung out of the conical flask; insert the bung and start the stopclock promptly as the magnesium is dropped into the acid. Demonstrate how to swirl the conical flask gently throughout the reaction to prevent bubbles of hydrogen from adhering to the magnesium. [O2]

- Groups could be encouraged to **collaborate** in order to collect duplicate results for each concentration tested. [O3]

- In order to determine the initial rates of reaction with each acid concentration, students need to plot graphs. This could be done by hand or using Graph plotter 5.14. If the graphs are to be plotted in Excel, demonstrate how to fit a trend line to the linear part of the graph, display the equation for the trend line and read off its gradient – which is a measure of the reaction rate in cm^3 of hydrogen per second. [O3]

Explain

- Groups describe how well their results **support** their hypotheses and **suggest** improvements that would make their evidence stronger. [O3]

Consolidate and apply

Low demand

- Students complete Practical sheet 5.14.2 to **practise** analysing graphs. [O2]

Standard and high demand

- Complete Practical sheet 5.14.1 to describe a quantitative relationship between the acid concentration and the rate of the reaction between magnesium and hydrochloric acid. [O3]

Extend

- Ask students able to progress further to **consider** whether or not the hypothesis they tested could be extended to cover other reactions. [O3]

Plenary suggestion

Display Presentation 5.14.2 'Evaluation' and challenge groups to **describe** what they did to improve the reliability of their results and the validity of their conclusions. [O3]

Answers to Practical sheet 5.14.1

1. When the acid concentration doubles the reaction rate quadruples; 2. The reaction rate is multiplied by nine; 3. About 0.11 cm Mg/s; 4. 0.108 cm Mg/s; 5. The temperature of the acid will rise; especially at higher concentrations; so the rate will be even higher

Answers to Practical sheet 5.14.2

1. Student plot; 2. At 40 °C the rate = 5 cm^3/s; at 30 °C the rate = 2.4 cm^3/s; at 20 °C the rate = 1.2 cm^3/s; 3. For this reaction the rate doubles approximately for every 10 °C rise in temperature

Lesson 15: Reversible reactions and energy changes

Lesson overview

OCR Specification reference

OCR C5.3

Learning objectives

- Investigate reversible reactions.
- Explore the energy changes in a reversible reaction.
- Find out how reaction conditions affect reversible reactions.

Learning outcomes

- Identify a reversible reaction. [O1]
- Explain the energy changes in reversible reactions. [O2]
- Consider changing the conditions of a reversible reaction. [O3]

Skills development

- WS 1.2 Use models in explanations.
- WS 1.2 Use models to make predictions.
- WS 4.1 Use scientific vocabulary, terminology and definitions.

Resources needed Equipment as listed in Technician's notes 5.15; Practical sheet 5.15; Worksheet 5.15

Digital resources Presentation 5.15 'A reversible reaction'

Key vocabulary backwards, exothermic, forwards, reversible

Teaching and learning

Engage

- Demonstrate the formation of a precipitate in the reaction between potassium iodide and lead nitrate. Emphasise that familiar reactions like this appear to be instantaneous and irreversible. However, many reactions are reversible. [O1]

Challenge and develop

- Demonstrate the reversible decomposition of ammonium chloride by heating it in a boiling tube with a piece of damp universal indicator paper halfway up the tube. The indicator paper should turn blue and then red as ammonium chloride decomposes to form ammonia and hydrogen chloride. These gases then recombine to deposit white ammonium chloride on the cooler parts of the tube. [O1]

- Pairs observe the reversible dehydration of blue copper(II) sulfate(VI)-5-water following the guidance on Practical sheet 5.15. [O1]

Explain

- Use a video clip to review the reversible decomposition of ammonium chloride [Google search string:] What are Reversible Reactions? | The Chemistry Journey | The Fuse School. [O1, O2]

Low demand

- Students complete the questions on Worksheet 5.15 to review what happens in each reversible reaction. [O1, O2]

Standard and high demand

- Students read 'Reversible reactions' in Section 5.15 of the Student Book and list the similarities between the two reversible changes they have observed. [O1, O2]

Consolidate and apply

- Use a video clip to extend students' understanding of reversible reactions and introduce the idea of equilibrium [Google search string:] If molecules were people - George Zaidan and Charles Morton. [O3]

- Use the 'Changing direction' section in Section 5.15 of the Student Book to introduce the idea that many important industrial reactions are reversible, so they appear to stop before all the reactants have turned into products. [O3]

Extend

- Ask students able to progress further to work in pairs to **formulate** an answer to question 3 in Section 5.15 of the Student Book. [O3]

Plenary suggestion

Display Presentation 5.15 'A reversible reaction' and ask groups to **generate** two questions about the reactions shown for another group to answer. Groups can then swap and answer each other's questions.

Answers to Worksheet 5.15

1. Ammonia; hydrogen chloride; 2. Ammonium chloride; 3. The ammonia and hydrogen chloride recombine in the cooler part of the tube; 4/5. The forward reaction should be labelled endothermic; and the reverse reaction exothermic

Lesson 16: Equilibrium

Lesson overview

OCR Specification reference

OCR C5.3

Learning objectives

- Recognise reactions that can reach equilibrium.
- Find out what happens to the reactants and products at equilibrium.
- Use Le Chatelier's principle to make predictions.

Learning outcomes

- Describe how reversible reactions reach equilibrium. [O1]
- Explain what happens to the forward and reverse reactions at equilibrium. [O2]
- Predict the effects of changes on systems at equilibrium. [O3]

Skills development

- WS 1.2 Use models in explanations.
- WS 1.2 Use models to make predictions.
- WS 4.1 Use scientific vocabulary, terminology and definitions.

Resources needed Equipment as listed in Technician's notes 5.16; Worksheets 5.16

Digital resources Presentation 5.16 'Equilibrium'

Key vocabulary equilibrium, counteract, reversible reaction, Le Chatelier's principle

Teaching and learning

Engage

- Display slide 1 of Presentation 5.16 'Equilibrium' and elicit students' ideas about what would happen if the water produced by the forward reaction could not escape. [O1]

Challenge and develop

- Use slide 2 of Presentation 5.16 'Equilibrium' to introduce the term 'equilibrium'. [O1]

- Demonstrate how dynamic equilibrium is reached using ball-pond balls. Model hydrated copper(II) sulfate using a tray of blue and white balls stuck together. Separate one pair at a time to simulate the forward reaction and put them on a separate product tray. Enlist a volunteer to begin sticking the balls back together and moving them back to the first tray. Stress that eventually the number of balls on each side stops changing – but there are rarely equal numbers on both sides at this point. [O1]

- Pairs of students could **model** a reversible reaction at equilibrium. They stand apart from each other on the left-hand side of the classroom and join hands to form pairs on the other side. Every time a pair joins up to move to the other side, an existing pair splits up and moves back.

- Use a video clip to reinforce the idea that at equilibrium the forward and backward reactions proceed at the same rates [Google search string:] What is Dynamic Equilibrium? | The Chemistry Journey | The Fuse School. [O2]

- Use an animation to reinforce the idea that the amounts of product and reactant present at equilibrium are unlikely to be equal, and show that the same equilibrium is established whether you start with the reactants or the products of the forward reaction [Google search string:] freezerray.com chemistry equilibrium 1. [O2]

Explain

Low demand

- Students complete the questions on Worksheet 5.16. [O1, O2]

Standard demand

- Students read 'Reaching equilibrium' and 'Equilibrium position' in Section 5.16 of the Student Book and answer questions 1 and 2. [O1, O2]

High demand

- Students **explain** why reactions that reach equilibrium could make it difficult to make pure products. [O2]

Consolidate and apply

- Use slide 4 of Presentation 5.16 'Equilibrium' to introduce the demonstration detailed in the Technician's notes 5.16. This shows how a reversible reaction at equilibrium can be disturbed by changing a concentration. Students should notice that adding extra chloride ions makes the solution more blue because it shifts the equilibrium to the right, but adding more water makes the solution more pink because it shifts the equilibrium to the left. Students may recognise the colour change because it is used in the cobalt chloride paper in testing for water. [O3]

- Students can read 'Le Chatelier's principle' in Section 5.16 of the Student Book and complete question 3. [O3]

Extend

- Ask students able to progress further to **suggest** what might happen if a solution containing cobalt ions in equilibrium was heated, given that the forward reaction is endothermic. [O3]

Plenary suggestion

Ask students to write one thing they know about reversible reactions, one thing they are unsure about and one thing they need to know more about. They could then **share** ideas and **agree** on a group list.

Answers to Worksheet 5.16

1. Diagram shows two HI molecules added to the right-hand box; 2. Diagram shows three to five HI molecules added to the right-hand box; a shorter forward arrow; and a longer reverse arrow; 3. Diagram shows more molecules in the right-hand box; the forward and reverse arrows are the same size; 4. None of the reactants or products can escape; nor can anything else enter; 5. The concentration of the reactants is lower than the concentration of the products

Lesson 17: Changing concentration and equilibrium

Lesson overview

OCR Specification reference

OCR C5.3

Learning objectives

- Distinguish between reactants and products.
- Explore how changing their concentrations affects reversible reactions.
- Use Le Chatelier's principle to make predictions about changing concentrations.

Learning outcomes

- Identify reactants and products in a reversible reaction. [O1]
- Explain how changing concentrations changes the position of equilibrium. [O2]
- Interpret data to predict the effect of a change in concentration. [O3]

Skills development

- WS 1.2 Use models in explanations.
- WS 1.2 Use models to make predictions.
- WS 4.1 Use scientific vocabulary, terminology and definitions.

Resources needed Equipment as listed in Technician's notes 5.17; Worksheet 5.17

Digital resources Presentation 5.17 'Disturbing equilibria'

Key vocabulary equilibrium position, concentration, dinitrogen tetroxide, reversible reaction

Teaching and learning

Engage

- Display slide 1 of Presentation 5.17 'Disturbing equilibria'. Ask groups to **decide** which one of the statements is true and **create** replacements for the incorrect ones. Only the second statement is true. Possible rewrites for the other two are: the concentrations of reactants and products do not have to be equal for an equilibrium to be reached; some, but not all, of the added reactant will be used up as the system reaches equilibrium.[O2]

Challenge and develop

- Display slide 2 of Presentation 5.17 'Disturbing equilibria' and ask students to **predict** the effect of adding hydrogen ions to the equilibrium shown, and of taking them away. Students should **predict** that adding extra hydrogen ions will make the solution more orange because it shifts the equilibrium to the right. Conversely, using alkali to remove hydrogen ions makes the solution more yellow because it shifts the equilibrium to the left.

- Demonstrate the effect of changing the hydrogen ion concentration following the guidance in Technician's notes 5.17.

Explain

- Students can be asked to read 'Removing products' in Section 5.17 of the Student Book and to complete question 1. [O3]

Consolidate and apply

- Display slide 3 of Presentation 5.17 'Disturbing equilibria' and introduce the term 'yield' to describe the percentage of ammonia present at equilibrium. Ask groups to **suggest** ways of maximising ammonia production.

- Use a video clip to introduce the industrial manufacture of ammonia and confirm that the ammonia must be constantly removed [Google search string:] Formation of ammonia using the Haber Process – BBC. [O2, O3]

Low demand

- Students complete the questions on Worksheet 5.17. [O1, O2, O3]

Standard and high demand

- Students read 'Changing concentrations' in Section 5.17 of the Student Book and answer questions 2 and 3. [O3]

Extend

- Ask students able to progress further to use ideas about reaction rates to **suggest** ways of speeding up both the forward and reverse reactions during ammonia manufacture. [O2]

Plenary suggestion

Ask students to write a question about reversible reactions on a coloured strip of paper and the answer on a strip of another colour. In groups of six to eight, hand out the strips so that everyone has a question and an answer. One student reads a question. Then the student with the right answer reads it out, followed by their own question.

Answers to Worksheet 5.17

1. Reading anticlockwise from the top left: the labels are 'nitrogen and hydrogen are added'; 'unused nitrogen and hydrogen are recycled'; 'ammonia is condensed and removed'; 2. A = nitrogen, B = hydrogen, C = ammonia; 3. $N_2 + 3H_2 \rightleftharpoons 2NH_3$; 4. Removing ammonia causes the equilibrium to shift to the right; and converts more nitrogen and hydrogen into ammonia

Lesson 18: Changing temperature and equilibrium

Lesson overview

OCR Specification reference

OCR C5.3

Learning objectives

- Distinguish between exothermic and endothermic forward reactions.
- Explore how changing the temperature affects reversible reactions.
- Use Le Chatelier's principle to make predictions about changing temperatures.

Learning outcomes

- Explain how exothermic reversible reactions respond to temperature changes. [O1]
- Explain how endothermic reversible reactions respond to temperature changes. [O2]
- Apply Le Chatelier's principle to the effect of temperature changes on reactions in equilibrium. [O3]

Skills development

- WS 1.2 Use models in explanations.
- WS 1.2 Use models to make predictions.
- WS 4.1 Use scientific vocabulary, terminology and definitions.

Resources needed Equipment as listed in Technician's notes; Worksheets 5.18.1 and 5.18.2; Technician's notes 5.18

Digital resources Presentation 5.18 'Changing temperatures'

Key vocabulary endothermic, equilibrium position, exothermic, Le Chatelier's principle

Teaching and learning

Engage

- Display slide 1 of Presentation 5.18 'Changing temperatures' and ask groups to **decide** how raising the temperature would affect the reaction shown. Students should **recognise** that adding heat energy will speed up both the forward and the reverse reactions but they should also expect the temperature change to affect the equilibrium. [O1, O2, O3]

Challenge and develop

- Demonstrate the shift in the position of equilibrium that takes place when you heat or cool an endothermic reversible reaction following the procedure detailed in Technician's notes 5.18. Students should notice that heating makes the solution more blue because it shifts the equilibrium to the right, but cooling makes the solution more pink because it shifts the equilibrium to the left. [O2]

- Use slide 2 of Presentation 5.18 'Changing temperatures' to stress that raising the temperature increases the equilibrium yield of endothermic reversible reactions. [O2]

- Slide 3 of Presentation 5.18 shows the exothermic reversible reaction used to make ammonia. An animation could be used to show that increasing the temperature has the opposite effect on this sort of reaction [Google search string:] freezerray.com chemistry equilibrium 3. [O1]

- Use slide 4 of Presentation 5.18 'Changing temperatures' to stress that raising the temperature lowers the equilibrium yield of exothermic reversible reactions. [O1]

Explain

- A video clip could be used to apply Le Chatelier's Principle to the equilibrium between nitrogen dioxide and dinitrogen tetroxide [Google search string:] Le Chatelier's Principle: Part 2 | The Chemistry Journey | The Fuse School. [O3]

Low and standard demand

- Students complete Worksheet 5.18.1 using Section 5.18 in the Student Book for reference. [O1, O2, O3]

High demand

- Ask students to read 'Changes with increasing temperature' in Section 5.18 in the Student Book and answer questions 1 and 2. [O3]

Consolidate and apply

- Use a video clip to introduce the idea that the conditions actually used industrially are often a compromise designed to give the most cost-effective combination of reaction rate and equilibrium yield [Google search string:] What is the Haber Process | The Chemistry Journey | The Fuse School. [O1]

Low and standard demand

- Students complete Worksheet 5.18.2. [O1]

High demand

- Students can be asked to read 'Example in industrial processes' in Section 5.18 of the Student Book and **find** three things that are the same about the Haber process and the Contact process. [O1]

Extend

- Ask students able to progress further to consider this statement: an endothermic reversible reaction has reached equilibrium; if the temperature is raised the rate of the forward reaction increases but the rate of the reverse reaction decreases. The statement is false – heating makes the rates of both the forward and reverse reactions increase. However, the rate of the forward reaction will increase more than that of the reverse reaction so the position of equilibrium shifts to the right. [O3]

Plenary suggestion

Use slide 5 of Presentation 5.18 'Changing temperatures' to present a problem. Ask each group to list three ideas and **explain** their reasoning.

Answers to Worksheet 5.18.1

1. It decreased; 2. To the left; 3. Endothermic; 4. The reverse reaction; 5. Reduce the temperature; add more reactants; 6. The relative amount of products at equilibrium decreases

Answers to Worksheet 5.18.2

1. 35%; 2. To the left; 3. The reaction rate would be very low; 4. The unreacted gases are recycled; 5.It increases

Lesson 19: Changing pressure and equilibrium

Lesson overview

OCR Specification reference

OCR C5.3

Learning objectives

- Recognise the number of product and reactant molecules in a reaction.
- Explore how changing the pressure affects reversible reactions.
- Use Le Chatelier's principle to make predictions about changing pressures.

Learning outcomes

- Predict the effects of changes in pressure. [O1]
- Explain why these effects occur. [O2]
- Interpret data to predict the effect of a change in pressure. [O3]

Skills development

- WS 1.2 Use models in explanations.
- WS 1.2 Use models to make predictions.
- WS 4.1 Use scientific vocabulary, terminology and definitions.

Resources needed Worksheet 5.19

Digital resources Presentation 5.19 'Changing pressures'

Key vocabulary compromise, equilibrium position, molecule, pressure

Teaching and learning

Engage

- Display slide 1 of Presentation 5.19 'Changing pressures' and ask groups to **suggest** whether high or low pressures would be best for the reversible reaction used in the Contact process. Students should **recognise** that increasing the pressure will speed up both the forward and the reverse reactions, but they should also expect pressure to affect the equilibrium. [O1, O2]

Challenge and develop

- Use slide 2 of Presentation 5.19 'Changing pressures' to confirm that both the reaction rate and the equilibrium yield increase with pressure, but high pressures are not used because the equipment needed to keep gases at high pressures is too expensive. [O1, O2]

- Use slide 3 of Presentation 5.19 'Changing pressures' to remind students about the equilibrium between brown nitrogen dioxide and colourless dinitrogen tetroxide and ask them to use Le Chatelier's principle to **predict** how the colour will change if the pressure is lowered by increasing the volume. [O1, O2]

- A video clip could be used to check their prediction [Google search string:] Volume Effect on Equilibrium – LeChatelier's Principle Lab Extension. [O1, O2]

Explain

Low demand

- Students complete questions 1–3 in Section 5.19 of the Student Book. [O1, O2, O3]

Standard and high demand

- Use examples from Section 5.19 in the Student Book to explain how equations allow students to predict the effect a pressure change will have on an equilibrium reaction. [O1, O2]

Consolidate and apply

- Students work in pairs to **practise** using Le Chatelier's principle to make predictions using Worksheet 5.19. [O3]

Extend

- Ask students able to progress further to answer question 4 in Section 5.19 of the Student Book. [O2]

Plenary suggestion

Pairs list things they know about changing the position of equilibrium. Then they join into larger groups to **share ideas** and **produce** an agreed list.

Answers to Worksheet 5.19

1. C; 2. A; 3. C; 4. C; 5. B; 6. C; 7. A

Lesson 20: Maths skills: Use the slope of a tangent as a measure of rate of change

Lesson overview

OCR Specification reference

OCR C5.3

Learning objectives

- Practice drawing graphs.
- Use graphs to compare reaction rates.
- Use tangents to measure rates that change.

Learning outcomes

- Draw graphs from numeric data. [O1]
- Draw tangents to the curve to observe how the slope changes. [O2]
- Calculate the slope of the tangent to identify the rate of reaction. [O3]

Skills development

- WS 3.1 Present data using appropriate methods.
- WS 3.2 Translate data from one form to another.
- WS 3.5 Recognise or describe patterns and trends in data.

Maths focus Plot two variables from experimental data; draw and use the slope of a tangent to a curve as a measure of the rate of reaction

Resources needed Graph paper; calculators; Worksheets 5.20.1 and 5.20.2

Digital resources Presentation 5.20.1 'Comparing rates' and Presentation 5.20.2 'Which graph?'

Key vocabulary tangent, rate, slope, graph

Teaching and learning

Engage

- Display Presentation 5.20.1 'Comparing rates' and challenge students to **describe** what is happening to the reaction. They should appreciate that it starts fast and gradually slows to a halt. [O2]

Challenge and develop

Low and standard demand

- Students use the data given in Section 5.20 of the Student Book to **plot** a graph and answer questions 1 and 2. [O1]

High demand

- Students sketch a graph to show the volume of a gas collected from a reaction that starts very slowly and gradually speeds up. [O1]

Explain

- Students read 'Finding the gradient' in Section 5.20 of the Student Book and complete questions 3 and 4 for practice. [O3]

Consolidate and apply

Low and standard demand

- Students **practice** using the slopes of tangents to measure the rate of reaction at different times using Worksheet 5.20.1. [O3]

High demand

- Students use Worksheet 5.20.2 to complete question 6 from the Student Book. [O2]

Extend

Ask students able to progress further to **explain** why the two reactions shown on Worksheet 5.20.2 produce the same volume of hydrogen.

Plenary suggestion

Use Presentation 5.20.2 'Which graph?' to review the shapes of graphs showing the amount of product formed, or reactant used up, against time.

Answers to Worksheet 5.20.1

1. The reaction is slowing down; 2. At the start the rate = $(60 - 0)$ cm^3 of gas ÷ $(12 - 0)$ s = 5.00 cm^3 of gas/s; after 30 s the rate = $(60 - 26)$ cm^3 of gas ÷ $(60 - 0)$ s = 34/60 cm^3 of gas/s = 0.57 cm^3 of gas/s

Answers to Worksheet 5.20.2

From the Student Book answers

When and how to use these pages: Check your progress, Worked example and End of chapter test

Check your progress

Check your progress is a summary of what students should know and be able to do when they have completed the chapter. Check your progress is organised in three columns to show how ideas and skills progress in sophistication. Students aiming for top grades need to have mastered all the skills and ideas articulated in the final column (shaded pink in the Student book).

Check your progress can be used for individual or class revision using any combination of the suggestions below:

- Ask students to construct a mind map linking the points in Check your progress
- Work through Check your progress as a class and note the points that need further discussion
- Ask the students to tick the boxes on the Check your progress worksheet (Teacher Pack CD). Any points they have not been confident to tick they should revisit in the Student Book.
- Ask students to do further research on the different points listed in Check your progress
- Students work in pairs and ask each other what points they think they can do and why they think they can do those, and not others

Worked example

The worked example talks students through a series of exam-style questions. Sample student answers are provided, which are annotated to show how they could be improved.

- Give students the Worked example worksheet (Teacher Pack CD). The annotation boxes on this are blank. Ask students to discuss and write their own improvements before reviewing the annotated Worked example in the Student Book. This can be done as an individual, group or class activity.

End of chapter test

The End of chapter test gives students the opportunity to practice answering the different types of questions that they will encounter in their final exams. You can use the Marking grid provided in this Teacher Pack or on the CD Rom to analyse results. This shows the Assessment Objective for each question, so you can review trends and see individual student and class performance in answering questions for the different Assessment Objectives and to highlight areas for improvement.

- Questions could be used as a test once you have completed the chapter
- Questions could be worked through as part of a revision lesson
- Ask Students to mark each other's work and then talk through the mark scheme provided
- As a class, make a list of questions that most students did not get right. Work through these as a class.

Marking Grid for Single Chemistry for End of Chapter 5 Test 3816

Student Name	Getting Started [Foundation Tier]							Going Further [Foundation and Higher Tiers]			More Challenging [Higher Tier]						Most demanding [Higher Tier]			Total marks	Percentage
	Q. 1 (AO1) 1 mark	Q. 2 (AO1) 1 mark	Q. 3 (AO1) 2 marks	Q. 4 (AO1) 1 mark	Q. 5 (AO2) 1 mark	Q. 6 (AO2) 2 marks	Q. 7 (AO1) 2 marks	Q. 8 (AO1) 1 mark	Q. 9 (AO2) 2 marks	Q. 10 (AO1) 2 marks	Q. 11 (AO2) 4 mark	Q. 12 (AO2) 2 marks	Q. 13 (AO2) 1 mark	Q. 14 (AO2) 1 mark	Q. 15 (AO2) 2 mark	Q. 16 (AO3) 4 marks	Q. 17 (AO3) 2 marks	Q. 18 (AO3) 2 marks	Q. 19 (AO3) 6 marks		

© HarperCollins*Publishers* Limited 2016

Check your progress

You should be able to:

recognise that when a reaction has stopped one of the reactants has been used up →	describe the reactant that is used up first in a reaction as the limiting reactant →	explain the effect of a limiting quantity of a reactant on the amount of products it is possible to obtain, using moles or grams
relate mass, volume and concentration →	calculate the mass of solute in a solution →	relate concentration in mol/dm³ to mass and volume
describe how to carry out titrations →	calculate the concentrations in titrations in mol/dm³ and in g/dm³ →	explain how the concentration of a solution in mol/dm³ is related to the mass of the solute and the volume of the solution
identify the balanced equation needed for calculating yields →	calculate the theoretical amount of products from the amounts of reactants →	calculate the percentage yield from the actual yield and the theoretical yield
identify the balanced equation for a reaction →	calculate the atom economy of a reaction to form a desired product →	explain why a particular reaction pathway is chosen to produce a product given the atom economy, yield, rate, equilibrium position and usefulness of by-products
explain that the same amount of any gas (in moles) occupies the same volume at room temperature and pressure (rtp) →	calculate the volume of a gas at rtp from its mass and relative formula mass →	calculate volumes of gases from a balanced equation and a given volume of a reactant or product
identify how to measure the amount of gas given off in a reaction →	calculate the mean rate of a reaction →	draw tangents to the curves as a measure of the rate of reaction
identify which factors affect the rate of reactions →	explain how the changes of surface area affect rates →	explain how rates are affected by different factors
analyse experimental data on rates of reaction →	predict the effects of changing conditions on rates of reactions →	explain the effects of changes of factor on rates of reaction using collision theory
identify catalysts in reactions →	explain catalytic action →	explain activation energy
identify a reversible reaction →	describe how equilibrium is reached →	predict the effects of changes on systems at equilibrium
identify reactants and products in a reversible reaction →	explain how changing concentrations changes equilibrium →	interpret data to predict the effect of a change in concentration
explain how exothermic reactions behave if the temperature of systems at equilibrium changes →	explain how endothermic reactions behave if the temperature changes →	interpret appropriate data to predict the effect of a change in temperature on reactions at equilibrium
predict the effects of changes in pressure →	explain why these effects of pressure change occur →	apply Le Chatelier's principle to reactions in equilibrium

© HarperCollins*Publishers* Limited 2016

Worked example

1 **Explain how catalysts work.**

They speed up a reaction because they lower the activation energy.

This answer is correct as the description of what catalysts do is linked to the explanation.

2 **Draw the equation symbol used to show a reversible reaction.**

The answer is correct

3 **Draw an energy profile to show the activation energy needed for a reaction to take place.**

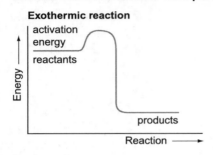

Exothermic reaction

activation
energy

reactants

Energy

products

Reaction

The profile is for a chemical reaction. The activation energy is the *difference between* the energy levels of the reactants and products not the initial energy level.

4 **Explain how increasing temperature increases the rate of reaction.**

The particles move faster so hit each other more likely.

This answer is worth one mark as a better word would be collide and an explanation is needed. A better answer is the particles move with more energy so there are more successful collisions

5 **Look at the table. Fill in the missing number and describe the pattern shown by this reaction at equilibrium. Predict the percentage of product probably formed at 150°C and suggest what the owner of the factory making this product should do.**

Temperature in °C	100	200	300	400
% reactants at equilibrium	22	37	52	59
% products at equilibrium	78	63	48	41

The higher the temperature the less products are made, so the owner should do this at a low temperature.

The data on the table is completed and the pattern is correctly described. The student has made a good suggestion to use a low temperature but needs to be careful to use the word lower not low. This reaction may not work at a low temperature and may not be fast enough. This time the 'benefit of doubt' is given. The answer is not complete though as they have missed the prediction. An answer of 60%–61% would give the fourth mark.

Global challenges: Introduction

When and how to use these pages

This unit builds on concepts students have acquired in previous lessons, particularly atomic structure, covalent bonding and chemical change.

Overview of the unit

Students learn about different aspects of sustainable development and fit their knowledge of recycling into a wider framework of life cycle assessments. They study recycling alongside reusing and reducing resource use. The unit offers a number of opportunities for students to evaluate the life style choices required for sustainable development and use data to relate the properties of substance to specific uses.

Students also consider methods used to control chemical reactions and reduce waste. They explore the composition of different alloys and how this affects their properties, find out how raw materials are used to make ceramics, polymers and composites with different properties and uses, revisit the Haber process and explore the use of ammonia in fertilisers. They also investigate voltaic cells and fuel cells.

They learn carbon chemistry and find out how fractional distillation and cracking are used to produce fuels and the petrochemicals used to produce polymers. They also study the structure of amino acids, proteins and DNA.

Students explore the composition of the atmosphere and use ideas and evidence to evaluate theories about its composition. In the process they will learn why the data needed to answer scientific questions may be uncertain, incomplete or unavailable. They will also consider the way the atmosphere has changed over geological time scales; evaluate the environmental implications of greenhouse gas emissions and other pollutants; explore the use of computer models to make predictions; evaluate the quality of evidence in reports about global climate change; and learn how peer review works. This will lead to an exploration of how greenhouse gas emissions could theoretically be reduced and why these reductions are difficult to achieve in practice.

This unit offers a number of opportunities for the students to use molecular modelling kits to represent organic structures. Students will improve their ability to work with symbols and equations. They will also use data and observations from demonstrations; make and record observations; describe patterns in data; use data to make predictions; and consider the extent to which data is consistent with a hypothesis.

There are several opportunities for students to use mathematics. They use fractions and percentages to describe the composition of the atmosphere; use powers of ten to locate times in Earth's history; translate data between graphical and numerical form; draw pie charts; plot variables against each other to describe trends and relationships; and interpret data presented in graphical form.

Obstacles to learning

Students may need extra guidance with the following terms and concepts.

- Students may need extra guidance working with ideas about the effects of temperature and pressure on the dynamic equilibrium involved in the Haber process.

- Students need to ensure they have shown the correct number of bonds to each atom in a displayed formula. A common mistake is showing carbon atoms in a C=C double bond with five bonds.

- There is little consistency between various texts and websites concerning boiling ranges and the number of carbon atoms in different fractions from the fractional distillation of crude oil. This is apparently because these vary from refinery to refinery and with different sources of crude oil. Students may need this explaining to prevent confusion.

- Students will find most of the common GCSE websites helpful but search for 'amino acids' and 'DNA' is more likely to produce A-level material and be confusing to many at this stage.

- Students may struggle to appreciate that the data needed to answer scientific questions may be uncertain, incomplete or unavailable and that theories often need to change as new evidence becomes available. They may assume that pollution is currently uncontrolled and be unaware of the constant monitoring that takes place and the number of controls that already limit pollution. They may not realise that emissions of most pollutants are currently decreasing in developed countries.

Practicals in this unit

In this unit students will do the following practical work:

- Reduce iron(III) oxide using carbon on a match head
- Use electrolysis and displacement to extract copper metal from solution.
- Prepare crystals of ammonium phosphate fertiliser
- Devise investigations concerning the conditions needed for rusting.
- Investigate rust prevention.
- Distinguishing between alkanes and alkenes
- Investigate the conditions required for fermentation
- Compare the chemical properties of ethanoic acid and hydrochloric acid
- Fractional distillation of a crude oil substitute
- Make voltaic cells and relate the voltages measured to the reactivity difference between the metals used
- Test salt solution for the presence of sodium ions and chloride ions and then remove the ions by distillation.

Students will observe:

- A copper ore and the compounds that can be extracted from it
- The reversible reaction when ammonium chloride is heated
- Making nylon.
- Cracking hydrocarbons
- A hydrogen fuel cell
- How the percentage of oxygen in the atmosphere can be measured by reacting it with copper.
- Evidence that plants release oxygen.
- Evidence that air contains very little carbon dioxide.
- Evidence that carbon dioxide is released when limestone is decomposed.
- A comparison of the soot released by blue and yellow Bunsen flames

	Lesson title	Overarching objectives
1	Extraction of metals	Explain how extraction methods depend on metal reactivity.
2	Using electrolysis to extract metals	Explain the process of the electrolysis of aluminium oxide
3	Alternative methods of metal extraction	Know how phytomining and bioleaching can be used to extract copper from low grade ores, and be able to evaluate alternative methods of metal extraction.
4	The Haber process	Explain how commercially used conditions in the Haber process are related to cost, and how to control the equilibrium position and rate.
5	Production and use of NPK fertilisers	Describe how fertilisers are produced both industrially and in the laboratory and compare the processes.
6	Life cycle assessment and recycling	Interpret and carry out simple life cycle assessments for objects such as shopping bags.

 ©HarperCollins*Publishers* Limited 2016

7	Ways of reducing the use of resources	Describe ways of recycling and reusing materials and explain why these are needed. Evaluate ways of reducing the use of a limited resource.
8	Alloys as useful materials	Describe and interpret the composition of alloys and evaluate their uses.
9	Corrosion and its prevention	Describe and interpret experiments on rusting and explain methods of preventing corrosion.
10	Ceramics, polymers and composites	Compare glass, ceramics, polymers, composites and metals and be able to select materials according to their properties and required use.
11	Maths skills: Translate between graphical and numerical form	Translate information between graphical and numerical form. Be able to draw and interpret pie charts, bar charts, histograms and line graphs.
12	Functional groups and homologous series	Compare the structure and properties of useful homologous series.
13	Structure and formulae of alkenes	Know the formulae and names for the first four alkenes and be able to describe why they are unsaturated.
14	Reactions of alkenes	Be able to use displayed formulae to describe the addition reactions of alkenes.
15	Alcohols	Know how alcohol is produced by fermentation and be able to describe some chemical reactions of the homologous series of alcohols.
16	Carboxylic acids	Recognise carboxylic acids as a homologous series and be able to describe some reactions of ethanoic acid, including the formation of an ester.
17	Addition polymerisation	Understand the formation of polymers by addition polymerisation of given alkenes and be able to write equations for the process using repeating units.
18	Condensation polymerisation	Understand the formation of a polymer by condensation polymerisation and be able to write equations for the process using repeating units.
19	Amino acids	Explain how amino acids combine to form polypeptides and proteins.
20	DNA and other naturally occurring polymers	Know the monomers used to make naturally occurring polymers, such as proteins, starches and cellulose, and DNA.
21	Fractional distillation and petrochemicals	Describe how and why crude oil can be separated into fractions by fractional distillation.
22	Crude oil, hydrocarbons and alkanes	Understand that crude oil is a source of alkanes and be able to describe the structure of the first four alkanes.
23	Properties of hydrocarbons	Identify the properties of different hydrocarbons and know how they influence their use.
24	Cracking and alkenes	Know that cracking produces more useful hydrocarbons, and be able to write word and balanced chemical equations for the process.
25	Cells and batteries	Relate the voltages measured in voltaic cells to the reactivity difference between the metals used.
26	Fuel cells	Describe how a hydrogen fuel cell works and evaluate their use in comparison with rechargeable cells and batteries
27	Key concept: Intermolecular forces	Understand the nature and effects of weak intermolecular forces.
28	Maths skills: Visualise and represent 3D models	Understand how 3D models are used to represent molecules.

29	Proportions of gases in the atmosphere	Recall the proportions of gases in the atmosphere and explain how these are maintained.
30	The Earth's early atmosphere	Use ideas and evidence about the Earth's early atmosphere to evaluate theories about its composition.
31	How oxygen increased	Explore the processes that allowed the percentage of oxygen in the atmosphere to rise to its present-day value.
32	Key concept: Greenhouse gases	Explain how greenhouse gases trap heat and consider the consequences of adding greenhouse gases to the atmosphere.
33	Human activities	Evaluate the quality of evidence in reports about global climate change and understand how peer review systems work.
34	Global climate change	Discuss the environmental implications of climate change.
35	Carbon footprint and its reduction	Describe how emissions of greenhouse gases could theoretically be reduced.
36	Limitations on carbon footprint reduction	Explain why emissions of greenhouse gases are difficult to reduce in practice.
37	Atmospheric pollutants from fuels	Use a fuel's composition, and the condition in which it is used, to predict which pollutants it will release.
38	Properties and effects of atmospheric pollutants	Explain the environmental implications of air pollution and evaluate its role in damaging human health.
39	Potable water	Distinguish between potable water and pure water and know how ground water and salty water are treated to produce potable water.
40	Waste water treatment	Know how waste water is treated to provide potable water.
41	Required practical: Analysis and purification of water samples from different sources, including pH, dissolved solids and distillation	Analysis and purification of water samples from different sources, including pH, dissolved solids and distillation.
42	Maths skills: Use ratios, fractions and percentages	Use fractions and percentages to describe the composition of mixtures and use ratios to determine the mass of products expected and calculate percentage yields.

Lesson 1: Extraction of metals

Lesson overview

OCR Specification reference

OCR C6.1

Learning objectives

- Find out where metals come from.
- Extract iron from its oxide using carbon.
- Consider how other metals are extracted from their ores.

Learning outcomes

- Identify substances reduced by loss of oxygen. [O1]
- Explain how extraction methods depend on metal reactivity. [O2]
- Interpret or evaluate information on specific metal extraction processes. [O3]

Skills development

- WS 1.4 Describe and explain a technological application of science.
- WS 2.3 Describe the technique that should be used for a particular purpose and explain why.
- WS 4.1 Use scientific vocabulary, terminology and definitions.

Maths focus: Recognise that gaining negative electrons converts positive ions into neutral atoms

Resources needed Equipment as listed in Technician's notes 6.1; Practical sheet 6.1; Worksheet 6.1

Digital resources Presentation 6.1 'Extracting metals'

Key vocabulary iron(III) oxide, reactivity, reduction, reduction with carbon

Teaching and learning

Engage

- Show slide 1 of Presentation 6.1 'Extracting metals' and point out that gold ore contains pieces of gold, copper ores do not. Instead they contain compounds such as green copper(II) carbonate. Challenge groups to **describe** how copper could be obtained from its ore.

Challenge and develop

- Check that students **appreciate** that copper carbonate can be separated from crushed rock but that a chemical reaction is needed to remove the copper. The thermal decomposition of green copper carbonate could be demonstrated to produce the black copper oxide from which copper was extracted in Lesson 4.1. [O1]

- Use a video clip to show students how copper was first extracted [Google search string:] BBC Bitesize – National 4 Chemistry – Extracting copper. [O1]

- Students reduce iron(III) oxide using carbon on a match head following the guidance in Practical sheet 6.1. [O1]

Explain

- Show slide 2 of Presentation 6.1 'Extracting metals' to link the reactivity of metals to the method used to extract them from ores. [O1, O2]

- Use a video clip to emphasise the difference between the two extraction techniques [Google search string:] The Extraction of Metals by Jon Dicks (first half only). [O1, O2]

Low demand

- Students read 'Reduction' and 'Reduction using carbon' from Section 6.1 in the Student Book and answer questions 1–3. [O1, O2]

Standard and high demand

- Students work in pairs to **compose** a page for a revision website about why different extraction techniques are needed for different metals. [O1, O2]

Consolidate and apply

Low demand

- Students **explore** the link between the discovery dates of metals and their reactivity by completing Worksheet 6.1. [O3]

Standard and higher demand

- Students read 'Ionic equations' in Section 6.1 in the Student Book and answer questions 4 and 5. [O3]

Extend

- Ask students able to progress further to watch a video clip of the thermit reaction [Google search string:] Iron –Periodic Table of Videos. They then write symbol and ionic equations for the reaction. [O3]

Plenary suggestion

Ask students to draw a large triangle with a smaller inverted triangle that just fits inside it (so they have four triangles). In the outer three ask them to write, something they've seen, something they've done and something they've discussed. Then add something they've learned in the central triangle.

Answers to Worksheet 6.1

1. It is unreactive/near the bottom of the reactivity series; so it does not form compounds; 2. Removing oxygen; 3. $2CuO + C \rightarrow 2Cu + CO_2$; 4. It can only be extracted by electrolysis; it couldn't be extracted until electricity had been discovered; 5. Aluminium oxide

Lesson 2: Using electrolysis to extract metals

Lesson overview

OCR Specification reference

OCR C6.1

Learning objectives

- Review the connection between the reactivity series and the ways metals are extracted.
- Consider how aluminium is extracted from aluminium oxide.
- Learn the oxidation and reduction reactions involved.

Learning outcomes

- Explain why some metals have to be extracted by electrolysis. [O1]
- Explain the process of the electrolysis of aluminium oxide. [O2]
- Write half equations for the reactions at the electrodes. [O3]

Skills development

- WS 1.2 Use models in explanations.
- WS 1.4 Describe and explain how aluminium is extracted.
- WS 4.1 Use scientific vocabulary, terminology and definitions.

Resources needed Worksheet 6.2

Digital resources Presentation 6.2.1 'Name that metal'; Presentation 6.2.2 'What was the question?'

Key vocabulary aluminium, bauxite, carbon anode, cryolite

Teaching and learning

Engage

- Display Presentation 6.2.1 'Name that metal' and challenge students to name the metal described in less than five clues. Clues appear on 'click' and become increasingly specific.

Challenge and develop

- Use a video clip to show the industrial scale of aluminium extraction [Google search string:] Aluminium extraction. [O1, O2]

Explain

Low demand

- Students complete Worksheet 6.2 to summarise the process of aluminium extraction using Section 6.2 of the Student Book as a reference. [O1, O2]

Standard and high demand

- Students produce diagrams of the equipment needed to electrolyse aluminium oxide and list the three main inputs the process requires. [O1, O2]

Consolidate and apply

Low demand

- Create an advert to encourage people to recycle aluminium cans. Explain why this could make aluminium cheaper and reduce pollution. [O2]

Standard and high demand

- Students read 'Electrode half equations' in Section 6.2 of the Student Book and add the relevant half equations to their diagrams of the electrolysis equipment. [O3]

Extend

- Ask students able to progress further to label the half equations at each electrode as oxidation or reduction. [O3]

Plenary suggestion

Challenge groups to **devise** questions to match the answers on Presentation 6.2.2 'What was the question?'

Answers to Worksheet 6.2

1. a) The container should be labelled 'carbon cathode where aluminium forms'; b) One of the rectangles should be labelled 'carbon anode where oxygen forms'; c) The space between the two should be labelled 'the electrolyte – aluminium oxide dissolved in molten cryolite'; 2. Aluminium is more reactive than carbon; so carbon cannot displace aluminium from its ores; 3 So that its ions are free to move; the cryolite has a lower melting point then aluminium oxide, so this saves energy; 4. The oxygen released at the anodes; reacts with the carbon to make carbon dioxide; 5. Aluminium ions are attracted to the negative cathode; where they gain electrons; and turn into atoms; oxide ions are attracted to the positive anode; where they lose electrons; to form atoms, which pair up to form oxygen molecules

Lesson 3: Alternative methods of metal extraction

Lesson overview

OCR Specification reference

OCR C6.1

Learning objectives

- Describe the process of phytomining.
- Describe the process of bioleaching.
- Evaluate alternative biological methods of metal extraction.

Learning outcomes

- Describe how some plants concentrate metal compounds in their leaves and shoots that can be harvested and burned to produce copper rich ash; copper can be extracted from the ash. [O1]
- Describe how some bacteria leach copper compounds from their ores, and how these can be used to produce copper metal. [O2]
- Compare the use of traditional mining methods and smelting with phytomining and bioleaching in terms of environmental and economic factors. [O3]

Skills development

- WS 1.4 Explain everyday and technological applications of science, evaluate associated personal, social, economic and environmental implications and make decisions based on the evaluation of evidence and argument.
- WS 2.4 Carry out experiments appropriately having due regard for the correct manipulation of apparatus, the accuracy of measurements and health and safety concerns.
- WS Use scientific vocabulary, terminology and definitions.

Resources needed Practical sheet 6.3; Worksheets 6.3.1 and 6.3.2; Technician's notes 6.3

Key vocabulary phytomining, bioleaching, leach, hyperaccumulators, toxic metals

Teaching and learning

Engage

- Ask students to **recall** how we use copper. Tell them that hundreds of years ago deposits of pure copper could be found in the Earth. These have now largely been exhausted, but we still use vast quantities of copper. Instead, copper ores are mined in opencast mines and chemical reactions are traditionally used to extract the copper metal from the ore. These involve heat energy. Ask students to work in pairs and list the drawbacks of using chemical reactions to extract copper from an ore. [O1, O2]

  Ask students to imagine they are working in the copper mining industry. What properties do they want in an ore when considering whether it is worth mining? Establish that ores with high copper percentages are better economically. Explain that these are called high grade ores and over time, and as high grade ores have been used up, we now rely on low grade ores to extract copper. Students can **suggest** how this affects their lists of extraction problems.

Challenge and develop

- High demand
- Explain that students are going to find out about two alternative methods of extracting copper from low grade ores. Students can answer question 1 on Worksheet 6.3.1. This sets a framework for the different treatments of different grades of ore. [O1, O2]

 © HarperCollins*Publishers* Limited 2016

Explain

- Students can read about phytomining in Section 6.3 of the Student Book. They can answer question 2 on Worksheet 6.3.1. Check their answers. [O1].

- Students can read about bioleaching in Section 6.3 of the Student Book. They can answer question 1 on Worksheet 6.3.2. Check that they **understand** the meaning of the term 'leach'. [O2]

- Remind students that both phytomining and bioleaching produce a solution of a copper compound from which copper is obtained by either electrolysis or displacement. They can carry out the experiment on Practical sheet 6.3. This is in two parts, electrolysis and displacement. They can answer the Evaluation section on Practical sheet 6.3. [O1, O2]

- Students can read 'Evaluating production' in Section 6.3 of the Student Book. They can answer question 2 on Worksheet 6.3.2. This compares the feasibility of traditional mining methods, phytomining and bioleaching. [O3]

Consolidate and apply

- Students can **discuss** the use of phytomining in removing metal compounds from contaminated land. [O3]

- Students can answer questions 3, 4, 5 and 6 in Section 6.3 of the Student Book. [O2, O3]

Extend

Ask students able to progress further to:

- **Research** the Chilean copper industry where 10% of the copper produced is produced by bioleaching. [O2]

- **Find out** how bioleaching is used in gold extraction. [O2]

Plenary suggestions

Students can form groups of three. Allocate each student in each group a type of mining – 'traditional mining', 'phytomining' or 'bioleaching'. Ask questions to which the answers are one of the methods used for copper extraction. Students can 'claim' their answers.

Ask students to write down three 'big ideas' they have **learnt** this lesson. They can **combine** their ideas in pairs, and then as a class to **produce** a master list.

Answers to Worksheet 6.3.1

1.a) A high grade ore has a higher percentage of copper; b) Large opencast mines; waste rock/ore; smelting uses energy; produces SO_2; c) Higher grade ores are being used up; d) $CuS + O_2 \rightarrow Cu + SO_2$; 2. a) Plants grown on rock containing copper compounds; absorbed through plant roots; accumulate in leaves and shoots; harvested; burned; copper extracted from ash; b) CO_2 absorbed in photosynthesis; is returned to the atmosphere during burning; c) A small percentage of copper compounds remain in the ore; d) A plant that accumulates a metal in its leaves and shoots

Answers to Worksheet 6.3.2

1. a) Bacteria are grown on low grade ores; bacteria change composition of ore; and copper compounds leach out; solutions of these are used to extract copper; b) Copper sulfide is insoluble; copper sulfate is soluble; 2. Mainly student response and opinion, but should include reference to decreasing availability of high grade ores; and cost

Answers to Practical sheet 6.3

1. a) Gas is given off; b) Red/brown metal (copper) is deposited; 2. Becomes paler blue; 3. Copper ions are removed from the solution; as copper is deposited; so the colour becomes paler; 4. $CuSO_4(aq) + Fe(s) \rightarrow Cu(s) + FeSO_4(aq)$

Lesson 4: The Haber process

Lesson overview

OCR Specification reference

OCR C6.1

Learning objectives

- Apply the principles of dynamic equilibrium to the Haber process.
- Use graphs to explain the trade-off between rate and equilibrium.
- Explain how commercially used conditions relate to cost.

Learning outcomes

- Explain that $N_2 + 3H_2 \rightleftharpoons 2NH_3$ is a reversible reaction in which the rate of the forward reaction equals the rate of the reverse reaction at equilibrium. [O1]
- Use graphs to explain that the reaction to produce ammonia favours a low temperature and high pressure, but low temperatures produce slow rates of reaction and high pressures are expensive, so compromise conditions are used. [O2]
- Explain how compromise pressures are used in the Haber process to reduce cost. [O3]

Skills development

- WS 1.4 Explain everyday and technological applications of science, evaluate associated personal, social, economic and environmental implications and make decisions based on the evaluation of evidence and arguments.
- WS 3.5 Interpret observations and other data (presented in verbal, diagrammatic, graphical, symbolic or numerical form), including identifying patterns and trends, making inferences and drawing conclusions.
- WS 3.8 Communicate the scientific rationale for investigations, methods used, findings and reasoned conclusions, through written and electronic reports and presentations using verbal, diagrammatic, graphical, numerical and symbolic forms.

Maths focus Recognise and use expressions in decimal form; use ratios, fractions and percentages.

Resources needed Class sets of molecular model kits (e.g. Molymods); Practical sheet 6.4; Worksheets 6.4.1 and 6.4.2; Technician's notes 6.4.

Key vocabulary Haber process, equilibrium, dynamic equilibrium, reversible reaction, optimum conditions

Teaching and learning

Engage

- Allow selected students to cautiously smell the stopper from a dilute ammonia solution bottle. Tell them this gas is ammonia and is the compound used in old-fashioned smelling salts – the bottles wafted under the noses of fainting ladies in old detective stories like Sherlock Holmes. Explain that ammonia has a far more important role – its compounds are used as fertilisers and are responsible for feeding half the world. [O1]
- **Use** Molymods to make a model of an ammonia molecule. Explain that it has covalent bonding and ask students to draw a dot-and-cross diagram to show its bonding. Check their diagrams. [O1]

Challenge and develop

High demand

- Explain that the ammonia used to make fertilisers is made in chemical plants using nitrogen from the air and hydrogen. The process is called the Haber process after the chemist who developed it. Students can use Figure 6.12 and read 'The manufacture of ammonia' in Section 6.4 of their Student Book to follow the process. There are several online PowerPoint presentations available that can be used to illustrate the process. Students can answer question 1 in Worksheet 6.4.1. [O1]

 © HarperCollins*Publishers* Limited 2016

- Explain the source of the raw materials. Students can read about this in 'Commercially used conditions' in Section 6.4 of the Student Book. They can answer questions 3 and 4 in the Student Book and/or question 2 in Worksheet 6.4.1. [O1]

Explain

- Explain that the reaction to produce ammonia is reversible. You can use Practical sheet 6.4 to demonstrate a reversible reaction by heating ammonium chloride in a glass vessel. Discuss the balanced symbol equation to produce ammonia and the use of the equilibrium sign, $\rightleftharpoons$. Explain the forward and reverse reactions. You can use Molymods to model the forward and reverse reactions between hydrogen gas, nitrogen gas and ammonia gas. Explain that, eventually, an equilibrium is reached where the concentration of each gas does not change, but reactions continue in both directions at the same rate. Explain that this is called a 'dynamic equilibrium'. Students can answer question 1 in Worksheet 6.4.2. [O1]

- Explain that nitrogen and hydrogen do not react under room conditions, but will if the temperature and pressure are changed. They can find out how changes in pressure and temperature affect the reaction by reading 'Equilibrium versus rate' in Section 6.4 of the Student book. Explain the problem of using a low temperature on the rate of the reaction. They can answer questions 5 and 6 in the Student Book and question 2 in Worksheet 6.4.2. [O2]

- Ask students to work in pairs and **identify** the costs associated with producing ammonia. These should include the energy costs of obtaining nitrogen from air and hydrogen from methane, as well as the costs of maintaining high pressures and temperatures. They should also include the cost and availability of the raw materials. Students can double up and **share** their lists. A class list of costs can be **compiled**. [O3]

Consolidate and apply

- Ask students to explain the fact that ammonia manufacturers opt to produce ammonia little and often rather than a lot slowly. [O2, O3]

- Ask students to imagine they are controlling the conditions being used to manufacture ammonia. Ask them to **identify** all the factors they need to take into consideration. [O2, O3]

Extend

Ask students able to progress further to:
- **Research** Fritz Haber and find out why and when he developed the Haber process. [O1]
- **Find out** what Le Chatelier's principle is and how it affects the manufacture of ammonia.

Plenary suggestions

Ask students to write down one thing from the lesson they are sure about, one thing they are unsure of and one thing they need to know more about. **Identify** the weak areas.

Ask students to write a question from the lesson and a mark scheme. Their question must have at least three marks, so three points. They can **try** their questions out on each other.

Answers to Worksheet 6.4.1

1. a) To produce fertilisers; b) Air and natural gas; c) To increase the rate of reaction; d) To give a larger surface area for the reaction to take place on; e) 450 °C and 200 atmospheres; f) It is recycled in the apparatus; g) When the mixture is cooled, ammonia liquefies at −33 °C, but hydrogen and nitrogen remain as gases; ammonia can be separated; 2. a) Nitrogen from the air and hydrogen from methane in natural gas; b) Air is cooled until it liquefies; then it is heated slowly and the fraction that boils off at −196 °C is collected; c) Methane + steam → carbon monoxide + hydrogen; $CH_4 + H_2O \rightarrow CO + 3H_2$

Answers to Worksheet 6.4.2

1. a) $N_2 + 3H_2 \rightleftharpoons 2NH_3$; b) The reaction proceeds in both directions; c) $N_2 + 3H_2 \rightarrow 2NH_3$; d) $2NH_3 \rightarrow N_2 + 3H_2$; e) The reaction is proceeding at the same rate in both directions; 2. a)(i) 65%; (ii) 20%; (iii) 30–35%; (iv) <10%; b) Low/350 °C; c) High/400 atmospheres; d) Slow rate of reaction; e) Difficult to maintain and expensive; f) Left; g) Increases the rate of reaction in both directions

Answers to Practical sheet 6.4

1. $NH_4Cl \rightleftharpoons NH_3 + HCl$; 2. $NH_4Cl \rightarrow NH_3 + HCl$; 3. $NH_3 + HCl \rightarrow NH_4Cl$

Lesson 5: Production and use of NPK fertilisers

Lesson overview

OCR Specification reference

OCR C6.1

Learning objectives

- Describe how to make a fertiliser in the laboratory.
- Explain how fertilisers are produced industrially.
- Compare the industrial production of fertilisers with laboratory preparation methods.

Learning outcomes

- Use a titration to make a sample of ammonium phosphate. [O1]
- Explain how fertilisers are produced industrially from ammonia made in the Haber process and from mining potassium chloride, potassium sulfate and phosphate rock. [O2]
- List the similarities and differences between the industrial process to produce ammonium phosphate and the laboratory preparation. [O3]

Skills development

- WS 1.4 Explain everyday and technological applications of science, evaluate associated personal, social, economic and environmental implications and make decisions based on the evaluation of evidence and arguments.
- WS 2.4 Carry out experiments appropriately having due regard for the correct manipulation of apparatus, the accuracy of measurements and health and safety considerations.
- WS 2.6 Make and record observations and measurements using a range of apparatus and methods.
- WS 2.7 Evaluate methods and suggest possible improvements and further investigations.

Maths focus Use ratios, fractions and percentages

Resources needed Practical sheet 6.5; Worksheet 6.5; Technician's notes 6.5

Digital resources Digital photographs of fertiliser bags with NPK labels and ratios

Key vocabulary fertiliser, nitrogen, phosphorus, potassium, NPK fertiliser

Teaching and learning

Engage

Low demand

- Remind students that the major use of ammonia is to make fertilisers. Tell them that modern farms have tractors with computers with GPS systems. They can detect exactly where fertiliser is needed in a field and deliver the correct amount to the correct area. Explain that the three major elements that need to be replaced are nitrogen (N), phosphorus (P) and potassium (K). Figure 6.16 in Section 6.5 of the Student Book shows a fertiliser bag with an NPK label. There are several images online at Google images. If students have already covered the nitrogen cycle in their Biology lessons, this can be linked to the use of fertilisers. [O1]

- Explain that the ratio of nitrogen to phosphorus to potassium is given on the fertiliser packaging – for example an NPK fertiliser labelled 25 – 4 – 2 contains 25% nitrogen, 4% phosphorus and 2% potassium. Discuss which compounds fertiliser manufacturers can use in fertilisers to provide these elements. [O1]

Challenge and develop

- Introduce students to Practical sheet 6.5 and explain the use of the apparatus as needed, reminding them how to use a burette tap and read a meniscus. You may prefer to fill the burettes before the lesson. Students may need to **practise** controlling the burette tap so that they can add a drop at a time. Allow them to carry out the practical in pairs. They are performing one titration using an indicator to measure reacting volumes and a second 'titration' without indicator. The ammonium phosphate solution will need to be kept until the next lesson to crystallise. Note that students will be **measuring** the mass of product obtained to use in Lesson 5.4 on percentage yield. Students can answer the questions in the Evaluation section of Practical sheet 6.5, and questions 1 and 2 in Section 6.5 of the Student Book. [O1]

Explain

Standard demand

- Explain that the industrial production of NPK fertilisers uses a variety of raw materials. Ammonium phosphate, supplying nitrogen and phosphorus, is made industrially using the same reaction as for the lab preparation. Students can read 'Industrial production of fertilisers' in Section 6.5 of the Student Book and complete the diagram on Worksheet 6.5 (questions 1 and 2). This can act as a revision sheet if required. [O2]

High demand

- Students can answer question 3 on Worksheet 6.5. They are **comparing** the industrial production of ammonium phosphate with the laboratory preparation. [O3]

- Students can answer questions 3, 4, 5 and 6 in Section 6.5 of the Student Book. [O3]

Consolidate and apply

- Ask students what a 25 – 0 – 0 NPK fertiliser contains and to **name** a suitable compound that could be used in the fertiliser.

- Ask students to work in pairs and **list** the steps needed to produce a sample of ammonium sulfate from dilute sulfuric acid and ammonia solution. Combine the lists into a class list. [O1]

- Discuss the environmental and social effects of producing and using fertilisers. Include energy consumption, the environmental impact of mining, potential pollution from chemical processes, effects of excess fertiliser on waterways and the need to feed a growing population. [O2]

Extend

Ask students able to progress further to:

- **Calculate** the percentage by mass of nitrogen in ammonium phosphate and of phosphorus in ammonium phosphate. They will need to **research** the method. [O2]

- **Calculate** the percentage by mass of nitrogen in ammonium sulfate and in ammonium nitrate, and hence **decide** which provides more nitrogen in a fertiliser. [O2]

Plenary suggestions

Ask students to draw a large triangle with a smaller inverted triangle inside. In the three outer triangles, they write something they have done, something they have discussed and something they have seen. In the centre triangle, they write something they have learnt.

Ask students to write down two 'big ideas' they have **learned** this lesson. They can **share** their ideas in groups and **compile** a master list in rank order.

Answers to Worksheet 6.5

1. *From left to right*: hydrogen from natural gas; potassium chloride; potassium sulfate; calcium nitrate; phosphoric acid; *Arrows should have labels*: Haber process; sulfuric acid; phosphoric acid; 2. An arrow from ammonia to a new box labelled 'nitric acid'; 3. *Differences*: scale; produced indirectly from phosphate rock; calcium nitrate also produced; *Similarities*: both are produced in a neutralisation reaction; both produced from ammonia and phosphoric acid

Answers to Practical sheet 6.5

1. a) To measure the volume of acid; b) To measure the volume of ammonia solution; c) To detect the end point/when the exact volume of acid has been added to neutralise the ammonia solution; 2. a) Student's suggestions; may include errors in apparatus measurements, especially the measuring cylinder; and detecting the end point; b) For example, use more accurate apparatus to measure the volume of ammonia solution; use a pH probe; 3. $6NH_3 + 2H_3PO_4 \rightarrow 2(NH_4)_3PO_4$

Lesson 6: Life cycle assessment and recycling

Lesson overview

OCR Specification reference

OCR C6.1

Learning objectives

- Describe the components of a life cycle assessment (LCA).
- Interpret LCAs of materials or products from information.
- Carry out a simple comparative LCA for shopping bags.

Learning outcomes

- Describe that a LCA involves assessment of the environmental impact of a product during the extraction and processing of raw materials, manufacturing and packaging, use and operation during its lifetime and disposal at the end of its life. [O1]
- Recognise that some LCA information is qualitative and dependent on the author, while other information is quantitative. [O2]
- Compare the environmental impact of different types of carrier bags at different stages of the LCA. [O3]

Skills development

- WS 1.3 Appreciate the power and limitations of science and consider any ethical issues that may arise.
- WS 1.4 Explain everyday and technological applications of science, evaluate associated personal, social, economic and environmental implications and make decisions based on the evaluation of evidence and argument.
- WS 1.5 Evaluate risks both in practical science and the wider societal context, including perception of risk in relation to data and consequences.

Maths focus Recognise and use expressions in decimal form; use ratios, fractions and percentages; make estimates of the results of simple calculations; use an appropriate number of significant figures; translate information between graphical and numerical forms

Resources needed Graph paper; Worksheets 6.6.1 and 6.6.2

Digital resources Computer access for the extended section

Key vocabulary life cycle assessment, extracting, manufacturing, disposal

Teaching and learning

Engage

- Show students a selection of supermarket carrier bags (or similar). Try to include the conventional lightweight bags, the thicker glossier bags, paper carriers and cotton bags. You could refer to the changes in law in the UK regarding payment for plastic carrier bags and the idea of bags for life. Ask them to work in pairs and decide which bag they think is the most environmentally friendly and why. They can refer to these at the end of the lesson. [O1]

Challenge and develop

Low demand

- Discuss the life of a carrier bag from raw materials to manufacturing to use and, finally, disposal. Students may wish to **amend** their ideas about which are the most environmentally friendly at this stage. Tell them that companies and governments make assessments of the impact of a product at each stage of its life and this is called a 'Life Cycle Assessment', or LCA. They can use Figure 6.17 in the Student Book. They can read 'Life cycle assessments' in Section 6.6 of the Student Book, which lists these stages. [O1]
- Students can answer question 1 on Worksheet 6.6.1. [O1]

Explain

Standard demand

- Ask students which factors in the life cycle stages can be measured, and numbers applied, and which are more difficult to assess. Establish that energy consumption, water consumption, resource consumption and some wastes can be measured. Discuss the difference between the terms 'qualitative' and 'quantitative'. Students can read 'Objectivity and subjectivity' in Section 6.6 of the Student Book. Emphasise that measurements are difficult to argue with, but subjective assessments depend on who makes them. Discuss how advertising can be misleading. They can answer question 2 on Worksheet 6.6.1. [O2]

High demand

- Either ask students to read 'Life cycle assessments' in Section 6.6 of the Student Book and answer questions 5 and 6, or introduce Worksheet 6.6.2. This contains an LCA for lightweight poly(ethene) bags, heavier glossy poly(ethene) bags (as in 'bags for life') and paper carrier bags. Objective data for some aspects is provided in a table. Students have to make value **judgements** to complete the exercise and produce an LCA, as well display information in graphical form. [O3]

Consolidate and apply

- Discuss who might carry out an LCA and how they might use them. This should include governments, companies and businesses. [O1]

- Ask students to list the factors an LCA should include for a smart phone. [O3]

Extend

Ask studetns able to progress further to:

- **Carry out** a [Google search string:] Levi 501 LCA and find out which information is objective and which information is subject to other factors. [O2]

- **Produce** an LCA to compare the use of drinks bottles made with plastic with those made with glass.

Plenary suggestions

Ask students to write a question and a mark scheme about something from the lesson. The question needs to include at least two points. Students can try their questions out on each other.

Ask students to write down the 'big idea' they have learned this lesson. They can pair up and **share** ideas. Ideas can be **shared** as a class.

Answers to Worksheet 6.6.1

1. *Guidelines*: a) amount of raw materials; energy requirements; transport; processing; pollution; waste; b) Energy requirements; transport; packaging material; pollution; c) How it is used; recycling; (d) Pollution, transport; 2. a) *Easily measured*: energy consumption, water consumption, resource consumption, some waste produced; *Subjective*: transport requirements, customer use, disposal; b) They can select information to fit their case

Answers to Worksheet 6.6.2

1. Quantitative; 2. Students' graphs; 3. Four; 4. a) Raw materials used; transport; b) Pollution – greenhouse gases, etc.; c) Reuse; recycling; d) Method of disposal; pollution (not exhaustive); 5. Students' response; 6. Students' response

Lesson 7: Ways of reducing the use of resources

Lesson overview

OCR Specification reference

OCR C6.1

Learning objectives

- Describe ways of recycling and reusing materials.
- Explain why recycling, reusing and reducing are needed.
- Evaluate ways of reducing the use of limited resources.

Learning outcomes

- Describe how some products, such as glass bottles, can be reused or recycled, but other products cannot be reused and are recycled for a different use. [O1]
- Explain how recycling, reusing and reducing reduce the use of limited resources, energy consumption, waste and environmental impacts. [O2]
- Evaluate reasons for the recycling of aluminium cans and other products from a given list of facts. [O3]

Skills development

- WS Explain everyday and technological applications of science, evaluate associated personal, social, economic and environmental implications and make decisions based on the evaluation of evidence and arguments.
- WS 3.7 Be objective, evaluate data in terms of accuracy, precision, repeatability and reproducibility and identify potential sources of random and systematic error.
- WS 3.8 Communicate the scientific rationale for investigations, methods used, findings and reasoned conclusions, through written and electronic reports and presentations using verbal, diagrammatic, graphical, numerical and symbolic forms.

Maths focus Recognise and use expressions in decimal form; recognise and use expressions in standard form; use ratios, fractions and percentages

Resources needed A bin liner of rubbish including a glass bottle, a plastic water bottle, drinks can made of aluminium, tin can, old mobile phone, denim jeans, vegetable/fruit waste and a newspaper; Worksheets 6.7.1 and 6.7.2

Key vocabulary recycling, limited resource, reduction of use, reuse of resources

Teaching and learning

Engage

- Show students a bin liner of rubbish, including a glass bottle, a plastic water bottle, aluminium drinks can, tin can, old mobile phone, denim jeans, vegetable/fruit waste and a newspaper. Discuss which of these can be recycled and/or which can be reused. The glass bottle and the denim jeans can be reused – the rest can be recycled, plus the glass bottle. [O1]
- Ask students to write down why they think we need to recycle and/or reuse waste. They can add to these ideas through the lesson and use them in the plenary. [O1]

Challenge and develop

Low demand

- Students can read the information in question 1 in Worksheet 6.7.1. This is about the reuse and recycling of glass containers. It can be applied to local recycling collections. Students can answer question 1 on Worksheet 6.7.1. [O1

Standard demand

- Ask students why it is necessary to recycle glass and **discuss** their ideas. They can refer to their initial lists and **develop** them if appropriate. They can read 'Recycling and reusing' and 'Reducing the use of finite resources' in Section 6.7 in the Student Book. They can **answer** question 2 in Worksheet 6.7.1. [O2]

Explain

- Emphasise that we recycle and reuse to reduce the use of limited resources, energy consumption, waste and the environmental impact. [O1, O2]

- Ask students how metal recycling differs from glass recycling. You could refer to local scrap yards as collection points for scrap vehicles. Establish that because we use many different types of metal, these need to be sorted by type first. Once sorted, these can be melted down before being made into different products. Explain that some scrap metals can also be added to new metals to reduce the amount of ore needed. Students can read question 1 on Worksheet 6.7.2. This explains how scrap iron is added to cast iron in the steel-making process, so reducing the amount of ore needed. Students can answer question 1 in Worksheet 6.7.2. [O1, O2].

High demand

- Students can answer question 2 in Worksheet 6.7.2. This requires a longer answer and some **evaluation** as to why we should recycle aluminium. [O3]

- Students can answer questions 5 and 6 in Section 6.7 in the Student Book.

Consolidate and apply

- Students can **assess** how recycling and reusing is applied in their locality. [O1]

- Remind students that they learned about some methods used for copper extraction in Lesson 6.3. Discuss why 41% of copper is recycled. [O2]

Extend

Ask students able to progress further to:

- **Find out** how paper and plastics are recycled. [O1]

- **Find out** why London's road dust contains 100 times more palladium, platinum and rhodium metals than their respective ores, and why some companies are considering collecting road dust to extract these metals.

Plenary suggestions

Students need to work in groups of four. Each student is assigned one photograph (from Figures 6.21, 6.22, 6.23 and 6.24) in Section 6.7 in the Student Book. They take turns to **explain** the relevance of their photograph to reducing the use of resources to the rest of their group.

Ask students to think ahead to the year 2100. They can **suggest** how ways of reducing the use of resources might change and what impact it might have on their lives and those of their children.

Answers to Worksheet 6.7.1

1. a) Unreactive; b) For recycling it needs to stay chemically the same; c) Sand and limestone are extracted from the Earth's crust; sodium carbonate is made from rocks from the Earth's crust; d) Collecting bottles and fulfilling hygiene requirements to reuse them is expensive and time consuming; recycling was not common 50 years ago; e) Students' response; 2. a)(i) Fewer/smaller quarries needed; (ii) Less carbon dioxide emitted; so less global warming; (iii) Reduced; (iv) Reduced; b) There is a finite amount available in the Earth's crust; c) Energy needed to melt the raw materials is often from fossil fuels; which burn to produce CO_2; CO_2 given off when raw materials are heated; d) 4.14×10^9 kJ

Answers to Worksheet 6.7.2

1. a) Reduces; b) $C + O_2 \rightarrow CO_2$; c) Different scrap metals will have different compositions; and the amounts will change with the percentage of scrap used; d) For example, reduces use of iron ore; reduces the environmental impact of mining; less waste to land fill; 2. Student response

Lesson 8: Alloys as useful materials

Lesson overview

OCR Specification reference

OCR C6.1

Learning objectives

- Describe the composition of common alloys.
- Interpret the composition of other alloys from data.
- Evaluate the uses of other alloys.

Learning outcomes

- Describe the composition of bronze, brass, gold used for jewellery, high and low carbon steels and stainless steel, plus other alloys. [O1]
- Use data about the composition of other alloys, such as solder, to determine their properties, such as hardness. [O2]
- Explain how different percentage compositions of alloys, such as pewter, allow for different uses. [O3]

Skills development

- WS 1.4 Explain everyday and technological applications of science, evaluate associated personal, social, economic and environmental implications and make decisions based on the evaluation of evidence and argument.
- WS 3.3 Carry out and represent mathematical and statistical analysis.
- WS 3.5 Interpret observations and other data (presented in verbal, diagrammatic, graphical, symbolic or numerical form), including identifying patterns and trends, making inferences and drawing conclusions.

Maths focus Recognise and use expressions in decimal form; use ratios, fractions and percentages

Resources needed Samples of copper foil and aluminium foil; Worksheets 6.8.1, 6.8.2 and 6.8.3

Digital resources Class internet access

Key vocabulary alloy, brass, bronze, steel

Teaching and learning

Engage

Show students samples of copper foil and aluminium foil and explain that the pure metals are too soft for many uses. Remind them that they learned about alloys in Section 2.20 of the Student Book and found out that a mixture of metals is often stronger and harder than the pure metal. They may **recall** that different sized metal ions in the structure stop the layers of metal ions sliding over each other. Explain that other properties, such as melting point and resistance to corrosion, can be changed by using alloys. Tell them that most of the metals we use are alloys, not pure metals. [O1]

Challenge and develop

Low demand

- Ask students to name as many alloys as they can. Allow them to **research** the composition of some alloys and to complete Worksheet 6.8.1. They can use Section 6.8 of the Student Book and will need internet access to complete the list of alloys. [O1]
- Remind students that the proportion of gold in gold alloys is measured in carats – 24 carat gold being pure gold, 18 carat being 75% gold and 25% of another metal (such as silver, copper or zinc) and 9 carat containing nine parts of gold out of 24. They can **calculate** the percentage of gold in 9 carat gold. This is part of Worksheet 2.11.1. [O1]

Explain

Standard demand

- Explain that the percentage of each metal or element in an alloy can be varied to produce different properties. Students can read 'Other alloys' in Section 6.8 of their Student Book and find out how the composition of solder varies according to its use. They can answer questions 3 and 4. [O2]

- Students can answer Worksheet 6.8.2 to find out about silver alloys and the results of recent research at Middlesex University. [O2]

High demand

- Remind students that when iron is extracted from its ore, cast iron is produced. Cast iron contains up to 4% carbon and is very brittle and difficult to shape. Remind them that they found out how, during the steel making process, scrap steel is added to cast iron and oxygen blown into the molten mixture to remove some of the carbon. Establish that different steels can be produced using different carbon contents, and that other metals can be added to give different properties. Students can answer Worksheet 6.8.3 to find out about high, medium and low carbon steels and stainless steel. [O3]

- Students can read how different compositions of pewter produce different properties in 'Evaluating uses' in Section 6.8 of the Student Book. They can answer questions 5 and 6. [O3]

Consolidate and apply

- Answer questions 1 and 2 in Section 6.8 of the Student book. [O2, O3]

- Ask students what properties the alloys used to make spacecraft have. They may be able to **suggest** some constituent metals, such as aluminium for lightness. [O1]

Extend

Ask students able to progress further to:

- **Find out** how the smart alloy nitinol is able to change shape. [O1]

- **Find out** what titanium steel is and how its properties differ from other steels. [O2]

Plenary suggestions

Organise students into groups of five or six. Each student chooses an alloy. In turn, they **describe** their alloy to the group. The other students have to **identify** it.

Organise students into groups of two or three. They have to create a freeze frame to depict an alloy. Students can **perform** their frames to the rest of the class, who have to name the alloy.

Answers to Worksheet 6.8.1

Bronze: Cu/Sn, harder, statues; *Brass*: Cu/Zn, harder, taps, fittings; *Gold*: Au/Cu/Ag/Zn, harder, jewellery; *High carbon steel*: Fe/C, harder, brittle, cutting tools; *Low carbon steel*: Fe/C, softer, malleable, car panels; *Stainless steel:* Fe/C/Cr/Ni, hard, resists corrosion, cutlery; *Amalgam*: Hg/Ag/Sn/Cu, strong, durable, dental fillings; *Nitinol*: Ni/Ti, shape memory, dental braces; *Solder*: Sn/Pb, lower melting point, joining metals

Answers to Worksheet 6.8.2

1. Increases hardness; 2. Reduces lustre/shine; 3. Tarnishing decreases; 4. a) Expose same size samples to sulfur atmosphere, skin and UV light for the same time; and measure tarnishing (amount of light reflected, compare colour, etc.); b) Sterling silver taken as standard at 100; and others compared to it; 5, 30%

Answers to Worksheet 6.8.3

1. Low carbon steel; 2. High carbon steel; 3. High carbon steel; 4. Stainless steel; 5. Low carbon steel

Lesson 9: Corrosion and its prevention

Lesson overview

OCR Specification reference

OCR C6.1

Learning objectives

- Show that air and water are needed for rusting.
- Describe experiments on rusting and interpret the results.
- Explain methods for preventing corrosion.

Learning outcomes

- Know that corrosion involves chemical reactions of materials with substances in the environment, and that both water and air are needed for iron to rust. [O1]
- Plan and carry out experiments to show that both water and air are needed for iron to rust, and interpret the results. [O2]
- Explain how corrosion can be prevented by using a barrier, such as paint, or a reactive coating, such as zinc in galvanised iron. [O3]

Skills development

- WS 2.2 Plan experiments or devise procedures to make observations, produce or characterise a substance, test hypotheses, check data and explore phenomena.
- WS 2.7 Evaluate methods and suggest possible improvements and further investigations.
- WS 3.5 Interpret observations and other data (presented in verbal, diagrammatic, graphical, symbolic or numerical form), including identifying patterns and trends, making inferences and drawing conclusions.

Resources needed Practical sheets 6.9.1 and 6.9.2; Worksheets 6.9; Technician's notes 6.9.1 and 6.9.2

Digital resources Computer and data projector (internet access)

Key vocabulary corrosion, sacrificial protection, galvanising, electroplating

Teaching and learning

Note that this lesson includes a practical that needs to be left for 3 days.

Engage

- Show students photos of metal corrosion, including rust and corroded copper. You may have cases of corrosion in a locality you can refer to. There are several on Google images [Google search string:] metal corrosion, rust and corroded copper. Discuss what has happened to the metals and establish that a chemical reaction has occurred between the metal and substances in the environment. Emphasise that the term 'corrosion' is applied to all metals and that when iron corrodes, the red/brown substance is called rust and it is referred to as rusting. Emphasise that the term 'rust' cannot be applied to other metals. [O1]

Challenge and develop

Low demand

- Ask students to suggest which substances iron reacts with in the environment when it rusts. They can answer question 1 in Worksheet 6.9. [O1]
- Introduce Practical sheet 6.9.1. Students are planning their own experiments to find out if both air and water are required for rusting. They can work in pairs to **produce** an annotated diagram showing their experiment. Check their plans and allow them to set up their experiments. Students having problems controlling variables can use Section 6.9 of the Student Book as a guide. Experiments need to be left for at least 3 days. [O1] (Schools in coastal areas may wish to include sodium chloride in the apparatus list to illustrate the effects of salt in speeding up corrosion, but this is not in the specification.)

Explain

Standard demand

- When the results are available, students can **record** them and answer the evaluation section in Practical sheet 6.9.1. They may need help defining 'systematic errors'. [O2]

High demand

- Explain that rust prevention is a big industry and represents one of the measures needed to conserve limited resources. Students can read about methods involving a coating that acts as a barrier in Section 6.9 of their Student Book (painting, oil and electroplating). They can answer question 2 in Worksheet 6.9. [O3]

- Explain that some coatings used on metals contain corrosion inhibitors or more reactive metals. Students can read 'Electrons in corrosion' in Section 6.9 of the Student Book to find out how sacrificial protection works. They can answer question 3 in Worksheet 6.9 and questions 5 and 6 in Section 6.9 of the Student Book. [O3]

- Introduce Practical sheet 6.9.2. This is intended to be set up prior to the lesson so that students can **observe** and **interpret** the results. Depending on the time available, this could be organised as a class practical, a demonstration or a planning exercise. Discuss the results with the students. [O3]

Consolidate and apply

- Students can work in pairs to **discuss** and list the problems that arise from corrosion on North Sea oil rigs. They can **suggest** possible methods that can be used to prevent or limit corrosion. [O3]

- Students can **identify** measures used by car manufacturers to prevent corrosion. [O3]

Extend

Ask students able to progress further to:

- **Find out** how salt affects the rate of corrosion and **suggest** the implications for driving on salted roads and for coastal areas. [O2]

- **Explain** why the chemistry of corrosion is the reverse of metal extraction for many metals. [O2]

Plenary suggestions

Organise students into groups of five or six. Each student requires two strips of paper. They write a question on one strip and the answer on the other. Shuffle and redistribute the strips so that each student has a question and an answer. One student asks their question, and the student holding the correct answer responds and asks the next question.

In groups of five or six, each student is allocated one method of preventing corrosion. They **prepare** a sentence describing their method and saying why it works. Students can **share** their sentences with the group.

Answers to Worksheet 6.9

1. a)(i) A reaction between a metal and the environment; (ii) The reaction between iron, water and oxygen; (b) Water and air/oxygen; (c) Hydrated iron(III) oxide; (d) No water or oxygen; 2. a)(i), (ii), (iii), (iv) All the methods provide a barrier to keep water; and oxygen away from the metal; b) The layer of aluminium oxide acts as a protective barrier; 3. a) They involve a chemical reaction and not just a barrier; b)(i) Magnesium is more reactive than iron; and corrodes first; (ii) The magnesium eventually corrodes completely; and the products dissolve in the seawater; (iii) The layer of zinc acts as a barrier – if the zinc is scratched it corrodes first; because zinc is more reactive than iron; c) Copper is less reactive than iron; d) For example, aluminium

Answers to Practical sheet 6.9.1

1. Should conclude that both air/oxygen and water are required for rusting; 2. Depends on student's results; 3. Depends on student's results; but composition of nails may confuse the results; 4. Student's suggestions

Answers to Practical sheet 6.9.2

1. Nails 2, 3, 4, 5 and unscratched galvanised nail; 2. Student's suggestions; further investigations could include investigating which metals protect sacrificially

Lesson 10: Ceramics, polymers and composites

Lesson overview

OCR Specification reference

OCR C6.1

Learning objectives

- Compare properties of materials quantitatively.
- Compare glass, ceramics, polymers, composites and metals.
- Select materials by relating their properties to uses.

Learning outcomes

- Recognise a material as being a ceramic, a polymer or a composite and describe what they consist of or how they were made. [O1]
- Describe the structure of thermoplastics and thermosets, and of composites, including composites containing carbon fibres or nanotubes. [O2]
- Compare the properties of glass and clay ceramics, polymers and composites quantitatively. [O3]

Skills development

- WS 1.4 Explain everyday and technological applications of science, evaluate associated personal, social, economic and environmental implications and make decisions based on the evaluation of evidence and argument.
- WS 3.5 Interpret observations and other data (presented in verbal, diagrammatic, graphical, symbolic or numerical form), including identifying patterns and trends, making inferences and drawing conclusions.
- WS 3.8 Communicate the scientific rationale for investigations, methods used, findings and reasoned conclusions, through written and electronic reports and presentations using verbal, diagrammatic, graphical, numerical and symbolic forms.

Maths focus Understand and use the symbols $=$, $<$, $\ll$, $\gg$, $>$, $\propto$, $\sim$

Resources needed A selection of ceramic, polymer and composite materials, such as pottery, china, a thermosetting plastic (electric plug), a thermoplastic (poly bag), soda-lime glass, borosilicate glass, wood, concrete and fibreglass; Molymods (or other molecular models kit); Worksheets 6.10.1, 6.10.2 and 6.10.3

Key vocabulary thermosoftening, thermosetting, composite, reinforcement

Teaching and learning

Engage

Show students a selection of ceramic, polymer and composite materials, such as pottery, china, a thermosetting plastic (electric plug), a thermoplastic (poly bag), soda-lime glass, borosilicate glass, wood, concrete and fibreglass. Divide the samples into ceramics, polymers and composites. Note that glass is considered to be a ceramic here. Ask students to suggest the characteristics of each group. Note that clay ceramics are made by shaping wet clay and then hardening at high temperatures, whereas glass ceramics are shaped when molten. [O1, O2]

Challenge and develop

Low demand

- Students can complete the table in Worksheet 6.10.1 to show what the different ceramics, polymers and composites are made of and some examples of how they are used. They can use Section 6.10 in the Student Book as a reference. You may wish to explain thermoplastic, thermoset and composite structure, as discussed below, before students complete all or part of this worksheet. [O1], [O2]

Standard demand

- Use molecular model kits, such as Molymods, to demonstrate that polymer molecules consist of long chains of molecules. Show how they are tangled in a thermoplastic polymer, but able to slide over each other. Explain that this means the plastic is soft. Now add some cross-links between the polymer chains to show the structure of a thermoset. Explain that this is a more rigid structure and does not soften when heated. In fact it decomposes before it melts, so it has no melting point. Students can answer question 1 in Worksheet 6.10.2. [O2]

- Describe the structure of a composite. You can use the diagram in question 2 in Worksheet 6.10.2. Students can answer question 2. [O2]

- Explain that some modern composites are made using either carbon fibres or carbon nanotubes in a polymer matrix, instead of glass fibres. This makes the composite very strong and lightweight. Students can answer question 3 in Worksheet 6.10.2. [O2]

Explain

High demand

- Explain that if a particular property is required for an object, properties of different materials can be compared and the best material chosen. Introduce students to Worksheet 6.10.3. It includes melting point and density data for some ceramics, polymers and composites. They can answer the questions and select the most appropriate materials for a specific use. [O3]

Consolidate and apply

- Remind students that fibreglass has fine glass fibres in a polymer matrix. The glass fibres are strong but very brittle, but the polymer matrix holds them together and protects them. Discuss how fibreglass has the combined properties of both glass and polymer. [O2]

- Composites can be designed to have particular properties. Ask students to suggest composites for building bridges, making aircraft panels and making tennis rackets. [O2, O3].

Extend

Ask students able to progress further to:

- **Find out** what 'wattle and daub' is and why it is one of the oldest man-made composites known. [O1]

- **Find out** why composites and polymers do not have exact melting points.

Plenary suggestions

Assign each student a term such as 'composite', 'ceramic' or 'polymer', or the name of a specific substance. Ask questions to which these terms are the answers. Students can 'claim' their answers.

Organise students in pairs. Ask each pair to **decide** on one 'big idea' they learned this lesson. They can double up and **share** their ideas and eventually **compile** a class list of 'big ideas'.

Answers to Worksheet 6.10.1

Soda-lime glass: heating sand, Na_2CO_3, $CaCO_3$, windows; *Borosilicate glass*: heating sand and BO_3, laboratory apparatus; *Clay ceramics*: moulding and heating, pottery; *Thermosets*: ethene with catalyst; electrical fittings; *Thermoplastics*: ethene with catalyst, poly bags; *Wood*: natural, building; *Concrete*: stones in cement, building; *Fibreglass*: glass fibres in polymer, thermal insulation

Answers to Worksheet 6.10.2

1. a)(i) Long tangled polymer chains; (ii) Long polymer chains with cross-links; b) Chains can slide over each other in a thermoplastic; but are held rigid in a thermoset; c) Long chains held rigid; d)(i) Thermoset, (ii) Thermoplastic; 2. a) Reinforce the matrix; b) Binder; c) *Wood*: cellulose fibres in lignin; *concrete*: stones in cement; *fibreglass*: glass fibres in polymer; 3. a) Student's diagrams; b) Strong and lightweight; c) Student's suggestion

Answers to Worksheet 6.10.3

1. Borosilicate glass; 2. Iron; 3. Decomposes on heating; 4. Thermoset; 5. HDPE; 6. LDPE or HDPE

 © HarperCollins*Publishers* Limited 2016

Lesson 11: Maths skills: Translate information between graphical and numerical form

Lesson overview

OCR Specification reference
OCR C6.1

Learning objectives
- Represent information from pie charts numerically.
- Represent information from graphs numerically.
- Represent numeric information graphically.

Learning outcomes
- Understand the information shown in a pie chart and be able to draw one from given information. [O1]
- Be able to extract information from a graph show how a dependent variable changes as the independent variable changes. [O2]
- Be able to draw a suitable graph from a given set of data. [O3]

Skills development
- WS 3.2 Translate data from one form to another.
- WS 3.3 Carry out and represent mathematical and statistical analysis.
- WS 3.5 Interpret observations and other data (presented in verbal, diagrammatic, graphical, symbolic or numerical form), including identifying patterns and trends, making inferences and drawing conclusions.

Maths focus Translate information between graphical and numerical form; plot two variables from experimental or other data

Resources needed Protractors, graph paper; Worksheets 6.11.1 and 6.11.2

Digital resources Computer and data projector

Key vocabulary graph, numerical scale, pie chart

Teaching and learning

Engage
- Show students bar charts or graphs of mean monthly temperatures and rainfall for a holiday destination. There are plenty on travel websites. Discuss why the travel companies have chosen to use bar charts/graphs to show this data. You could repeat this exercise using a hospital temperature chart and asking why a graph is preferable. [O2]
- Ask students to **explain** the difference between a pie chart, a bar chart, a histogram and a line graph. Gauge their existing understanding. [O1, O2]

Challenge and develop
Low demand
- Ask students what they understand from Figure 6.38 in Section 6.11 of the Student Book. Establish the features of a pie chart, in that they are useful for showing proportions of different types of data. Students can read 'Representing information with pie charts' in Section 6.11 of the Student Book and answer Worksheet 6.11.1. They are required to interpret a pie chart and to construct one. [O1]

Explain

Standard demand

- Ask students to name different types of graph and explain that the rest of the lesson will concentrate on line graphs. You may wish to set this in context by explaining that bar graphs have bars of equal width, each with a label and a scale on the *y*-axis. Histograms are drawn from a frequency table and the *y*-axis is frequency. A line graph shows the relationship between two variables – usually an independent variable and a dependent variable. [O2]

- Students can complete question 1 in Worksheet 6.11.2. They are interpreting a line graph of ammonia production from 1947 to 2007. You may need to explain the significance of '1000 tonnes' on the *y*-axis. [O2]

- Students can read 'Reading from graphs' in Section 6.11 of the Student Book and answer questions 3 and 4. [O2]

High demand

- Remind students how to draw a line graph. Establish that the independent variable is on the *x*-axis and the dependent variable on the *y*-axis. Discuss choosing suitable scales and labelling axes. They can read 'Drawing graphs from data' in Section 6.11 of the Student Book and answer question 6. They can also answer question 2 on Worksheet 6.11.2. [O3]

Consolidate and apply

- Students can **work** in pairs to **decide** the types of graph they would use to show the results of a survey about the reasons why people access the internet, the results of an experiment to measure how the rate of a reaction varies with temperature and the results of a survey to find the numbers of students of each shoe size in a school. [O1, O2]

- Students can **work** in pairs to **produce** a guide to drawing a line graph of the changes in an independent variable and a dependent variable. [O3]

Extend

Ask students able to progress further to **calculate** the rate of reaction from the graph drawn in question 2b on Worksheet 6.11.2 at the beginning of the experiment and after 30 seconds. [O3]

Plenary suggestions

Ask students to write down one thing they are sure off, one thing they are unsure off and one thing they need to know more about. They can **share** their lists with the class. Identify any areas that need revisiting.

Ask students to write a two-mark question on the lesson's content, with a mark scheme. They can try their questions out on each other.

Answers to Worksheet 6.11.1

1. a) Toilet flushing; b) Having a shower; c) Washing clothes; d) Pie charts are drawn clockwise from 12 o'clock in decreasing segment size; e) 108°; 2. Angles clockwise; should be 108°, 72°, 72°, 36°, 36°, 18°, 18°

Answers to Worksheet 6.11.2

1.a) 'Year' is the independent variable; 'production of ammonia' is the dependent variable; b)(i) 10 000 000 tonnes; (ii) 90 000 000 tonnes; (c) 100 000 000 tonnes; (d) 87 000 tonnes

2. a)

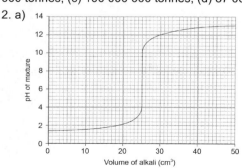

b) Time (s) should be on the *x*-axis; volume of hydrogen (cm^3) on the *y*-axis; points should be joined by a line of best fit

Lesson 12: Functional groups and homologous series

Lesson overview

OCR Specification reference

OCR C6.2a, C6.2b

Learning objectives

- Identify the first four hydrocarbons in the alkane series.
- Name the first four compounds in homologous series.
- Identify the functional group of a series.

Learning outcomes

- Recognise the molecular or displayed formulae of methane, ethane, propane or butane. [O1]
- Name the first four members of an homologous series. [O2]
- Recognise that each homologous series has a different functional group and use the general formula of the alkanes to make predictions. [O3]

Skills development

- WS 1.1b Use a variety of models to make predictions and to develop scientific explanations.
- WS 1.4a Use scientific vocabulary, terminology and definitions.
- WS 1.4c Use IUPAC chemical nomenclature unless inappropriate.

Maths focus Visualise and represent two-dimensional and three-dimensional forms including 2D representations of 3D objects.

Resources needed Class sets of Molymod molecular models (or similar) containing at least four carbon atoms and ten hydrogen atoms; Worksheet 6.12.1.

Key vocabulary alkane, functional group, hydrocarbon, homologous

Teaching and learning

Engage

- Show displayed and molecular formulae of methanol, methane and propane and ask students to suggest which molecule is the odd one out. [O1]
- Build on students suggestions to introduce the idea that carbon compounds can be grouped into homologous series of compounds with identical functional groups. [O1, O2]

Challenge and develop

Low demand

- Explain that hydrocarbons are molecules containing only carbon atoms and hydrogen atoms and that alkanes are an homologous series with a backbone of carbon atoms joined by single bonds. [O1]

250 © HarperCollins*Publishers* Limited 2016

Explain

Standard demand

- Introduce students to Molymod kits (or similar). Explain that these are used to make models of molecules and that they will be using the black carbons and white hydrogens. Ask students to **make** a model of CH_4. Discuss what represents the bonds between the carbon and hydrogen atoms. Explain that this is a hydrocarbon called *methane*. It is one of a *homologous series* of similar molecules called alkanes. Show students how a 2D-displayed formula is used to represent the 3D model. [O1, O2]

Consolidate and apply

Standard demand

- Students complete Worksheet 6.12.1. They will be **making** models of ethane, propane and butane. Check their displayed formulae and discuss the origin of the alkane names (parts b and c). [O1, O2]

High demand

- Explain that a family of chemicals like the alkanes is called a *homologous series*. The chemicals in a homologous series have the same general formula. The general formula for alkanes is C_nH_{2n+2}. Check that students can use this general formula to write the molecular formula of alkanes containing more than four carbon atoms. [O3]

Extend

Ask students able to progress further to:

- Use the molecular model kits to make two different structures for butane. Ask them to **draw** their displayed formulae. Explain that these are called *isomers*. Students can **investigate** the isomers of alkanes containing five and six carbon atoms. They can **discuss** why some displayed formulae may look different, but they are actually the same isomer. Students can then **discuss** why displayed formulae are frequently used to describe carbon compounds. [O1, O2]

Plenary suggestions

Ask students to draw a large triangle with a smaller inverted triangle inside that just fits. In the outer three triangles they write something they have seen, something they have done and something they have discussed. In the centre triangle, they write something they have learned.

Answers to Worksheet 6.12.1

1. a)

b) Similar composition, structure

c) Two, three and four carbon atoms respectively

d) Easier to draw

e) Does not show the 3D structure

Lesson 13: Structure and formulae of alkenes

Lesson overview

AQA Specification reference

AQA 4.7.2.1

Learning objectives

- Describe the difference between an alkane and an alkene.
- Draw the displayed structural formulae for the first four members of the alkenes.
- Explain why alkenes are called unsaturated molecules.

Learning outcomes

- Know that alkenes have a double covalent bond between two of the carbon atoms in their molecules, whereas all the bonds in alkanes are single covalent bonds. [O1]
- Be able to write the formula for ethene, propene, butene and pentene and draw their displayed formulae. [O2]
- Know that alkenes are unsaturated because they contain fewer hydrogen atoms than the alkane with the same number of carbon atoms. Be able to write the formula of any alkene from its general formula. [O3]

Skills development

- WS 1.2 Use a variety of models, such as representational, spatial, descriptive, computational and mathematical, to develop scientific explanations and understanding of familiar and unfamiliar facts.
- WS 4.1 Use scientific vocabulary, terminology and definitions.
- WS 4.3 Use SI units and IUPAC nomenclature unless inappropriate.

Maths focus Visualise and represent two-dimensional (2D) and 3D forms, including 2D representation of 3D objects.

Resources needed Class sets of Molymods (or similar); Worksheets 6.13.1 and 6.13.2

Key vocabulary alkenes, double bond, homologous series, unsaturated

Teaching and learning

Engage

- Students can work in pairs or small groups. Each group will need a set of Molymods. Ask students to make a model of a propane molecule. They can **write** the (molecular) formula for the molecule and its displayed formula. If needed, revisit the two different types of formulae and check that the students know these are single covalent bonds. Now ask them to **make** a molecule with the formula C_3H_6 (propene). Explain that a double bond is needed between two of the carbon atoms so that each carbon atom can make four bonds and each hydrogen atom can make one bond. Explain that this is an example of an alkene. Students can answer question 1 in Worksheet 6.13.1. Note that some students may make cyclopropane instead of propene. [O1]

Challenge and develop

Low demand

- Remind students that alkanes are called a *homologous series* and tell them that alkenes are another homologous series. Students can continue working in pairs. Ask them to use their model kits to **make** models of ethene, propene, butene and pentene. They can **peer assess** their models. Note that for butene and pentene, students may make different positional isomers of the alkenes. You may need to tell them that, although these are valid structures, in this exercise they should always put the double bond on the first carbon in the chain. [O2]

Explain

Standard demand

- Show students how displayed formulae of the alkenes are drawn, reminding them that these are 2D representations of a 3D structure. If necessary, you could revisit the displayed formulae for the alkanes. Students can complete question 2 in Worksheet 6.13.1. They can check their (molecular) formulae and displayed formulae in Section 6.1 of the Student Book. [O2]

High demand

- **Allow** students to work in pairs. They can use Section 6.13 of the Student Book. One student **researches** saturated and unsaturated hydrocarbons. The other **researches** the general formulae for alkanes and alkenes. Allow them five minutes to carry out the task. They can **report back** to each other. [O3]

- Students can answer questions 1–5 in Worksheet 6.13.2. Use directed questioning to check that they understand why the term 'saturated' is applied to some hydrocarbons, that single and double bonds can be present in hydrocarbons and how the general formulae for alkanes and alkenes differ. **Check** that students **realise** that if there are three or more carbon atoms in an alkene, not all the carbon atoms make double bonds. [O3]

Consolidate and apply

- Ask students to work in pairs choosing an alkene and using the molymod kits to make a model of it. They can decide on its formula and displayed formula, its name and the type of bonds. Each pair of students can then **quiz** another pair about the alkene. [O1, O2]

- Ask students where they have come across saturated and unsaturated fats and identify their knowledge. Tell them that part of all fat molecules consists of long chains of hydrocarbons, like those in alkanes and alkenes. Students can answer question 6 in Worksheet 6.13.2. [O3]

Extend

- Ask students able to progress further to draw the displayed formulae of isomers (positional) of butene and pentene. Tell them that they are drawing the formulae for the isomers showing different positions of the double bond. They can use Molymod kits if needed. They can also explain why propene does not have this type of isomer. [O2]

Plenary suggestions

List the following words on the board – alkane, alkene, hydrocarbon, single covalent bond, double covalent bond, saturated, unsaturated, C_nH_{2n+2} and C_nH_{2n}. Use traffic lights or a show of hands to identify students' understanding of each as: 'confident I know what it means'; 'unsure'; 'not a clue'. Identify areas of weakness to be revisited in future lessons.

Answers to Worksheet 6.13.1

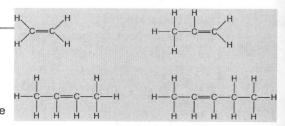

1. a)(i) Single covalent bonds; (ii) C_3H_8; b)(i) Single covalent bonds;
(ii) Double covalent bonds; (c) Propane is the alkane; propene is the alkene
2. a) C_2H_4, C_3H_6, C_4H_8 and C_5H_{10}; and see right;
b) Number and type of atoms in a molecule;
c) Number and type of atoms and bond ; how they are arranged in a molecule

Answers to Worksheet 6.13.2

1. a) Four; b) Single covalent bonds; 2. a) Some make four; some three; b) Single and double covalent bonds; 3. Every carbon atom makes four single covalent bonds; 4. At least two carbon atoms make one double covalent bond; and two single; 5. a) Shows the types of atoms in a molecule; and the ratio of the different atoms; b) C_nH_{2n+2}; c) C_nH_{2n}; d) Alkane formulae C_6H_{14}, C_7H_{16}, C_8H_{18}, C_9H_{20}, $C_{10}H_{22}$, $C_{12}H_{26}$; alkene formulae C_6H_{12}, C_7H_{14}, C_8H_{16}, C_9H_{18}, $C_{10}H_{20}$, $C_{12}H_{24}$; 6. a)(i) Single covalent bonds; (ii) Some double covalent bonds; b) A monounsaturated fat contains one double bond in each hydrocarbon chain; a polyunsaturated fat contains more than one double bond in each hydrocarbon chain

Lesson 14: Reactions of alkenes

Lesson overview

OCR Specification reference

OCR C6.2

Learning objectives

- Describe the addition reactions of alkenes.
- Draw the full displayed structural formulae of the products alkenes make.
- Explain how alkenes react with hydrogen, water and the halogens.

Learning outcomes

- Know that an alkene plus hydrogen, (or water or a halogen) react together to make a new compound. The double bond in alkenes makes them reactive. [O1]
- Draw the displayed formulae for the addition products when alkenes react with either hydrogen, water or a halogen. [O2]
- Be able to explain that the double bond in alkenes becomes a single bond when they form addition products; know the conditions under which alkenes react with hydrogen and water. [O3]

Skills development

- WS 1.2 Use a variety of models, such as representational, spatial, descriptive, computational and mathematical, to solve problems, make predictions and develop scientific explanations and understanding of familiar and unfamiliar facts.
- WS 1.5 Explain everyday and technological applications of science.
- WS 4.1 Use scientific vocabulary, terminology and definitions.
- WS 4.3 Use SI units and IUPAC nomenclature unless inappropriate.

Maths focus Visualise and represent two-dimensional (2D) and 3D forms including 2D representations of 3D objects.

Resources needed Molymods; Practical sheet 6.14; Worksheets 6.14.1 and 6.14.2; Technician's notes 6.14

Key vocabulary addition, carbon–carbon double bond, finely divided nickel catalyst, incomplete combustion

Teaching and learning

Engage

- Demonstrate the addition of a few drops of bromine water to a few cm^3 of cyclohexane and also of cyclohexene. This is covered on Practical sheet 6.14 (in step 1). Remind students that alkenes are more reactive than alkanes. [O1]
- Ask students to suggest reasons why alkenes are more reactive than alkanes. [O1]

Challenge and develop

Low standard

- Recap that alkenes are a homologous series with a C=C functional group, and it is the functional group that is responsible for the similar reactions of all alkenes. [O1]
- Ask students to **predict** what will happen when bromine water is added to test tubes of ethene and methane. Carry out the procedure in step 2 on Practical sheet 6.14. Deduce that alkenes react with bromine water to produce a colourless product, but alkanes do not. Explain that bromine and the alkene add together to make the new compound. [O1]

Explain

Students can use Molymods to **make** a model of an ethene molecule and suggest the structure of the product formed when it reacts with bromine water. Ask students to **draw** the displayed formula of the new substance and establish that it is called dibromoethane. Students can complete question 1 in Worksheet 6.14.1. Emphasise that chlorine and iodine react in a similar way with ethene to produce dichloroethane and diiodoethane, respectively. [O2]

- Ask students to **predict** the products formed when ethene reacts with hydrogen and with water. They can use Molymods if needed and **draw** the displayed formulae of the products. They need to realise that the addition of hydrogen results in an alkane being formed, and that the addition of water results in an alcohol being formed. They will find out more about alcohols in the next lesson. Explain that these reactions are typical of all alkenes. Students can answer question 2 in Worksheet 6.14.1. [O2]

- Ask students how they expect ethene to react with oxygen. Carry out step 3 in Practical sheet 6.14 to demonstrate that alkenes produce more smoke (carbon/soot) when they burn than alkanes do. Remind them that this is incomplete combustion. Students can answer question 3 in Worksheet 6.14.1. [O2]

- Ask students to suggest what happens to the bonds in ethene when it reacts with other substances. Establish that the double bond becomes a single bond and that each carbon atom is now able to form a new single bond; other atoms can add to it. Students can answer question 4 in Worksheet 6.14.1. [O3]

Consolidate and apply

- Students can answer Worksheet 6.14.2. They will need to look at Section 6.14 in the Student Book to find out about the conditions required for the addition reactions. [O1, O2, O3]

- Students can carry out an online search to find out what old-fashioned margarine has to do with the addition reactions of alkenes and unsaturated fats. [O3]

Extend

- Ask students able to progress further to **draw** the displayed formulae for the two possible position isomers of butene (but-1-ene and but-2-ene). Ask them to **draw** the displayed formulae for the possible products when butene reacts with bromine water. [O2]

Plenary suggestions

Ask students to draw a large triangle with a smaller inverted triangle inside, as big as possible. In the three outer triangles, ask them to write something they have seen, something they have done and something they have predicted. In the centre triangle, they can write something they have learnt.

Write the words 'ethane', 'dibromoethane' and 'ethanol' on the board. Ask students to name the substances that react with ethene to produce each substance.

Answers to Worksheet 6.14.1

1. a) i, iv and v; b)(i) See below; (ii) See below; (iii) It is saturated/ has no double bonds; (iv) Dibromoethane has all single bonds; ethene has a double bond; c) See below; diiodoethane

2. a)(i) See below; (ii) Ethane, (iii) Alkanes; b)(i) See below; (ii) Ethanol; 3. a) Combustion is incomplete; b) Ethene + oxygen → carbon + water; $C_2H_4 + O_2 → 2C + 2H_2O$. c) Propene + oxygen → carbon monoxide + water, $C_3H_6 + 3O_2 → 3CO + 3H_2O$; 4. Depends on students' response

1. b)(i)   (ii)   c) 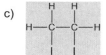   2. a)(i)   b)(i)

Answers to Worksheet 6.14.2

Depends on student's response

Answers to Practical sheet 6.14

1. Alkenes decolourise bromine water, alkanes do not; 2. Alkenes burn with a more sooty flame than alkanes

Lesson 15: Alcohols

Lesson overview

OCR Specification reference

OCR 6.2

Learning objectives

- Recognise alcohols from their name or from given formulae.
- Describe the conditions used for the fermentation of sugar using yeast.
- Write balanced chemical equations for the combustion of alcohols.

Learning outcomes

- Recognise that alcohol names end in '-ol' and that their formulae contain the –OH functional group. [O1]
- Know that fermentation occurs when sugar (glucose), water and the enzymes in yeast are present at a temperature between 25°C and 50°C; ethanol and carbon dioxide are produced. [O2]
- Recall how alcohols dissolve in water, react with sodium, oxidise and combust; be able to write balanced chemical equations for the combustion reactions. [O3]

Skills development

- WS 1.2 Use a variety of models, such as representational, spatial, descriptive, computational and mathematical, to solve problems, make predictions and develop scientific explanations and understanding of familiar and unfamiliar facts.
- WS 2.2 Plan experiments or devise procedures to make observations and test hypothesis.
- WS 2.4 Carry out experiments appropriately having due regard for the correct manipulation of apparatus, the accuracy of measurements and health and safety concerns.
- WS 4.3 Use SI units and IUPAC chemical nomenclature unless inappropriate.

Maths focus Visualise and represent two-dimensional (2D) and 3D forms, including 2D representations of 3D objects.

Resources needed Class sets of Molymods (or similar); Practical sheets 6.15.1 and 6.15.2; Worksheets 6.15.1 and 6.15.2; Technician's notes 6.15

Digital resources [Google string] RSC Fermentation of glucose

Key vocabulary alcohol, fermentation, oxidation, enzymes

Teaching and learning

Engage

- Remind students that in the previous lesson, they found out how water can add to ethene to make an alcohol. Ask them to use the Molymod kits to make a model of the alcohol produced. Explain that this is called ethanol and is one of many different alcohols. Tell students that ethanol is the alcohol in alcoholic drinks such as beer and wine. Alcohols are a homologous series of organic compounds. [O1]

Challenge and develop

Low demand

- Students can work in pairs. Deduce the (molecular) formula and the displayed formula for ethanol. Explain that the formula for ethanol can also be shown as CH_3CH_2OH and that this is called a *structural formula*. Ask them to use the Molymod kits to make models of propanol and butanol. You will have to explain that for their specification, they must attach the OH group to the first carbon atom. Students can answer questions 1–3 in Worksheet 6.15.1. They can use Section 6.15 in the Student Book to check their answers. [O1]

Standard demand

Explain that the ethanol made when water and ethene react together is used as a fuel and in the chemical industry as a solvent. It is not suitable for drinking because of impurities. Fermentation has been used for centuries to produce alcohol solutions fit to drink. Establish students' prior knowledge of fermentation, concentrating on the conditions required. Introduce Practical sheet 6.15.1. Students are going to **devise** their own experiments to investigate the conditions required for fermentation to occur. Allow students to carry out their experiments in pairs. You will need to check their plans. The experiments need to be left for several hours, depending on the yeast used, or until the next lesson. Alternatively, if time is short, you could set up a demonstration of fermentation using [Google string] RSC Fermentation of glucose. [O2]

Explain

High demand

- Explain that students are now going to observe some chemical reactions of alcohols, using ethanol as the example. Use Practical sheet 6.15.2 to demonstrate the reactions of ethanol with water and with sodium, and in oxidation and combustion reactions. Students can answer the Evaluation section of Practical sheet 6.15.2. [O3]

- Students can now answer questions 4 and 5 in Worksheet 6.15.1. [O3]

Consolidate and apply

- Students can complete Worksheet 6.15.2 and **construct** a summary chart to show the methods used to produce alcohols and their reactions. O1, O2, O3]

Extend

- Ask students able to progress further to write word and balanced symbol equations for the reactions between ethanol and sodium, and between propanol and sodium. [O3]

- Students **research** the formulae and structure of carboxylic acids in preparation for next lesson.

Plenary suggestions

Ask students to compile a list of three big ideas they have learnt in this lesson. They can work in pairs, then **share** their lists with other groups, **compiling** a master list of ideas with the most important at the top of the list.

Answers to Worksheet 6.15.1

1. CH_3OH, CH_3OH; C_2H_5OH, CH_3CH_2OH; C_3H_7OH, $CH_3CH_2CH_2OH$; C_4H_9OH, $CH_3CH_2CH_2OH$

2. Its name ends in '-ol'; 3; It contains the OH group; 4. a) Methanol + oxygen → carbon dioxide + water, $CH_3OH + O_2 → CO_2 + 2H_2O$; ethanol + oxygen → carbon dioxide + water; $C_2H_5OH + 3O_2 → 2CO_2 + 3H_2O$; propanol + oxygen → carbon dioxide + water, $2C_3H_7OH + 9O_2 → 6CO_2 + 8H_2O$; b) Add limewater; it turns milky; c) Add blue colbalt chloride paper; it turns pink; 5. a)(i) Sodium metal moves around, fizzes, gas given off; (ii) Burns with a blue flame; (iii) Turns green, smells of vinegar/ethanoic acid; (iv) Methanol dissolves/mixes, universal indicator remains green

Answers to Worksheet 6.15.2

A summary of alcohol reactions; depends on student's response

Answers to Practical sheet 6.15.1

A summary of alcohol reactions; depends on student's response

Answers to Practical sheet 6.15.2

1. a) Yes; b) 7; 2. a) Fizzes gently; b) Hydrogen; 3. Pale blue flame; carbon dioxide and water; 4. pH <7; reacts with some metals, metal oxides, metal hydroxides, carbonates

Lesson 16: Carboxylic acids

Lesson overview

OCR Specification reference

OCR C6.2

Learning objectives

- Describe the reactions of carboxylic acids.
- Recognise carboxylic acids from their formulae.
- Explain the reaction of ethanoic acid with an alcohol.

Learning outcomes

- Know that carboxylic acids form acid solutions in water and, like all acids, react with carbonates to produce carbon dioxide. [O1]
- Be able to write and to recognise formulae and displayed formulae for methanoic acid, ethanoic acid, propanoic acid and butanoic acid. [O2]
- Know that ethanoic acid reacts with ethanol in the presence of an acid catalyst to produce the ester ethyl ethanoate and water. [O3]

Skills development

- WS 1.2 Use a variety of models, such as representational, spatial, descriptive, computational and mathematical, to solve problems, make predictions and develop scientific explanations and understanding of familiar and unfamiliar facts.
- WS 1.4 Explain everyday and technological applications of science.
- WS 2.4 Carry out experiments appropriately having due regard for the correct manipulation of apparatus, the accuracy of measurements and health and safety concerns.
- WS 4.3 Use SI units and IUPAC chemical nomenclature unless inappropriate.

Maths focus Visualise and represent two-dimensional (2D) and 3D forms, including 2D representations of 3D objects

Resources needed Class set of Molymods; Practical sheet 6.16; Worksheets 6.16.1 and 6.16.2; Technician's notes 6.16

Key vocabulary ionises, incomplete ionisation, ester, acid catalyst

Teaching and learning

Engage

- Explain how vinegar is made from different types of wine and cider to produce wine vinegar and cider vinegar. Ask students to **recall** the reaction between ethanol and the oxidising agent acidified sodium dichromate, when vapours with a vinegary smell were given off. Explain that the vinegar industry uses bacteria (*Acetobacter*) to turn the wine to vinegar, not an oxidising agent like acidified sodium dichromate solution, and that this lesson is about the chemical responsible for the vinegary smell, ethanoic acid. [O1]
- Tell students that ethanoic acid belongs to the homologous series of *carboxylic acids*. These are organic acids. [O1]

Challenge and develop

Low demand

- Ask students to list as many properties of acids as they can. Explain that they are going to find out if carboxylic acids behave in the same way. Introduce Practical sheet 6.16. Students are going to carry out some test tube reactions to investigate the pH of ethanoic acid, and its reaction with carbonates and with ethanol. Students can work in pairs to carry out parts 1 and 2 of the practical. [O1]

Students can read Section 6.16 in the Student Book to find out why hydrochloric acid has a pH of 1, but ethanoic acid has a pH of 3. They can answer question 1 in Worksheet 6.16.1 [O1]

Explain

Standard demand

- Students can work in pairs. Write the displayed formula of ethanoic acid on the board and ask students to use Molymods to **make** a model of the molecule. You could use the model to reinforce just which hydrogen is involved when ethanoic acid is ionised. Ask students to **make** models of other carboxylic acids – methanoic, propanoic and butanoic acids. They can complete question 2 in Worksheet 6.16.1. Students can use the Student Book to check their formulae. [O2]

High demand

- Tell students they are now going to find out how ethanoic acid reacts with ethanol. This is part 3 in Practical sheet 6.16. You may wish to make this a demonstration because it involves students handling concentrated sulfuric acid and ethanoic acid. Either allow students to **carry out** the preparation or demonstrate it. Explain that the products are ethyl ethanoate and water. Ethyl ethanoate has a distinctive pear drop/glue smell and is an example of an ester. They may **recognise** it as nail polish remover for acrylic nails. Explain that it is also used in the food-flavouring industry and to make decaffeinated coffee. [O3]

- Students can answer question 3 in Worksheet 6.16.1. [O3]

Consolidate and apply

- Students could **construct** a flow diagram, similar to those in Worksheets 7.7.2 and 7.8.2 to **summarise** the reactions of carboxylic acids. [O1, O2, O3]

- They could do an Internet search to find products made using carboxylic acids. [O1]

Extend

- Ask students able to progress further to use Molymods kits to make models of ethanoic acid and ethanol. Use these to make models of the products when they react to produce ethyl ethanoate and water. They can write a balanced chemical equation for the reaction. [O3]

Plenary suggestions

Students could use the cards in Worksheet 6.16.2 to match the names with the formulae.

Answers to Worksheet 6.16.1

1. a) Molecules break-up to produce two charged ions; b) Hydrochloric acid ionises almost completely into hydrogen ions and chloride ions; c) It only ionises partially; d) Hydrogen ions are responsible for the acid properties; hydrochloric acid is almost completely ionised, ethanoic acid is only partially ionised; e) Test tube 1; because hydrochloric acid is a stronger acid; 2. $HCOOH$; C_2H_5COOH; C_2H_5COOH; C_3H_7COOH

3. a) Ethyl ethanoate; water; b) Heat; with a concentrated acid catalyst; c) Ester; d) Sweet/pleasant/fruit smell; liquid at room temperature; e) Carboxylic acid + alcohol → ester + water

Answers to Practical sheet 6.16

1. Ethanoic acid pH 3; hydrochloric acid pH 1; 2. Similar – both give off carbon dioxide; different – hydrochloric acid reacts faster; 3. Volatile, pear drop/glue-like small; 4. Catalyst

Lesson 17: Addition polymerisation

Lesson overview

OCR Specification reference

OCR C6.2

Learning objectives

- Recognise addition polymers and monomers from diagrams.
- Draw diagrams of the formation of a polymer from an alkene.
- Relate the repeating unit of the polymer to the monomer.

Learning outcomes

- Recognise monomers from their displayed formulae and the presence of a C=C double bond; recognise polymers from their displayed formula. [O1]
- Use displayed formulae to show the formation of poly(ethene) from ethene and of poly(propene) from propene. [O2]
- Know that the repeating unit has the same number of atoms as the monomer in addition polymerisation because no other molecule is formed in the reaction. [O3]

Skills development

- WS 1.2 Use a variety of models, such as representational, spatial, descriptive, computational and mathematical, to solve problems, make predictions and develop scientific explanations and understanding of familiar and unfamiliar facts.
- WS 1.4 Explain everyday and technological applications of science.
- WS 4.1 Use scientific vocabulary, terminology and definitions.
- WS 4.3 Use SI units and IUPAC chemical nomenclature unless inappropriate.

Maths focus Visualise and represent two-dimensional (2D) and 3D forms, including 2D representations of 3D objects

Resources needed Class sets of Molymods (or similar); Worksheets 6.17.1 and 6.17.2

Key vocabulary monomer, polymer, polymerisation, repeating unit

Teaching and learning

Engage

- Allow students to work in pairs. Ask them to use the Molymods to **make** as many ethene molecules as the sets allow. Bring the class together and ask them to open the double bonds and to join up the ethene molecules with other ethene molecules to make a long molecule. Explain that the long molecule they have made is poly(ethene) – 'poly' meaning many, so 'many ethenes'. It is commonly called polythene. Explain that molecules are usually at least 10 000 carbon atoms long. [O1]
- Explain that poly(ethene) is just one example of many materials commonly called plastics. Ask students to **estimate** the percentage of their clothes and belongings with them today that are made from plastics. You might want to review their estimates later. [O1]

Challenge and develop

Low demand

- Explain that in the example they have just modelled, ethene is called the monomer and poly(ethene) is called the polymer. Monomers are the small molecules from which polymers are made. The process is called *addition polymerisation* because no other molecule is formed in the reaction. Tell students there is another type of polymerisation they will find out about next lesson. Alternatively, ask students to use Section 6.17 of the Student Book to find definitions for monomer, polymer and addition polymerisation. Students can answer question 1 on Worksheet 6.17.1. [O1]

Explain

Standard demand

- Show students how diagrams using displayed formulae and repeating units are used to show polymerisation reactions using poly(ethene) as the example. Students can answer question 2 in Worksheet 6.17.1. They may need Molymods to model the addition polymerisation reaction to produce poly(propene). Either check their answers or let them **peer assess** them. [O2]

- Students can answer question 3 in Worksheet 6.17.1 to **practice** drawing polymers from given monomers. [O2]

High demand

- Ask students to work in pairs to compare the monomers they have learned about. Emphasise that they have the same number of atoms because no other molecules have been formed. The difference is that the monomer molecules have a double covalent bond and the polymer molecules have no double covalent bond. [O3]

Consolidate and apply

Low demand

- Students can read how poly(ethene) was discovered using Worksheet 6.17.2 and answer question 1. [O1]

High demand

- Students can answer question 2 in Worksheet 6.17.2 to **practice** drawing monomers given the repeating unit. [O3]

Extend

Ask students able to progress further to:

- **Research** the names of other common polymers and where/how they are used. Students could revisit their initial estimates of the percentage of plastic/polymer material in their clothing and belongings. [O1]

- **Research** High- and low-density poly(ethene) (HDPE and LDPE) and find out how their structure and properties differ. [O1]

Plenary suggestions

Ask students to make up a sentence containing the words 'monomer', 'polymer' and 'polymerisation'. They can **construct** a second sentence containing the words 'ethene', 'poly(ethene)', 'repeating unit' and 'monomer'. Students can **share** their sentences with the rest of the class.

Answers to Worksheet 6.17.1

1. a)(i) Ethene; (ii) Poly(ethene); (iii) Polymerisation; b) A double C=C bond; c) i and iv; 2. a) A large number; b)(i) To show that this is the section that is repeated; (ii) The monomer has a double C=C bond; the polymer has single C–C bonds; c)

3. a) poly(chloroethene):

b) poly(styrene):

Answers to Worksheet 6.17.2

1. a) Ethene; b) Many monomers join up to make a polymer; no other molecules are made; c) Oxygen was needed as the catalyst

2. (a)   (b)   (c)

Lesson 18: Condensation polymerisation

Lesson overview

OCR Specification reference

OCR C6.2

Learning objectives

- Explain the basic principles of condensation polymerisation.
- Explain the role of functional groups in producing a condensation polymer.
- Explain the structure of the repeating units in a condensation polymer.

Learning outcomes

- Explain that condensation polymerisation produces a polymer and another molecule. [O1]
- Recall that condensation polymerisation involves monomers with two functional groups; the functional groups react together to produce the polymer and another molecule. [O2]
- Explain that a condensation polymer formed from two different monomers will have a repeating unit made from the two monomers with a small molecule eliminated. [O3]

Skills development

- WS 1.2 Use a variety of models, such as representational, spatial, descriptive, computational and mathematical, to solve problems, make predictions and develop scientific explanations and understanding of familiar and unfamiliar facts.
- WS 1.4 Explain everyday and technological applications of science.
- WS 4.1 Use scientific vocabulary, terminology and definitions.

Maths focus Visualise and represent two-dimensional (2D) and 3D forms, including 2D representations of 3D objects

Resources needed Class sets of Molymods (or similar); Practical sheet 6.18; Worksheets 6.18.1 and 6.18.2; Technician's notes 6.18

Key vocabulary condensation, polyester, ethanediol, hexanedioic acid

Teaching and learning

Engage

- Show students some objects made from nylon and polyester (e.g. clothing, bags). Identify the polymer present and tell students that these are made in a different type of polymerisation reaction called *condensation polymerisation*. [O1]
- Demonstrate making nylon using Practical sheet 6.18.

Challenge and develop

High demand

- Ask students to work in small groups. They are going to make models of two monomers used to make a condensation polymer. They will need a set of Molymods. Half the groups can make models of ethanediol, $HO-CH_2-CH_2-OH$. The other half can make models of hexanedioic acid, $HOOC-CH_2-CH_2-CH_2-CH_2-COOH$. Discuss the names of these compounds and that they each have two functional groups, one at each end of the molecule. Students can answer question 1 in Worksheet 6.18.1. [O2]
- Bring the class together and arrange them in a circle or a line. Ask them to join the OH group of ethanediol to the COOH group of hexanedioic acid by eliminating a water molecule. They should produce a model of a condensation polymer. Explain that this is called a polyester. Some students might make the link with the reaction of a carboxylic acid with an alcohol to produce an ester. Ask students to answer question 2 in Worksheet 6.18.1. They can **peer assess** their answers. [O1, O2]

Explain

High demand

- Explain that, like addition polymers, condensation polymers have a repeating unit. In the case of polyester, the repeating unit must contain both monomers. Explain how the repeating unit is derived for polyester. Students can answer question 3 in Worksheet 6.18.1. [O3]

Consolidate and apply

- Explain that nylon is also a condensation polymer, but is made from different monomers. Students can answer question 1 in Worksheet 6.18.2. They may need Molymods to model the reaction, or you could make it a class activity, as with polyester. Students will need to know that NH_2 is a functional group. [O3]

Extend

- Ask students able to progress further to carry out an online search to find out which monomers are used to make Kevlar. They can answer question 2 in Worksheet 6.18.2. They may need help in **interpreting** the benzene rings. [O3]

Plenary suggestions

Ask students to work in pairs and make a list of the similarities and differences between addition polymerisation and condensation polymerisation. One pair can **share** their lists with the rest of the class and other pairs can add any other similarities or differences from their lists.

Ask each student to **compose** a question on anything from today's lesson. They need to have at least two points in their questions. Students can pair up and try their questions out on each other.

Answers to Worksheet 6.18.1

1. a) OH/alcohol; b) Two; c) Two carbon atoms, so ethane; two OH groups so diol; d) COOH/carboxylic acid; e) Two; f) Two carboxylic acid groups so dioic acid; six carbon atoms so hexane; 2. (a) $-CH_2-CH_2-$; (b) $-CH_2-CH_2-CH_2-CH_2-$;

c) $nHO-\boxed{}-OH + nHOOC-\boxed{}-COOH \longrightarrow \left(\boxed{}-OOC-\boxed{}-COO\right)_n + 2nH_2O$; (d) $-COO-$;

3. a) $-CH_2-CH_2-$; b) $-CH_2-CH_2-CH_2-CH_2-$; c) An H from a COOH group and the OH have been removed to make water; – COO– is left

Answers to Worksheet 6.18.2

1. a) (i) COOH/carboxylic acid; (ii) NH_2/amine; (iii) The two COOH groups contain two carbons; (iv) $HOOC-\boxed{}-COOH$, $H_2N-\boxed{}-NH_2$; b $nHOOC-\boxed{}-COOH + nH_2N-\boxed{}-NH_2 \longrightarrow \left(OC-\boxed{}-COHN-\boxed{}-NH\right)_n + 2nH_2O$

(ii) $\left(OC-\boxed{}-COHN-\boxed{}-NH\right)_n$

(c) There are six carbon atoms in each monomer molecule; 2. a) It is exceptionally strong; b) Benzene-1,4-dicarboxylic acid; 1,4-diaminobenzene;

c) ; d) ;

e) Water is also produced

Answers to Practical sheet 6.18

1. 1,6-diaminohexane; decanedioyl dichloride

© HarperCollins*Publishers* Limited 2016

Lesson 19: Amino acids

Lesson overview

OCR Specification reference

OCR C6.2

Learning objectives

- Describe the functional group of an amine.
- Identify the two functional groups of an amino acid.
- Explain how amino acids build proteins.

Learning outcomes

- Be able to recognise NH_2 as the functional group of an amine. [O1]
- Know that amino acids have an NH_2/amine functional group and a COOH/carboxylic acid functional group. [O2]
- Be able to explain that amino acids react by condensation polymerisation to produce polypeptides, and that different amino acids can be combined in the same chain to form proteins. [O3]

Skills development

- WS 1.2 Use a variety of models, such as representational, spatial, descriptive, computational and mathematical, to solve problems, make predictions and develop scientific explanations and understanding of familiar and unfamiliar facts.
- WS 4.1 Use scientific vocabulary, terminology and definitions.
- WS 4.3 Use SI units and IUPAC nomenclature unless inappropriate.

Maths focus Visualise and represent two-dimensional (2D) and 3D forms, including 2D representations of 3D objects

Resources needed Class sets of Molymods; Worksheet 6.19

Key vocabulary amino acid, glycine, peptide, protein

Teaching and learning

Engage

- Write the terms 'DNA', 'starch', 'cellulose' and 'protein' on the board. Ask students to list as many scientific facts about them as they can **recall**. Most of these will probably be from their Biology lessons, but they will hopefully make the connection with foods, digestion and cells. Explain that these are examples of natural polymers and that they will find out about some of these natural polymers and how they are made. [O1]

Challenge and develop

High demand

- Remind students that the monomers they have learned about so far either contain a double C=C bond and have addition polymerisation reactions, or contain two functional groups. The functional groups react together in a condensation polymerisation reaction. Explain that the monomers that make up proteins are called *amino acids* and that proteins are formed by condensation polymerisation. Tell students that the difference between the monomers used to make polyester and nylon and the amino acid monomers used to make proteins is that amino acids have two different functional groups. [O2]

Explain

- Students can work in pairs and need a set of Molymods. Ask students to make a model of a molecule of ethanoic acid, CH_3COOH. Ask them to identify the functional group. Explain that this is one of the functional groups in an amino acid, and hence the name 'acid'. Explain that the other functional group is called an amine group and has the formula $-NH_2$.

264 © HarperCollins*Publishers* Limited 2016

- Students may **recall** the amine functional group in one of the monomers used to make nylon. They can read more about amines in Section 6.19 in the Student Book. Ask them to replace one of the hydrogens in the 'CH$_3$' end of their model with an amine group. Explain that this is an amino acid called glycine and is the simplest amino acid possible. Students can complete question 1 in Worksheet 6.19. [O2]

- Ask students to make as many models of glycine as their Molymod kits allow. Students can form a circle or a line and join the monomers to make a condensation polymer. Water molecules should also be formed. Explain that they have just made a model of a polypeptide. Students can answer question 2 in Worksheet 6.19. [O3]

- Explain that the condensation polymers made from up to 50 amino acids are called **polypeptides** and that polypeptides can be made from many amino acids. (There are up to 500 different amino acids.) When different amino acids are combined in the same chain of more than 50 amino acid monomers, the compound is called a **protein**. Students can answer question 3 in Worksheet 6.19. [O3]

Consolidate and apply

- Write the formula

$$N_2H - \overset{\displaystyle H}{\underset{\displaystyle R}{\overset{|}{\underset{|}{C}}}} - COOH$$

on the board.

- Ask students to identify the functional groups and explain what 'R' represents. They can identify 'R' in glycine and alanine. Explain that a peptide link is the remains of the two functional groups in the monomers. Students can identify the peptide links when glycine polymerises and when alanine polymerises.

Extend

Ask students able to progress further to:

- Find out what a dipeptide and a tripeptide are.

- **Research** the structure of proteins and find out how they have a twisted structure. [O3]

Plenary suggestions

Write the words 'amino acid', 'condensation polymerisation', 'polypeptide', 'protein' and 'glycine' on the board. Ask students to work in pairs and write one or two sentences to include all the words. They can **share** their sentences with the rest of the class.

Ask students to draw a large triangle with a smaller inverted triangle inside it, so that they have four triangles. In the three outer triangles, they can write one thing they have revisited, one thing they have **modelled** and one thing they have **described**. In the middle triangle, they can write the most important thing they have learned.

Answers to Worksheet 6.19

1. a) Polypeptides; b) Condensation polymerisation; c) Carboxylic acid group/COOH; amine group/NH$_2$; d)(i) The NH$_2$ group and the COOH group should be circled; (ii) NH$_2$CH$_2$COOH; e)

2. a) nH$_2$NCH$_2$COOH $\rightarrow$ (−HNCH$_2$COO−)$_n$ + nH$_2$O; (b) H$_2$NC$_2$H$_4$COOH; (c) nH$_2$NC$_2$H$_4$COOH $\rightarrow$ (−HNC$_2$H$_4$COO−)$_n$ + nH$_2$O;
3. a) NH$_2$ and COOH groups react to produce a water molecule; b) Protein; c) Essential in all living cells

Lesson 20: DNA and other naturally occurring polymers

Lesson overview

OCR Specification reference

OCR C6.2

Learning objectives

- Describe the components of natural polymers.
- Explain the structure of proteins and carbohydrates.
- Explain how a molecule of DNA is constructed.

Learning outcomes

- Know that sugars, starch and cellulose are carbohydrates containing carbon, hydrogen and oxygen only. [O1]
- Know that proteins are polymers of amino acids and starch and cellulose are polymers of sugars. [O2]
- Know that DNA molecules consist of two polymer chains made from monomers called nucleotides. [O3]

Skills development

- WS 1.2 Use a variety of models, such as representational, spatial, descriptive, computational and mathematical, to solve problems, make predictions and develop scientific explanations and understanding of familiar and unfamiliar facts.
- WS 4.1 Use scientific vocabulary, terminology and definitions.

Maths focus Use ratios, fractions and percentages; visualise and represent two-dimensional (2D) and 3D forms, including 2D representations of 3D objects

Resources needed Class sets of Molymods (or similar); molecular model of a section of DNA if available; Worksheets 6.20.1 and 6.20.2

Key vocabulary: cellulose, deoxyribonucleic acid, DNA bases, protein

Teaching and learning

Engage

- Either: Remind students about the scientific facts they listed at the start of the previous lesson on DNA, starch, cellulose and protein. Note that this was a higher tier lesson only. Remind students that proteins are polymers of amino acids and ask the question 'What monomers make up starch and cellulose?' Some may **make the link** that because starch is broken down into sugars during digestion, then simple sugars are the monomers that make starch. [O2]

- Or: Write the terms 'DNA', 'starch', 'cellulose' and 'protein' on the board. Ask students to list as many scientific facts about them as they can **recall**. Most of these will probably be from their Biology lessons, but they will hopefully make the connection with foods, digestion and cells. Explain that these are examples of natural polymers and that they will find out about some of these natural polymers and how they are made. [O2]

Challenge and develop

Low demand

- Write the formulae of glucose ($C_6H_{12}O_6$) and sucrose ($C_{12}H_{22}O_{11}$) on the board. Ask students what they have in common and establish that they are both sugars. Now add some alkane formulae and deduce the difference between the terms 'hydrocarbon' and 'carbohydrate'. Explain that sugars, starch and cellulose are carbohydrates, containing carbon, hydrogen and oxygen only. [O1]

- Students can answer question 1 in Worksheet 6.20.1. They will find Section 6.20 in the Student Book helpful. [O1]

Explain

Standard demand

- Confirm that there are several different types of sugar molecule. Explain that the smallest sugar molecules are called *simple sugars* and that these polymerise to make starches and cellulose. You can make the link between glucose as a digestion product of starch and as a simple sugar that makes the polymer starch. Note that they do not need to know the difference between starch and cellulose, but knowing that the different way the monomers are arranged in the polymer may help prevent confusion. You can refer students to Figure 6.47 in the Student Book. [O2]

- Students can answer question 2 in Worksheet 6.20.1. They can reinforce their understanding by using Molymods (or similar) to make molecular models of parts of a starch molecule. These can be linked together to make a larger polymer model, as in previous lessons. [O2]

- Students who have not completed the higher tier Lesson 14 will need to answer question 3 in Worksheet 6.20.1. Explain that proteins are also polymers, but this time the monomers are amino acids.

High demand

- Write the term 'deoxyribonucleic acid' on the board and explain why it is referred to as 'DNA'. Explain that DNA contains natural polymers. Students can use Figure 6.48 in the Student Book and/or molecular models of a section of DNA. They can use Section 6.20 of their Student Book to answer question 1 in Worksheet 6.20.2. You may wish students to work in pairs since some of these questions are quite challenging. [O3]

Consolidate and apply

- Write the following sentence on the board: 'Nature builds polymers from simple molecules in plants and we break the polymers down in digestion'. Allow students to work in pairs to **decide** on an explanation of the sentence. Pairs of students can join to make fours and **decide** on a joint explanation. These can be **presented** to the class. [O1, O2, O3]

Extend

Ask students able to progress further to:

- Find out the structure of glucose.

- Find out the difference between starch and cellulose.

- **Research** the discovery of DNA and the awarding of Nobel prizes. Some students may be interested in researching the controversy that still exists over who is accredited with its discovery. [O3]

Plenary suggestions

List the following words on the board: 'sugar', 'starch', 'cellulose', 'protein', 'DNA', 'deoxyribonucleic acid', 'nucleotide' and 'double helix' (the list can continue). These words are the answers to questions that students are to **compose**. They can be **shared** with the class.

Answers to Worksheet 6.20.1

1. a) Sugar, starch, cellulose; b) Carbon, hydrogen, oxygen;
c) Hydrocarbons contain carbon and hydrogen only; carbohydrates also contain oxygen;
d)(i) $C_6H_{12}O_6$, (ii) $C_{12}H_{22}O_{11}$; (e) $C_6H_{12}O_6 + 6O_2 \rightarrow 6CO_2 + 6H_2O$;
2. a) Sugar with starch and cellulose; amino acids with proteins;
b) Glucose, fructose; c) A monosaccharide contains one simple sugar only;
a polysaccharide contains many; d) See right;
3. a) COOH; NH_2; b) Different amino acids can join in different orders; to make different proteins

Answers to Worksheet 6.20.2

1. Sensible label placing; 2. Two; 3. Nucleotides; 4. The shape made by two polymer chains; twisted around each other; 5. Sugar, phosphate, base; 6. Four different nucleotides; they are different because there are four different bases; 7. The bases in the polymer chains in DNA always pair with each other in the same way; so information can be passed on when the cell divides

Lesson 21: Crude oil, hydrocarbons and alkanes

Lesson overview

OCR Specification reference

OCR C6.2

Learning objectives

- Describe why crude oil is a finite resource.
- Identify the hydrocarbons in the series of alkanes.
- Explain the structure and formulae of the alkanes.

Learning outcomes

- Describe why crude oil is a finite resource in terms of rates of formation and use, and changes in conditions. [O1]
- Name the first four alkanes and write their molecular and displayed formulae. [O2]
- Use the general formula to calculate the formula of other alkanes; draw electronic structure diagrams to describe why carbon forms four single covalent bonds and hydrogen forms one single covalent bond in alkanes. [O3]

Skills development

- WS 1.2 Use a variety of models to make predictions and to develop scientific explanations.
- WS 4.1 Use scientific vocabulary, terminology and definition.
- WS 4.3 Use SI units and IUPAC chemical nomenclature unless inappropriate.

Maths focus Visualise and represent two-dimensional (2D) and 3D forms, including 2D representations of 3D objects

Resources needed Class sets of Molymods (or similar) containing at least four carbon atoms and ten hydrogen atoms; Worksheets 6.21.1 and 6.21.2

Key vocabulary alkane, crude oil, finite resource, hydrocarbon

Teaching and learning

Engage

- Ask students what crude oil is. You may need to explain that petroleum is an alternative name for crude oil, but the petrol we buy at service stations is obtained from crude oil. [O1]
- Determine student's **understanding** of how crude oil formed. [O1]
- Explain why crude oil is important by identifying the range of materials we obtain from it, including pharmaceuticals, polymers and toiletries. Students' belongings can be used to illustrate the range of products obtained from crude oil. [O1]

Challenge and develop

Low demand

- Students can use pair talk. Ask them to read the information in question 1 in Worksheet 6.21.1 and list the reasons why oil is a finite resource. Discuss their ideas and establish that crude oil takes millions of years to form and it is being consumed rapidly. Also, conditions on Earth have changed. You could extend this by discussing oil-saving strategies. [O1]
- Explain that hydrocarbons are molecules containing only carbon atoms and hydrogen atoms. You may need to differentiate between hydrocarbons and carbohydrates. [O2]

Explain

Standard demand

- Introduce students to Molymod kits (or similar). Explain that these are used to make models of molecules and that they will be using the black carbons and white hydrogens. Ask students to **make** a model of CH_4. Discuss what represents the bonds between the carbon and hydrogen atoms. Explain that this is a hydrocarbon called *methane*. It is one of a family of similar molecules called alkanes. Show students how a 2D-displayed formula is used to represent the 3D model. [O2]

- Students can answer question 2 in Worksheet 6.21.1. They will be **making** models of ethane, propane and butane. Check their displayed formulae and discuss the origin of the alkane names (parts b and c). [O2]

High demand

- Explain that a family of chemicals like the alkanes is called a *homologous series*. The chemicals in a homologous series have the same general formula. The general formula for alkanes is C_nH_{2n+2}. Students can answer question 1 in Worksheet 6.21.2. [O3]

Consolidate and apply

High demand

- Ask students to **draw** the electronic structure of a carbon atom and a hydrogen atom. Check their diagrams. Students can complete questions 2 and 3 in Worksheet 6.21.2. [O3]

Extend

Ask students able to progress further to:

- Use the molecular model kits to make two different structures for butane. Ask them to **draw** their displayed formulae. Explain that these are called *isomers*. Students can **investigate** the isomers of alkanes containing five and six carbon atoms. They can **discuss** why some displayed formulae may look different, but they are actually the same isomer. Students can then **discuss** why displayed formulae are frequently used to describe carbon compounds. [O2]

- Students can refer back to Worksheet 6.21.1, question 1, and **calculate** the volume (dm^3) of crude oil consumed daily worldwide and the volume estimated to be left in the Earth. They can then **calculate** how long this will last at the present rate of consumption. [O1]

Plenary suggestions

Ask students to draw a large triangle with a smaller inverted triangle inside that just fits. In the outer three triangles they write something they have seen, something they have done and something they have discussed. In the centre triangle, they write something they have learned.

Petrol contains the alkane called octane. Octane molecules have eight carbon atoms each. Ask students to write its molecular formula and displayed formula, and to **describe** where it formed naturally and why this source is described as finite.

Answers to Worksheet 6.21.1

1. a) Should include that oil forms over millions of years but is being consumed quickly; conditions have changed; c) Depends on student's experience
2. a) See right; b) Similar composition, structure;
c) Two, three and four carbon atoms respectively; d) Easier to draw;
e) Does not show the 3D structure

Answers to Worksheet 6.21.2

1. a) The number of carbon atoms in a molecule; b)(i) C_5H_{10}; (ii) $C_{10}H_{22}$; (iii) C_7H_{16}; (iv) $C_{20}H_{42}$; (c) 20; 2. a)(i) See below; (ii) See below; (b) Carbon has four electrons in the outer shell so can accomodate four more; (c) Hydrogen has one electron in the outer shell and so can accomodate one more; 3. (a) Covalent bonds; (b) (i) See below; (ii) See below; (iii) See below

2. a)(i)   (ii)   3. b)(i)   (ii)   (iii)

Lesson 22: Fractional distillation and petrochemicals

Lesson overview

OCR Specification reference

OCR C6.2

Learning objectives

- Describe how crude oil is used to provide modern materials.
- Explain how crude oil is separated by fractional distillation.
- Explain why the boiling points of the fractions are different

Learning outcomes

- Name some fuels and petrochemicals that are produced from crude oil. [O1]
- Describe how fractional distillation is used to separate the hydrocarbons in crude oil into fractions according to their boiling points; carry out or observe a laboratory demonstration of synthetic crude oil. [O2]
- Use ideas about weak forces between molecules to explain why longer hydrocarbon chain compounds have higher boiling points. [O3]

Skills development

- WS 1.2: Use a variety of models, such as representational, spatial, descriptive, computational and mathematical, to develop scientific explanations.
- WS 1.4: Explain everyday and technological applications of science.
- WS 2.4: Carry out experiments appropriately having due regard for the correct manipulation of apparatus, the accuracy of measurements and health and safety considerations.
- WS 4.3 Use SI units and IUPAC nomenclature unless inappropriate.

Maths focus Plot two variables from experimental or other data

Resources needed Molymods (optional): Practical sheet 6.22; Worksheets 6.22.1 and 6.22.2; Technician's notes 6.22

Digital resources Computer access: resources.schoolscience.co.uk/exxonmobil/index.html

Key vocabulary boiling point, fractional distillation, hydrocarbon, liquid petroleum gas

Teaching and learning

Engage

- Either show or allow students computer access to **view** the virtual refinery tour at Fawley refinery on Southampton Water (resources.schoolscience.co.uk/exxonmobil/index.html). **Explain** that this is a chemical factory or plant in which crude oil is processed into useful fuels such as heating oil and petrol, and into chemicals used to make materials like fabrics, detergents (hair shampoo), toiletries (eye shadow) and many more. These substances are called *petrochemicals*. [O1]

- Alternatively, allow students to **research** the boiling points of different hydrocarbons and **suggest** how boiling points could be used to separate hydrocarbons. [O2]

Challenge and develop

Low demand

- Emphasise that crude oil is not very useful until we separate the mixture into hydrocarbons with similar numbers of carbon atoms. These are called *fractions*. Explain that the process is called *fractional distillation* and is carried out in a refinery. [O2]

Standard demand

- Introduce Practical sheet 6.22. Explain the procedure and either demonstrate the experiment or allow able students to **carry out** the fractional distillation. Students will need to work in pairs

Some groups may be able to carry out the experiment more successfully if it is demonstrated first. Students should find that the fractions become darker and more viscous as the boiling point increases, and also more difficult to ignite, burning with more sooty flames. Students can answer the Evaluation section on Practical Sheet 6.22. [O2]

Explain

Standard demand

- Students can answer question 1 in Worksheet 6.22.1. They will find out about the trend in boiling points of the alkanes. Explain that they have just used this trend in the fractional distillation experiment. Describe how the hydrocarbons in crude oil are separated into fractions by their different boiling points; first in the class experiment and then industrially. Students can use Section 6.22 in the Student Book to find out how the different fractions are used. Explain that these fractions are processed further to provide fuels and chemicals for the petrochemical industry. [O2]

- Students can answer question 2 in Worksheet 6.22.1. [O2]

High demand

- Recap the covalent bonding between atoms in a hydrocarbon molecule. Explain that hydrocarbon molecules are also attracted to each other, but that relatively weak forces exist between different hydrocarbon molecules. These are about one hundredth the strength of a typical covalent bond, but are important because they determine many of the properties of hydrocarbons. You can use Molymods or similar to illustrate the weak forces (intermolecular forces). Explain that when a hydrocarbon boils, these weak forces are overcome, so the longer the hydrocarbon chain, more weak forces have to be overcome and more energy is needed. Longer hydrocarbons have higher boiling points. [O3]

- Students can answer question 1 in Worksheet 6.22.2.

Consolidate and apply

- Students can answer question 2 in Worksheet 6.22.2. They find out about liquid petroleum gas and **apply** some of the ideas covered in this lesson. [O2]

Extend

Ask students able to progress further to:

- **Carry out** an internet search to find the formulae of the alkanes in each of the fractions obtained from the fractionating tower. [O2]

- **Explain** why liquid petroleum gases do not condense in the fractionating tower. [O3]

Plenary suggestions

Students can work in pairs to list three ways in which the class fractional distillation experiment is similar to the industrial process, and three ways in which it is different.

Each pair of students is allocated a word from the vocabulary list above. This is their answer to the question they have to **compose**. They can test their questions on other pairs of students.

Answers to Worksheet 6.22.1

1. a) Student graph; b) Methane, ethane, propane, butane; c) Pentane, hexane, heptane; 2. a) As in Figure 6.53 in the Student Book ; b) Hottest at the bottom; coolest at the top; c) LPG; d) Bitumen

Answers to Worksheet 6.22.2

1. a) A: propane, C_3H_8, B: hexane, C_6H_{14}; b) Covalent; c) Wavy lines should be between the two displayed formulae; d) Covalent bonds are stronger than weak forces; weak forces are overcome when alkane boils; covalent bonds are not broken; e) More weak forces between adjacent molecuels require more energy to overcome; 2. a) For example, extraction; transportation to refinery; heated in fractionating column; LPG rises to the top of the column; and is collected; b) Diesel; petrol; c) Butane has a higher boiling point than propane; so more propane is added in winter to prevent the fuel freezing

Answers to Practical sheet 6.22

1. Depends on student observations, but fractions should become thicker, darker and burn less easily as boiling point rises; 2. Thickens and darkens; 3. Fractions burn less easily as the boiling point rises; 4. Dark tar-like substance

Lesson 23: Properties of hydrocarbons

Lesson overview

OCR Specification reference

OCR C6.2

Learning objectives

- Describe how different hydrocarbon fuels have different properties.
- Identify the properties that influence the use of fuels.
- Explain how the properties are related to the size of the molecules.

Learning outcomes

- Name some fuels obtained from crude oil, such as petrol and LPG, and explain that because they have different properties, they have different uses. [O1]
- Describe how boiling point, viscosity and flammability influence the use of a fuel. [O2]
- Explain why boiling point and viscosity increase with increasing molecular size of the hydrocarbon while flammability decreases. [O3]

Skills development

- WS 1.2: Use a variety of models, such as representational, spatial, descriptive and mathematical, to develop scientific explanations.
- WS 1.4: Explain everyday and technological applications of science.
- WS 4.1 Use scientific vocabulary, terminology and definitions.

Resources needed Molymods (optional); Worksheets 6.23.1 and 6.23.2

Key vocabulary flammability, ignite, viscosity, volatile

Teaching and learning

Engage

- Identify students' understanding of what a fuel is. Establish that it must burn in oxygen and transfer heat energy to its surroundings. Remind students of the work they did on exothermic reactions in Chapters 3 and 5. [O1]

- Students can work in pairs. Tell them they are setting out on a Duke of Edinburgh's Gold Award expedition/camping trip/climbing Everest. They need to take cooking fuel with them. Ask students to list the properties the fuel must have. Students can **compare** their lists with another pair of students. [O1]

- Remind students of the practical carried out in the previous lesson and ask them to recall the different properties of the fractions they obtained. Ask which of these fractions would make the best cooking fuel. [O2]

Challenge and develop

- Remind students that petrol, diesel and LPG can be used as fuels for motor vehicles. Students can answer question 1 in Worksheet 6.23.1. Discuss how freezing point can determine a fuel's use. You could point out that most diesel engines today are adapted to keep the fuel above its freezing point. [O2]

- Discuss what is required of a fuel in order to release energy. It must combust, which requires it to vaporise and then ignite. Emphasise that it is the vapours from liquid fuels that ignite, not the liquid fuel. You can use a burning candle to illustrate hydrocarbon combustion. More viscous fuels do not vaporise as easily as less viscous fuels. If they do not vaporise easily, they do not ignite easily. Students can answer question 2 in Worksheet 6.23.1. [O2]

Explain

Standard demand

- Allow students to work in pairs. Each pair writes three sentences to **describe** the trends between boiling point and molecular size, viscosity and molecular size, and flammability and molecular size. Pairs of students can double up to **check** each other's sentences. [O2]

High demand

- If students have covered the high demand section in Lesson 16, they can be reminded of the weak forces that exist between adjacent hydrocarbon molecules. For the other students, you can use Molymods to explain the weak forces (intermolecular forces) that exist between molecules. These forces must be overcome before the hydrocarbon can boil. Explain the boiling point trend. Similarly, these forces must be overcome before the hydrocarbon can vaporise. Explain the viscosity trend and hence why flammability decreases with increasing molecular size. [O3]

- Students can answer questions 1 and 2 in Worksheet 6.23.2. [O2, O3]

Consolidate and apply

- Tell students that jet aircraft use kerosene fuel. They can refer back to Section 6.23 in the Student Book to look up out some kerosene facts. Less able students can use Molymods to make kerosene molecules and **discuss** their properties **compared** to other fuels. Students can answer question 3 in Worksheet 6.23.2. Check their answers. [O2, O3].

Extend

- Ask students to carry out an internet search to find the displayed formula of biodiesel and **compare** it to diesel produced from crude oil. They can **identify** similarities and differences in the molecules and **find out** about any problems caused by using biodiesel in modern diesel engines. [O2]

Plenary suggestions

Ask students to write down one thing about today's lesson they are sure about, one thing they are unsure of and one thing they need to know more about. Students can form groups of three to four and **share** their points to produce a common list. Identify any topics that need revisiting.

Answers to Worksheet 6.23.1

1. a) For example, all are hydrocarbons; all are produced from crude oil; b) LPG is a gas at room temperature, petrol and diesel are liquids at room temperature; c) LPG 3–4, less than 30; petrol: 6–10, 80–120 °C; diesel: 9–16, 110–190 °C; d) Diesel freezes at –6 to –18 °C, so fuel could freeze; 2. a) Gas; b) Become more viscous as molecular size increases; c) Diesel; d) Petrol; e) Too viscous, do not vaporise easily so less flammable; f) Molecular size increased

Answers to Worksheet 6.23.2

1. a)(i) Thickness, ease of flowing, (ii) Ability to ignite; b) (i) Increase, (ii) Increase, (iii) Decrease; c)(i) Diesel, (ii) Kerosene, (iii) Heavy fuel oil; 2. a)(i) Larger molecules have more of the weak forces between them so more energy is needed to overcome them; (ii) Diesel molecules are larger than petrol molecules so have more of the weak forces between them which prevent them sliding over each other as easily; (iii) Heavy fuel oil has larger molecules than kerosene and more of the weak forces between them so they do not vaporise as easily and cannot ignite as easily; (iv) LPG has small molecules with fewer of the weak forces between them so the boiling point is lower; (b) More viscous hydrocarbons do not vaporise as easily so cannot ignite as easily; 3. a) 12–19; b) Kerosene has a higher boiling point, is more viscous and less flammable; c) Kerosene does not vaporise too easily in a hot jet engine.

© HarperCollins*Publishers* Limited 2016

Lesson 24: Cracking and alkenes

Lesson overview

OCR Specification reference

OCR C6.2

Learning objectives

- Describe the usefulness of cracking.
- Balance chemical equations as examples of cracking.
- Explain why modern life depends on the uses of hydrocarbons.

Learning outcomes

- Know that cracking produces more useful hydrocarbon molecules from longer chained alkanes by passing their vapours over a hot catalyst or mixing them with steam at a very high temperature. [O1]
- Observe an experiment to demonstrate cracking and know that alkenes are more reactive than alkanes and react with bromine water; balance chemical equations to show that cracking produces shorter chained alkanes and alkenes. [O2]
- Explain that the demand for products from the fractional distillation of crude oil does not match those obtained from the process and cracking is a method used to help redress this balance. [O3]

Skills development

- WS 1.2 Use a variety of models, such as representational, spatial, descriptive, computational and mathematical, to develop scientific explanations and understanding of familiar and unfamiliar facts.
- WS 1.4 Explain everyday and technological applications of science.
- WS 4.1 Use scientific vocabulary, terminology and definitions.
- WS 4.3 Use SI units and IUPAC chemical nomenclature unless inappropriate.

Maths focus Use ratios, fractions and percentages

Resources needed Class sets of Molymods (or similar); Practical sheet 6.24; Worksheet 6.24; Technician's notes 6.24

Key vocabulary alkene, bromine, catalyst, cracking, supply and demand

Teaching and learning

Engage

- Set students a puzzle – they can work in groups as appropriate. Ask students to use Molymods (or similar) to make a molecular model of decane ($C_{10}H_{22}$). Ask them to break a C–C bond so that one section contains eight carbon atoms and the other contains two carbon atoms. Ask students about the possible products. If Molymods or similar are not available, students can **draw** the displayed formulae. [O1]

Challenge and develop

Low demand

- Introduce the idea of a double bond and show students how octane and ethene can be produced from decane. Explain that ethene is from a family of another type of hydrocarbon called *alkenes* and students will find out more about them in the next lesson. Explain that this models a very useful reaction because octane is one of the hydrocarbons in petrol and also ethene is used in the petrochemical industry to make polymers (plastics) and other useful materials. The process is called *cracking*. Note that the structure and formulae of alkenes is dealt with earlier in this chapter. Students can answer question 1 in Worksheet 6.24. [O1]

Standard demand

- Introduce Practical sheet 6.24. This is a demonstration to crack liquid paraffin. You will need a second pair of hands to help collect the gas over water, either a technician or capable student. The gas collected is tested with bromine water and with a lighted splint.

- Confirm that alkanes do not react with bromine water but alkenes do by adding a few cm^3 of cyclohexane and cyclohexene (a typical alkane and a typical alkene) separately to bromine water. Emphasise that alkenes are more reactive than alkanes and that's what makes them so useful for making materials we use every day. Students can answer the Evaluation section of Practical sheet 6.24 and question 2 in Worksheet 6.24. [O2]

Explain

Standard demand

- Ask students to write both word and balanced chemical equations for the cracking of decane to produce octane and ethene. They can **peer assess** their equations. Explain that many cracking reactions are possible, but the products are always a mixture of alkanes and alkenes. Students can answer question 3 in Worksheet 6.24. [O2]

High demand

- Ask students why we need to crack long-chain hydrocarbon molecules. Identify the mismatch between the supply of fractions from the fractional distillation of crude oil and our demand for these. Students can **study** the data in question 4 in Worksheet 6.24 and in Section 6.24 of the Student Book. They can answer question 4 in Worksheet 6.24. [O3]

Consolidate and apply

Standard demand

- Give students the statement 'Modern life depends on hydrocarbons'. In pairs/small groups, students can list examples of how modern life depends on hydrocarbons. One group can give feedback to the class and others can add any other examples. [O2]

Extend

Ask students able to progress further to:

- **Find out** how zeolites are used as catalysts in modern cracking plants. Section 6.24 of the Student Book describes zeolites as having a 'cage-like' structure. Students can search online to find out how they work. [O1]

- In preparation for the next lesson, students **find out** about the structure and formulae of alkenes. [O2]

Plenary suggestions

Either as a class, in groups or in pairs, students write 'Crude oil' in the centre of a sheet of paper and use arrows and labels to list all the products we use from crude oil. They can label the arrows to show the processes. This could be added to in subsequent lessons.

Or students can **draw** a time line from the alkane chains in crude oil containing 15 carbon atoms down to ethene, including the processes and any relevant equations.

Answers to Worksheet 6.24

1. a) Smaller alkane molecules; alkenes; b) Fuels; c) Petrochemical industry; 2. a) Alkene; b) Alkenes react with bromine water and turn it colourless, alkanes do not react; c) To make new materials; 3. a)(i) $C_{10}H_{22} \rightarrow C_8H_{18} + C_2H_4$; (ii) $C_{16}H_{34} \rightarrow C_8H_{18} + 2C_3H_6 + C_2H_4$; (iii) $C_{12}H_{26} \rightarrow C_7H_{16} + C_3H_6 + C_2H_4$; b) Shorter alkane + alkene(s); 4. a) Petrol and diesel; b) Other fractions are produced; c) Kerosene; heavy fuel oils and bitumen

Answers to Practical sheet 6.24

1. Decolorised, 2. a) No change, (b) decolorised

Lesson 25: Cells and batteries

Lesson overview

OCR Specification reference

OCR **C6.2**

Learning objectives

- Make simple cells and measure their voltages.
- Consider the importance of cells and batteries.
- Find out how larger voltages can be produced.

Learning outcomes

- Explain how a voltage can be produced by metals in an electrolyte. [O1]
- Evaluate the uses of cells. [O2]
- Interpret data in terms of the relative reactivity of different metals. [O3]

Skills development

- WS 1.4 Describe and explain a technological application of science.
- WS 1.4 Make decisions based on the evaluation of evidence.
- WS 3.5 Use data to make predictions.

Maths focus Substitute numerical values into equations

Resources needed Equipment as listed in Technician's notes 6.25; Practical sheet 6.25; Worksheets 6.25.1 and 6.25.2

Key vocabulary cell, electrolyte, electrode, rechargeable

Teaching and learning

Engage

- Display a selection of cells and batteries and elicit students' ideas about their similarities and differences. [O1]

Challenge and develop

- Pairs follow Practical sheet 6.25 to **investigate** the relationship between the voltage of a simple cell and the reactivity of the metals used to make it. [O1]

- A virtual voltaic cell could be used as an alternative [Google search string:] Voltaic Cell Virtual lab with Concentrations.

Explain

Low demand

- Students complete Worksheet 6.25.1 to practice **interpreting** data from simple cells in terms of the relative reactivity of different metals. [O3]

Standard and high demand

Ask students to read 'Simple cells' in Section 6.25 in the Student Book and to answer question 1. [O1]

Consolidate and apply

- Show a video clip to reinforce ideas about how batteries work [Google search string:] How Do Batteries Work? : Ri Channel or [Google search string:] How batteries work – Adam Jacobson. [O1]

Low demand

- Students complete Worksheet 6.25.2 to **evaluate** data about rechargeable and non-rechargeable batteries. [O2]

Standard and high demand

- Ask students to read 'Cells and batteries' and 'Comparing voltages' in Section 6.25 in the Student Book and answer questions 2–4. [O2, O3]

- Ask students to **research** the main advantages and disadvantages of non-rechargeable (*primary*) batteries compared to rechargeable (*secondary*) batteries and list the main uses for each type [Google search string:] Battery University. [O2]

Extend

- Ask students able to progress further to find out what advantages lead–acid car batteries have over other types of rechargeable batteries. Could these account for the fact that lead–acid car batteries are still used even though lead is toxic? [O2]

Plenary suggestion

Ask groups to list as many differences between rechargeable and non-rechargeable batteries as they can in 30 seconds.

Answers to Worksheet 6.25.1

1. The further apart; the higher the voltage; 2. Silver and magnesium; 3. 1.60 V; 4. An electrolyte; 5. −2.70 V

Answers to Worksheet 6.25.2

1. Rechargeable batteries maintain a constant output until just before they need recharging; non-rechargeables gradually reduce their output as the chemicals in them get used up; 2. Rechargeable; 3. Non-rechargeable; 4. Rechargeable; 5. Non-rechargeable; 6. They can cause environmental damage if not recycled

Answers to Practical sheet 6.25

A higher voltage should be recorded when the two metals are further apart in the reactivity series

Lesson 26: Fuel cells

Lesson overview

OCR Specification reference

OCR C6.2

Learning objectives

- Find out how fuel cells work.
- Compare and contrast the uses of hydrogen fuel cells, batteries and rechargeable cells.
- Learn what reactions take place inside hydrogen fuel cells.

Learning outcomes

- Describe how a fuel cell works. [O1]
- Evaluate the use of hydrogen fuel cells in comparison with rechargeable cells and batteries. [O2]
- Write half equations for the electrode reactions in the hydrogen fuel cell. [O3]

Skills development

- WS 1.4 Describe and explain a technological application of science.
- WS 1.4 Make decisions based on the evaluation of evidence.
- WS 3.8 Use half equations.

Maths focus Substitute numerical values into equations

Resources needed Equipment as listed in Technician's notes 6.26; Worksheet 6.26

Digital resources Presentation 6.26.1 'Fuel cells' and Presentation 6.26.2 'True or false'

Key vocabulary fuel cell, less pollution, oxidation, oxygen

Teaching and learning

Engage

- Display Presentation 6.26.1 'Fuel cells' and ask groups to **decide** why the hydrogen-fuel-cell-powered car is so good for the environment. [O2]

Challenge and develop

- Demonstrate a hydrogen-fuel-cell-powered model car or show a video of one being operated [Google search string:] Hydrogen-fuel-cell model car. [O1]

- Challenge groups to **create** an advert for a hydrogen-fuel-cell-powered car using the information in Section 6.26 of the Student Book. Their advert should **explain** why their car is less polluting than a battery-powered electric car. [O1, O2]

Explain

Low demand

- Students complete Worksheet 6.26 to **consider** how fuel cells work and **review** their advantages and disadvantages. [O1, O2]

Standard and high demand

- Students answer questions 1 and 2 in Section 6.26 of the Student Book. [O1]

Consolidate and apply

- Show a video to explain in more detail how a fuel cell works [Google search string:] C.6 Three Fuels Cells (plus equations) HL. Stop the video when it starts to consider methanol fuel cells. [O3]

Standard and high demand

- Students use the equations for the reactions at each electrode to answer question 3 in Section 6.26 of the Student Book. [O3]

Extend

- Ask students able to progress further to **research** methanol fuel cells and explain their similarities to hydrogen fuel cells and any differences. [O3]

Plenary suggestion

- Run a whole class quiz using Presentation 6.26.2 'True or false'. [O1, O2]

Answers to Worksheet 6.26

1. Hydrogen and oxygen; 2. A, C, F and H are advantages (circled); B, D, E and G are disadvantages (boxed)

Lesson 27: Key concept: Intermolecular forces

Lesson overview

OCR Specification reference

OCR **C6.2**

Learning objectives

- Identify the bonds within a molecule and the forces between molecules.
- Explain changes of state.
- Explain how polymer structure determines its ability to stretch.

Learning outcomes

- Know that the atoms in a molecule are held together by strong covalent bonds and that weaker intermolecular forces form between molecules [O1]
- Explain how the size of molecules affects the intermolecular forces and boiling points. [O2]
- Explain that polymers with intermolecular forces between their chains can stretch, but those with covalently bonded cross-links do not. [O3]

Skills development

- WS 1.2 Use a variety of models, such as representational, spatial, descriptive, computational and mathematical, to solve problems, make predictions and develop scientific explanations and understanding of familiar and unfamiliar facts.
- WS 4.1 Use scientific vocabulary, terminology and definitions.
- WS 4.3 use SI units and IUPAC chemical nomenclature unless inappropriate.

Maths focus Visualise and represent two-dimensional (2D) and 3D forms, including 2D representations of 3D objects

Resources needed Class sets of Molymods (or similar); Worksheets 6.27.1 and 6.27.2

Key vocabulary intermolecular, covalent, force of attraction, intramolecular

Teaching and learning

Engage

- Ask students to work in pairs and use the Molymod kits to make a model of an octane molecule (depending on the number of 'carbons' in the class Molymod kits). They can double up with another pair and sit their models side by side. Ask students to describe all the bonds and forces represented by the models of two octane molecules lying side-by-side. [O1]
- Gather class ideas and deduce that covalent bonds join the carbon and hydrogen atoms together; and that weaker intermolecular forces exist between the octane molecules. Students can revisit Sections 6.16 and 6.17 the Student Book to refresh ideas about covalent bonds and intermolecular forces in alkanes. Those needing more help with covalent bonding can revisit Chapter 2. [O1]

Challenge and develop

- Remind students that one carbon atom can make four single covalent bonds. Ask them to list five examples of single covalent bonds formed by carbon, such as C–H. They can check back in this chapter if help is needed. Students can answer question 1(a) in Worksheet 6.27.1. This revisits the covalent bonds formed by carbon in the homologous series of alkanes, alkenes, alcohols and carboxylic acids. [O1]
- Ask students what has to happen to a compound to change its strong covalent bonds. Deduce that a chemical reaction, such as combustion, is needed. Ask what has to happen to overcome the intermolecular forces. Establish that these are weaker and can be overcome by heating (when a compound melts or boils) and by a force (in the case of stretchy poly(ethene)). Alternatively students can read Section 6.27 in the Student Book. They can answer questions 1(b) and (c) in Worksheet 6.27.1. [O1]

Explain

- Refer back to the student's octane models. Explain that octane is a liquid at room temperature (boiling point 125 °C) and ask them to describe what happens to the covalent bonds and the intermolecular forces as octane is heated. Establish that the octane molecules gain kinetic energy/move faster and get further apart; the strength of the intermolecular forces decreases; and at the boiling point, octane molecules have enough energy to leave the surface of the liquid and move freely as a gas. The intermolecular forces between the molecules in gaseous octane are relatively weak. Ensure that students understand that the covalent bonds within the octane molecules are strong bonds and do not break when octane boils. Students can answer question 2 in Worksheet 6.27.1. [O2]

- Remind students about the structure of polymers. They can refer to Section 6.10 in the Student Book. Ask them to work in pairs and **discuss** reasons why poly(ethene) film makes good cling film. [Note that poly(chloroethene) is also used to make cling film, but is being slowly replaced by low-density poly(ethene) because of concerns over the plasticiser content of poly(chloroethene) films. However, poly(ethene) cling film also contains additives to improve stretch and cling properties.] Students can **share** their ideas in a class list and then answer question 1 in Worksheet 6.27.2. [O3]

Consolidate and apply

- Explain that octane is one of the alkanes in petrol. In a car engine, octane is first evaporated in the carburettor and mixed with air. The fuel/air mix is ignited in the cylinders and combustion reactions occur to release energy to turn the wheels. Ask students to answer question 2 in Worksheet 6.27.2. [O2]

- 'The strong *intramolecular* bonds give the molecule its identity and its chemistry, but the weak *intermolecular* forces often affect the properties of the material made from these molecules.' Allow students to work in pairs to list four examples to illustrate this sentence. Note that the term 'intramolecular' is used in Section 6.27 of the Student Book and may need explaining. [O1, O2, O3]

Extend

- Ask students able to progress further to carry out an online search to find out why chemicals called plasticisers are added to poly(chloroethene) when it is used to make cling film. They could contrast this with the use of poly(chloroethene) to make drainpipes and guttering. [O3]

Plenary suggestions

Ask students to write down one thing they are sure of, one thing they are unsure of and one thing they need to know more about. **Identify** areas that need revisiting.

Divide the class into two groups – one group is 'covalent bonds', the other 'intermolecular forces'. Quickly ask questions to which the answers are either 'covalent bonds' or 'intermolecular forces'. Students have to claim the correct questions.

Answers to Worksheet 6.27.1

1. a) Ethane, C_2H_6, single covalent bonds; ethene, C_2H_4, double C=C covalent bond, single C–H covalent bonds; ethanol, C_2H_5OH, single covalent bonds; ethanoic acid, CH_3COOH, double C=O bond, all others single covalent bonds; b)(i) A chemical reaction; (ii) For example, $2C_8H_{18} + 25O_2 \rightarrow 16CO_2 + 18H_2O$; c) Heat; a force; 2. a) Students need to make the following points: liquid at 20 °C; weak intermolecular forces hold molecules together in liquid; strength of intermolecular forces decreases as temperature rises and energy increases; at 125 °C molecules have enough energy to overcome intermolecular forces and leave the liquid to move freely; covalent bonds do not break; b) Octane molecules are larger than methane molecules; and have stronger intermolcular forces; which need more energy to be overcome

Answers to Worksheet 6.27.2

1. a) Poly(ethene) molecules are large covalent molecules; with strong intermolecular forces; b)(i) Covalent; (ii) intermolecular forces; c) The intermolecular forces can be overcome easily when a force is applied; so poly(ethene) chains can slide over each other; the polymer stretches; d) The cross-links are strong covalent bonds that cannot be broken when a force is applied; so do not stretch easily; 2. a) Intermolecular forces are overcome as the temperature rises; so petrol vaporises; covalent bonds do not change; b) Intermolecular forces are relatively weak in gases; covalent bonds are broken and then new covalent bonds are made; in a combustion reaction

Lesson 28: Maths skills: Visualise and represent 3D models

Lesson overview

OCR Specification reference

OCR C6.2

Learning objectives

- Use three-dimensional (3D) models to represent hydrocarbons, polymers and large biological molecules.

Learning outcomes

- Identify 3D models of alkanes and draw 2D displayed formulae of them. [O1]
- Identify 3D models of polymers and be able to draw 2D displayed formulae of simple polymers and their monomers. [O2]
- Explain why 3D models are used to represent large biological molecules such as DNA. [O3]

Skills development

- WS 1.2 Use a variety of models, such as representational, spatial, descriptive, computational, and mathematical, to solve problems, make predictions and develop scientific explanations and understanding of familiar and unfamiliar facts.
- WS 4.1 Use scientific vocabulary, terminology and definitions.

Maths focus Visualise and represent 2D and 3D forms, including 2D representations of 3D objects

Resources needed Class sets of Molymods (or similar); model of DNA if available; Worksheets 6.28.1 and 6.28.2

Key vocabulary 3D model, representations, tetrahedron, helix

Teaching and learning

Engage

- Ask students why we use kits such as Molymods in lessons to represent atoms and molecules. Establish that atoms and molecules are very small and that using models helps to explain how they behave and predict how they might behave in other situations. You could make analogies with the use of model buildings by architects, body-part models in Biology lessons, etc. [O1]

- Alternatively, show students a range of molecular models that includes alkanes, alkenes, alcohols and part of a polymer. Discuss what the models represent to assess their current understanding. [O1]

Challenge and develop

Low demand

- Remind students that hydrocarbons are compounds containing carbon and hydrogen. They have learned about alkanes and alkenes. Remind them of examples of each if necessary. Students can work in pairs and need a set of Molymods (or similar). They can answer questions 1(a) and (b) in Worksheet 6.28.1. Check their diagrams and discuss the advantages and disadvantages of using 3D models versus 2D diagrams. Students can answer questions 1(c) and (d) in Worksheet 6.28.1. [O1]

- Ask students to **predict** the shape of a molecule of propene. They can make a model of propene to **check** their prediction. You could repeat this exercise with other hydrocarbons, such as butane, etc. [O2]

Explain

Standard demand

- Explain that 3D models are also used to show the structure of larger molecules like polymers, but in this case only part of the structure is usually shown. Students could make ethene molecules and join them as a reminder of the structure of poly(ethene). Remind them how 2D diagrams of a small section of a polymer chain are drawn showing atoms and bonds. Discuss the advantages and disadvantages of using 3D models and 2D diagrams to show a polymer chain. Remind students that the number of monomers in a polymer chain varies and that '*n*' is used in the displayed formula to reflect this fact. Students can read 'Models of larger structures' in Section 6.28 of the Student Book and answer questions 3 and 4. [O2]

- Students can answer question 2 in Worksheet 6.28.1 [O2]

High demand

- Students can read 'Representing large biological molecules' in Section 6.28 of the Student Book. Explain that shape is very important for large biological molecules because many of their functions depend on their shape, as well as the functional groups present. Students can answer question 1 in Worksheet 6.28.2.

Consolidate and apply

- Students can answer question 2 in Worksheet 6.28.2. They are **comparing** the use of 3D models to represent small hydrocarbon molecules, larger polymer molecules and DNA. Encourage students to write a longer essay-type answer containing at least six points. [O1, O2, O3].

Extend

- Ask students able to progress further to use Molymod kits to investigate how the presence of double bonds in larger hydrocarbon chains changes the shape of the molecule. [O1]

- They can also find out how 3D models help to explain the link between DNA and RNA.

Plenary suggestions

Ask students to draw a large triangle with a smaller inverted triangle inside. In the outer three triangles, they write one thing they have done, one thing they have discussed and one thing they have seen. In the middle triangle, they write one thing they have learned.

Answers to Worksheet 6.28.1

1. a)(i) Student's model; see below; (ii) Student's model; see below; (iii) Student's model; see below; (iv) Student's model; see below; b) Carbon forms four bonds in a tetrahedral shape in ethane; carbon form three bonds with 120° angles in ethene giving a flat shape; c) Shows the shape/direction of the bonds; d) Can be used in drawing diagrams; 2. a)(i) Poly(ethene) chains vary in length; (ii) Small hydrocarbon molecules have a definite number of carbon atoms in their molecules; (iii) see below; b) *3D models*: ethane shows whole molecules, can work out the molecular formula and shows shape; poly(ethene) shows a section of the molecule, cannot work out the molecular formula and shows the shape of a section; *2D diagrams*: ethane shows all of the molecule, can work out molecular formula and does not show shape; poly(ethene) shows a section of the molecule, cannot work out the molecular formula and does not show shape

Answers to Worksheet 6.28.2

1. a) Answer to include: sugar molecules / phosphate groups / polymer chains / base pairs / four bases / double helix shape; b) *2D model*: yes, no, partly, no; *3D model*: yes, yes, yes, no; 2. (Answers will vary) Six points to include: small hydrocarbon models show the whole molecule, plus one named feature; models of large hydrocarbon compounds show a section, plus one feature; models of biological molecules show a typical section (order of base pairs varies), plus one feature

Lesson 29: Proportions of gases in the atmosphere

Lesson overview

OCR Specification reference

OCR C6.3

Learning objectives

- Review the composition of the atmosphere.
- Measure the percentage of oxygen in the atmosphere.
- Consider why it stays the same.

Learning outcomes

- Identify the gases of the atmosphere. [O1]
- Recall the proportions of gases in the atmosphere. [O2]
- Explain how the balance of the gases is maintained. [O3]

Skills development

- WS 3.1 Draw pie charts.
- WS 3.2 Translate data from one form to another.
- WS 4.1 Use scientific vocabulary.

Maths focus Use fractions and percentages; draw an accurate pie chart

Resources needed Equipment as listed in the Technician's notes; Worksheets 6.29.1 and 6.29.2; Technician's notes 6.29.

Key vocabulary atmosphere, carbon dioxide, nitrogen, oxygen

Teaching and learning

Engage

- Challenge groups to name the gases in the atmosphere and the percentage of each gas present. [O1, O2]

Challenge and develop

- Correct any misconceptions about the composition of the atmosphere and ask how we know how much of each gas there is. [O2]
- Explain that we can measure the percentage of oxygen in air by using a chemical reaction to remove it and seeing how much 'air' is left. Then demonstrate this using the method outlined in Technician's notes 6.29. [O2]

Explain

Low and standard demand

- Students draw a pie chart to show the composition of the atmosphere using the templates cut from Worksheet 6.29.1 [O1, O2]

High demand

- Students draw pie charts to show the composition of the atmosphere using a compass and a protractor. [O1, O2]

 © HarperCollins*Publishers* Limited 2016

Consolidate and apply

- Demonstrate that the gas released by photosynthesising pondweed is oxygen by using it to relight a glowing splint. Then ask two students (wearing eye protection) to breathe out through limewater to demonstrate that we add carbon dioxide to the atmosphere because, like other living things, we respire. The third demonstration (see Technician's notes 6.29) shows that the carbon dioxide concentration in normal air is much lower than in air breathed out. [O3]

Low demand

- Students complete Worksheet 6.29.2 to review the demonstrations. [O3]

Standard demand

- Students read 'Air as a mixture' and 'Proportions' in Section 6.29 of the Student Book and complete questions 1–3. [O3]

Standard and high demand

- Students read 'Balancing the proportions' in Section 6.29 of the Student Book and answer questions 4 and 5. [O3]

Extend

Ask students able to progress further to answer question 6 in Section 6.29 of the Student Book. [O3]

Plenary suggestion

Ask students to draw a large triangle with a smaller inverted triangle that just fits inside it (so they have four triangles). In the outer three ask them to write something they've seen, something they've done and something they've discussed. Then add in the central triangle something they've learned.

Answers to Worksheet 6.29.2

1. It relights; 2. Oxygen; 3. Plants need carbon dioxide for photosynthesis; 4. It contains carbon dioxide; 5. It contains less carbon dioxide; 6. It releases carbon dioxide; 7. 1 is combustion; 2 is animal respiration; 3 is plant respiration; 4 is photosynthesis; 8. This time arrows 1–3 point downwards and 4 upwards; so 1 is combustion, 2 is animal respiration, 3 is plant respiration and 4 is photosynthesis

© HarperCollins*Publishers* Limited 2016

Lesson 30: The Earth's early atmosphere

Lesson overview

OCR Specification reference

OCR C6.3

Learning objectives

- Explore the origins of the Earth's atmosphere.
- Consider the evidence that ideas about the early atmosphere are based on.
- Consider the strength of the evidence these ideas are based on.

Learning outcomes

- Describe ideas about the Earth's early atmosphere. [O1]
- Interpret evidence about the Earth's early atmosphere. [O2]
- Evaluate different theories about the Earth's early atmosphere. [O3]

Skills development

- WS 1.1 Decide whether or not given data support a particular theory.
- WS 1.3 Explain why data needed to answer scientific questions may be uncertain, incomplete or unavailable.
- WS 1.6 Recognise the importance of peer review.

Resources needed Worksheet 6.30

Key vocabulary atmosphere, oceans, sediments, volcanic

Teaching and learning

Engage

- Show a video clip of gases being emitted from a volcano [Google search string:] BBC Bitesize Volcanoes and gases video, and elicit students' ideas about how common volcanoes were in the past. [O1]

Challenge and develop

- Use Section 6.30 of the Student Book to introduce the theory that the Earth's original atmosphere was produced by volcanic activity. [O1]

Explain

Low and standard demand

- Students complete Worksheet 6.30 to **summarise** the theories about the early atmosphere outlined in Section 6.30 of the Student Book. [O1]

High demand

- Students **create** a flow chart to outline the steps by which volcanic activity, temperature changes and chemical reactions could have produced an atmosphere rich in nitrogen on the early Earth. [O1]

Consolidate and apply

Low demand

- Show a video clip to summarise the first billion years of Earth's history [Google search string:] How the Earth was made – oceans. [O1]

Standard and high demand

- Use 'Evaluating the theories' in Section 6.30 of the Student Book to introduce some of the things scientists need to consider when evaluating evidence about the formation of the atmosphere. [O2, O3]

- Standard and high demand: Show a video clip to summarise the theory that most of Earth's water came from meteorites [Google search string:] BBC Bitesize The arrival of water on Earth. Challenge groups to **summarise** the evidence presented in the video and **consider** whether it is enough to convince them that the theory is correct. [O2, O3]

Extend

Ask students able to progress further to answer question 4 in Section 6.30 of the Student Book. [O3]

Plenary suggestion

Ask students to write down three things they learnt about the early atmosphere. Then ask them to **share** their facts in groups and **compile** a master list, with the most important facts at the top.

Answers to Worksheet 6.30

1. Carbon dioxide, water vapour, nitrogen and small amounts of methane and ammonia

2. A flowchart with the steps: 1 – no gases present; 2 – mainly carbon dioxide and water vapour with some nitrogen, methane and ammonia; 3 – mostly carbon dioxide (with some nitrogen); 4 – mostly nitrogen with some carbon dioxide; The steps should be linked by the: 1 – degassing of volcanoes; 2 – ocean formation; 3 – carbon dioxide dissolving (and producing sediments containing carbonates)

Lesson 31: How oxygen increased

AQA Specification reference

AQA 4.9.1.3

Lesson overview

Learning objectives

- Explore the processes that changed the oxygen concentration in the atmosphere.
- Consider the role of algae.
- Consider why oxygen levels in the atmosphere didn't rise when oxygen was first produced.

Learning outcomes

- Identify the processes allowing oxygen levels to increase. [O1]
- Explain the role of algae in the composition of the atmosphere. [O2]
- Recall the equation for photosynthesis. [O3]

Skills development

- WS 1.2 Use descriptive models.
- WS 1.3 Explain why data are needed to answer scientific questions.
- WS 3.2 Translate data between graphical and numeric forms.

Maths focus Plot a line graph using a reversed *x*-axis.

Resources needed Worksheet 6.31

Digital resources Presentation 6.31 'One of these is not like the others'

Key vocabulary algae, evolve, oxygen, photosynthesis

Teaching and learning

Engage

- Ask pairs to **suggest** evidence we could look for to help determine when oxygen first appeared in Earth's atmosphere. Then ask the pairs to join together in groups of four and then six to eight to **compile** an **agreed** list of evidence.

Challenge and develop

- Ask one person from each group to **report back** to the class and collate their ideas. Students should appreciate that we can either look for fossil evidence of the living things that released the oxygen or look for evidence for the presence of oxygen – that is, evidence that oxidation reactions have taken place. [O1]

- Use a video clip to review evidence for the earliest appearance of oxygen [Google search string:] Early life: oxygen enters the atmosphere BBC. [O1]

Explain

- Students complete Worksheet 6.31 to outline how the atmosphere has changed over the last 4.5 billion years. [O1]

- Students read 'The early production of oxygen' in Section 6.31 of the Student Book and annotate their graphs to show when each oxygen-producing organism first appears in the fossil record. [O1, O2]

 © HarperCollins*Publishers* Limited 2016

Consolidate and apply

- Low demand: Students read 'Algae and photosynthetic production of oxygen' in Section 6.31 of the Student Book and answer questions 3–5. [O3]

- Standard and high demand: Pairs read 'The Great Oxygenation event' in Section 6.31 of the Student Book and answer question 6. [O2, O3]

Extend

Ask students able to progress further to find out more about microbial mats and why NASA is studying them. [Google search string:] Microbial Mats – Space Science Division – Nasa. [O1]

Plenary suggestion

Use Presentation 6.31 'One of these is not like the others' to display the following sets of words and ask groups to **identify** the odd one out giving reasons: carbon dioxide, nitrogen and oxygen; water vapour, oceans and carbon dioxide; respiration, volcanic degassing and photosynthesis; microbial mats, banded iron formations, algae. Encourage them to **find** as many alternative answers as possible for each set – for example, for the first set: oxygen is not emitted by volcanoes/was not present in the first atmosphere; nitrogen is unreactive/not used in photosynthesis or respiration; carbon dioxide is the only one with a lower percentage in the atmosphere now than billions of years ago.

Answers to Worksheet 6.31

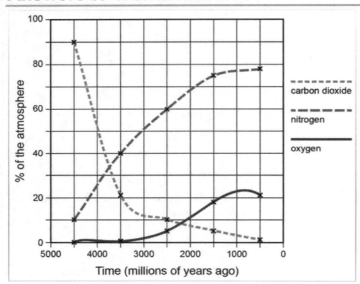

Lesson 32: Key concept: Greenhouse gases

Lesson overview

OCR Specification reference

OCR C6.3

Learning objectives

- Review the greenhouse effect.
- Explain how greenhouse gases trap heat.
- Consider the consequences of adding greenhouse gases to the atmosphere.

Learning outcomes

- Describe the greenhouse gases. [O1]
- Explain the greenhouse effect. [O2]
- Explain these processes as interaction of radiation with matter. [O3]

Skills development

- WS 1.2 Use models in explanations.
- WS 1.2 Recognise, draw and interpret diagrams.
- WS 1.4 Evaluate environmental implications of greenhouse gas emissions.

Resources needed Worksheet 6.32

Digital resources Presentation 6.32 'The greenhouse effect'

Key vocabulary wavelength, radiation, greenhouse, absorb

Teaching and learning

Engage

- Display slide 1 of Presentation 6.32 'The greenhouse effect' to prompt pairs to **discuss** what they know about the greenhouse effect. They should **recognise** that carbon dioxide, methane and water vapour are greenhouse gases that somehow trap solar energy. [O1, O2]

Challenge and develop

- Students use the data on Worksheet 6.32 to **construct** an annotated diagram that explains the greenhouse effect. [O2]

Explain

- Use a video clip to summarise the mechanism of the greenhouse effect and allow students to check their work [Google search string:] Green house Effect and Global Warming.

Low demand

- Students read 'Why are greenhouse gases important?' in Section 6.32 of the Student Book and complete questions 1 and 2. [O1]

Low and standard demand

- Students read 'How does the greenhouse effect work?' in Section 6.32 of the Student Book and complete questions 3 and 4. [O2]

Standard and high demand

- Students Read 'What types of radiation are involved?' in Section 6.32 of the Student Book and complete questions 5 and 6. [O3]

Consolidate and apply

- Use a video clip to introduce some of the complexities of the greenhouse effect and some of the consequences attributed to an increasing greenhouse effect [Google search string:] Global Warming – A video by NASA. [O1, O2, O3]

Extend

Ask students able to progress further to **explore** NASA resources related to atmospheric carbon dioxide [Google search string:] Carbon Dioxide – Global Climate Change – NASA. [O1, O2]

Plenary suggestion

Ask students to write down one thing about the greenhouse effect they are sure of, one thing they are unsure of and one thing they need to know more about. Ask them to work in groups of 5–6 to **agree** on group lists. Then take feedback and **agree** as a class what they are confident about, what they are less sure of and what things they want to know more about.

Answers to Worksheet 6.32

Students should add labels 1–7 to their diagrams at approximately the positions shown

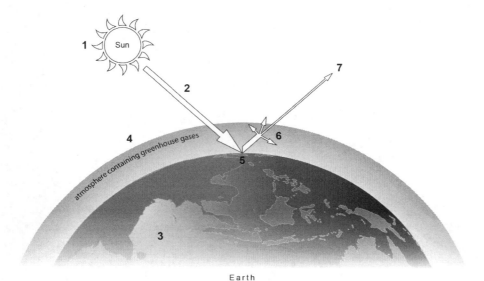

Lesson 33: Human activities

Lesson overview

OCR Specification reference

OCR C6.3

Learning objectives

- Consider the factors that affect the quality of scientific reports.
- Consider the reliability of computer models.
- Find out what a peer review involves.

Learning outcomes

- Evaluate the quality of evidence in a report about global climate change. [O1]
- Describe uncertainties in an evidence base. [O2]
- Recognise the importance of peer review and communicating results to a wide range of audiences. [O3]

Skills development

- WS 1.2 Recognise that computational models need to be tested by observation or experiment.
- WS 1.3 Recognise why data may be uncertain, incomplete or unavailable.
- WS 1.6 Recognise the importance of peer review.

Resources needed Worksheet 6.33

Digital resources Presentation 6.33.1 'Changing the atmosphere'; Presentation 6.33.2 'One of these is not like the others'

Key vocabulary correlation, deforestation, peer review, speculation

Teaching and learning

Engage

- Display Presentation 6.33.1 'Changing the atmosphere' and elicit students' ideas about how we add methane and carbon dioxide to the atmosphere. [O2]

Challenge and develop

- Check that students **recognise** that burning fossil fuels adds carbon dioxide directly and that deforestation adds carbon dioxide indirectly by reducing photosynthesis and increasing decay, but we cannot be certain how much is being added. At the same time, increasing numbers of farm animals, the area of rice cultivation and the amount of rubbish decomposing in landfill sites adds methane to the atmosphere. [O2]

- Use a video clip to summarise the processes that add carbon dioxide to the atmosphere and those that remove it [Google search string:] A Breathing Planet Off Balance. [O1]

- Explain that most scientists agree that the Earth's climate is changing, but not everyone agrees that humans are responsible for these changes and they do not all agree about the seriousness of these changes. To predict what will happen in the future, scientists use computer models. If a model replicates what has happened in the past accurately, scientists assume that it will predict future change accurately. However, models only give accurate predictions if they take account of everything that affects climate change, and we don't have all that information. Slide 2 of Presentation 6.33.2 'Changing the atmosphere' shows how different their predictions can be. We rely on peer review to maintain the quality of the ideas and evidence scientists report. [O2]

- Use a video clip to show how peer review works [Google search string:] Peer Review in 3 Minutes. [O3]

 © HarperCollins*Publishers* Limited 2016

Explain

Low demand

- Students read 'Human activities' and 'Evidence of human activity impact' in Section 6.33 in the Student Book and answer questions 1–4. [O3]

Standard and high demand

- Students read 'Evidence of human activity impact' and 'Modelling climate change' in Section 6.33 in the Student Book and answer questions 3–6. [O2, O3]

Consolidate and apply

- Pairs use Worksheet 6.33 to practise **evaluating** scientific reports. They then join up with another group to **explain** and **compare** ideas. [O1, O2, O3]

Extend

Ask students able to progress further to view a video clip about other factors that affect Earth's climate [Google search string:] Discovering the origins of ice ages BBC Bitesize. [O2]

Plenary suggestion

Use Presentation 6.33.2 'One of these is not like the others' to show groups of three items and ask groups to say why one of them is not like the others. Encourage groups to find as many answers as they can.

Answers to Worksheet 6.33

Panorama current affairs program: this satisfies criteria **A**, **B** and **C**, so it is quite strong. One possible question is 'Where did the evidence they examined come from?'

A peer-reviewed paper in a respected journal: this satisfies criteria **A**, **C** and **D**, so it is quite strong. One possible question is 'Were those who agreed or disagreed using the same evidence?'

The NASA website: this satisfies criteria **A**, **C** and **D**, so it is quite strong. One possible question is 'How do we know it is human-made carbon dioxide?'

Lesson 34: Global climate change

Lesson overview

OCR Specification reference

OCR C6.3

Learning objectives

- Explore the consequences of climate change.
- Consider the risks to human health.
- Judge the seriousness of these consequences.

Learning outcomes

- Discuss the scale of global climate change. [O1]
- Discuss the risk of climate change. [O2]
- Discuss the environmental implications of climate change. [O3]

Skills development

- WS 1.3 Explain why data may be uncertain.
- WS 1.4 Evaluate the environmental implications.
- WS 3.5 Interpret data presented in graphical form.

Resources needed Worksheet 6.34

Digital resources Presentation 6.34 'Extremes'

Key vocabulary average global temperature, distribution, food producing capacity, erosion

Teaching and learning

Engage

- Display Presentation 6.34 'Extremes' and challenge groups to **explain** what the graph shows. [O1]

Challenge and develop

- Check that students recognise that even a small rise in the mean temperature greatly increases the probability that we will experience more very hot periods, but we cannot predict when these will occur. Another variable is that different regions will be affected in different ways. [O2]

- Groups of four to six allocate one consequence of climate change from Worksheet 6.34 to each student. They regroup into 'expert' groups to **research** each topic using the internet before returning to their home groups. [O1, O2]

Explain

- Allow students 30 seconds to **report back** to their home group. Then challenge groups to **rank** the consequences of climate change in order of seriousness and to **justify** their decisions. [O1, O2, O3]

Consolidate and apply

Low and standard demand

- Students read 'The risks of global climate change' in Section 6.34 in the Student Book and answer questions 3 and 4. [O2, O3]

High demand

- Pairs read 'Environmental concerns of climate change' in Section 6.34 in the Student Book and **construct** a joint response to question 5. [O2, O3]

Extend

Ask students able to progress further to **explore** the UK Met Office's predictions of the global consequences of climate change [Google search string:] Human dynamics of climate change. [O1, O2, O3]

Plenary suggestion

Ask students to write down three things they know about climate change. Then get into groups to **compile** a master list with the most important at the top. The most important can then be **shared** with the rest of the class.

Lesson 35: Carbon footprint and its reduction

Lesson overview

OCR Specification reference

OCR C6.3

Learning objectives

- Find out what a 'carbon footprint' is.
- Consider factors that contribute to our carbon footprints.
- Explore ways of reducing our carbon footprints.

Learning outcomes

- Describe what a 'carbon footprint' is. [O1]
- Describe how emissions of carbon dioxide can be reduced. [O2]
- Describe how emissions of methane can be reduced. [O3]

Skills development

- WS 1.4 Describe ways of reducing human impacts on the environment.
- WS 1.5 Give examples of hazards associated with technology.
- WS 4.1 Use scientific vocabulary, terminology and definitions.

Resources needed Internet access; Worksheet 6.35

Digital resources Presentation 6.35 'Carbon footprints'

Key vocabulary alternative energy, carbon capture, carbon off-setting, carbon neutrality

Teaching and learning

Engage

- Display slide 1 of Presentation 6.35 'Carbon footprints' and ask groups to **suggest** a definition of the term. Their ideas could be compared with the one on slide 2. [O1]

Challenge and develop

- Use a video clip to clarify the concept of a carbon footprint [Google search string:] Simpleshow Explains The Carbon Footprint. [O1]
- Use a carbon-footprint calculator to learn which lifestyle choices contribute to a larger environmental footprint [Google search string:] WWF Footprint Calculator. [O1]
- Stress that it is important to cut methane emissions as well as carbon dioxide emissions. For example, cutting our consumption of meat reduces methane emissions from animal farming. [O1, O3]

Explain

- Groups **collaborate** to **construct** a list of lifestyle choices that increase the size of our carbon footprint and **rank** them in order of importance. Then each student is given a number and those with the same number join up to **share** ideas and **refine** their lists. [O2]
- Display slide 3 of Presentation 6.35 'Carbon footprints' and compare the major sources of greenhouse gases with the lists groups have agreed. [O2]

Consolidate and apply

- Introduce the idea that we can reduce our environmental footprints by either releasing fewer greenhouse gases or by taking more greenhouse gases out of the atmosphere. [O2]

Low and standard demand

- Students **read** Section 6.35 of the Student Book and complete Worksheet 6.35 to **summarise** the methods governments can use to reduce the UK's carbon footprint. [O2, O3]

High demand

- Students **read** Section 6.35 of the Student Book and **create** a concept map to **summarise** the methods used by individuals and governments to reduce our carbon footprints. [O2, O3]

Extend

Ask students able to progress further to **review** what steps the UK government is taking to manage greenhouse gas emissions and climate change [Google search string:] UK Climate Services booklet. [O2, O3]

Plenary suggestion

Ask students to draw a large triangle with a smaller inverted triangle that just fits inside it (so they have four triangles). In the outer three ask them to write something they've seen, something they've done and something they've **discussed**. Then add in the central triangle something they've **learned** about carbon footprints.

Answers to Worksheet 6.35

A3; B6; C1; D7; E5; F4; G2

Lesson 36: Limitations on carbon footprint reduction

Lesson overview

OCR Specification reference

OCR C6.3

Learning objectives

- Review the uncertainties about carbon emissions.
- Consider factors that limit our ability to reduce our carbon footprints.
- Decide which factors are most important.

Learning outcomes

- Recognise that carbon footprints are difficult to reduce. [O1]
- Give reasons why actions on reductions may be limited. [O2]
- Give reasons why methane emissions are difficult to reduce. [O3]

Skills development

- WS 1.3 Recognise that scientific data may be uncertain or incomplete.
- WS 1.4 Evaluate methods that can be used to limit greenhouse gas emissions.
- WS 4.1 Use scientific vocabulary, terminology and definitions.

Resources needed Worksheets 6.36.1 and 6.36.2

Digital resources Presentation 6.36 'Reducing emissions'

Key vocabulary economic considerations, international co-operation, population rise, scientific disagreement

Teaching and learning

Engage

- Display Presentation 6.36 'Reducing emissions' and elicit students' ideas about why greenhouse gas emissions are still rising even though we know how to reduce them. [O1]

Challenge and develop

- Use a video clip to remind students that there are often uncertainties in science [Google search string:] Carbon Emissions: Uncertain Choices – Science Museum. [O2]
- Introduce the task to consider some of the factors that make it difficult to reduce our carbon footprint. Then give each group a set of cards cut from Worksheet 6.36.1 and a copy of Worksheet 6.36.2. They take it in turns to take a card from the pile, read it out and **decide** which of the categories on Worksheet 6.36.2 it fits into. [O1, O2, O3]

Explain

Low and standard demand

- Students **list** the five limiting factors that can prevent us from reducing greenhouse gas emissions and give an example to illustrate each factor. [O1, O2]

High demand

- Students work in pairs to **construct** concept maps to summarise the factors that limit our ability to cut greenhouse gas emissions. [O2]

Consolidate and apply

Low demand

- Students answer question 1 in Section 6.36 of the Student Book. [O2, O3]

Standard demand

- Students questions 2 and 3 in Section 6.36 of the Student Book. [O2, O3]

High demand

- Students question 4 in Section 6.36 of the Student Book. [O1]

Extend

Ask students able to progress further to explore the advice of the Committee on Climate Change on reducing the UK's carbon footprint [Google search string:] Reducing the UK's Carbon Footprint. [O1]

Plenary suggestion

Ask students to write down two things that limit our ability to reduce our carbon footprints. Then get groups to **compile** a master list with the most important at the top. The most important can then be **shared** with the rest of the class.

Answers to Worksheet 6.36.1

Lack of public information and education – C, L; *Scientific disagreement* – D, F, I, O; *Incomplete international cooperation* – P, R; *Lifestyle changes* – B, E, H, N, Q; *Economic considerations* – A, G, I, J, K, M

© HarperCollins*Publishers* Limited 2016

Lesson 37: Atmospheric pollutants from fuels

Lesson overview

OCR Specification reference

OCR C6.3

Learning objectives

- Explore the products formed when fuels burn.
- Distinguish between complete and incomplete combustion.
- Write equations for complete and incomplete combustion.

Learning outcomes

- Describe some common unwanted products of combustion. [O1]
- Predict the products of combustion of a fuel knowing the composition of the fuel. [O2]
- Predict the products of combustion of a fuel knowing the conditions in which it is used. [O3]

Skills development

- WS 1.2 Use models in explanations.
- WS 1.4 Explain the environmental implications of burning fuels.
- WS 1.4 Describe methods that can tackle human impacts on the environment.

Resources needed Equipment as listed in the Technician's notes; sets of cards cut from Worksheets 6.37.1, 6.37.2, 6.37.3 and 6.37.4; Technician's notes 6.37

Digital resources Presentation 6.37 'Burning fuels'

Key vocabulary sulfur dioxide, oxides of nitrogen, particulates, hydrocarbons

Teaching and learning

Engage

- Display Presentation 6.37 'Burning fuels' and elicit students' ideas about the exhaust gases from car engines. They should **recognise** that they contain carbon dioxide and water vapour from burning fuels and solid particles of soot. [O1]

Challenge and develop

- Demonstrate that a beaker of water heated with a yellow flame takes longer to boil and accumulates a layer of soot. Describe this as evidence of incomplete combustion. When the Bunsen's air supply is reduced by closing the air hole some of the carbon in the fuel is not converted to carbon dioxide.

- Show students a carbon monoxide detector and stress that every home with gas central heating needs to have one. If the oxygen supply to their gas boiler was accidently cut for any reason it could release poisonous carbon monoxide, as well as soot, instead of carbon dioxide.

- Use a video clip to introduce symbol equations for the combustion of hydrocarbon fuels [Google search string:] What is Combustion? | The Chemistry Journey | The Fuse School. [O1, O2, O3]

Explain

Low and standard demand

- Pairs use Worksheet 6.37.2, and cards cut from Worksheet 6.37.1 to **construct** symbol equations for complete and incomplete combustion. [O1, O2, O3]

Higher demand

- Pairs use Worksheet 6.37.4 to **construct** symbol equations for complete and incomplete combustion. [O1, O2, O3]

Consolidate and apply

Low demand

- **Students** read 'Common atmospheric pollutants' and 'Gaseous pollutants' in Section 6.37 in the Student Book and answer questions 1–4. [O1]

Standard and high demand

- Students complete Worksheet 6.37.3 to **explain** how catalytic converters work. [O2]

- Students **explain** that catalytic converters cannot remove the fine particles of unburned carbon, called particulates, that are emitted in exhaust fumes – especially by diesel engines. Ask students to read 'Pollution by particulates' in Section 6.37 in the Student Book and answer questions 5 and 6. [O3]

Extend

Ask students able to progress further to **investigate** air pollution controls in their nearest city.

Plenary suggestion

Ask pairs to **compose** a question about burning fuels and a mark scheme for the answer. Then swap questions with other pairs.

Answers to Worksheet 6.37.2

1. $CH_4 + 2O_2 \rightarrow CO_2 + 2H_2O$; 2. $CH_4 + O_2 \rightarrow C + 2H_2O$; 3. $CH_4 + 1\frac{1}{2}O_2 \rightarrow CO + 2H_2O$; 4. $S + O_2 \rightarrow SO_2$; 5. $N + O_2 \rightarrow NO_2$

Answers to Worksheet 6.37.3

1.(i) NO; (ii) NO_2); (iii) N_2O; 2. ; 3.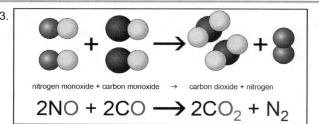

nitrogen monoxide + carbon monoxide → carbon dioxide + nitrogen

$$2NO + 2CO \rightarrow 2CO_2 + N_2$$

;

4. They use carbon monoxide to remove the oxygen from nitrogen monoxide to make harmless nitrogen; or use nitric acid to add oxygen to carbon monoxide to make less harmful carbon dioxide

Answers to Worksheet 6.37.4

1. $CH_4 + 2O_2 \rightarrow CO_2 + 2H_2O$; 2. $CH_4 + O_2 \rightarrow C + 2H_2O$; 3. $C_3H_8 + 2O_2 \rightarrow 3C + 4H_2O$; 4. $C_3H_8 + 3O_2 \rightarrow C + 2CO + 4H_2O$;
5. $C_2H_6 + 2O_2 \rightarrow C + CO + 3H_2O$; 6. $N + O_2 \rightarrow NO_2$; 7. $S + O_2 \rightarrow SO_2$

Lesson 38: Properties and effects of atmospheric pollutants

Lesson overview

OCR Specification reference

OCR C6.3

Learning objectives

- Review the hazards associated with air pollutants.
- Investigate correlations between pollutant emissions and deaths from asthma.
- Consider whether these support the hypothesis that air pollution makes asthma worse.

Learning outcomes

- Explain the problems caused by air pollutants. [O1]
- Explain the effects of acid rain. [O2]
- Evaluate the role of particulates in damaging human health. [O3]

Skills development

- WS 1.4 Explain the environmental implications of burning fuels.
- WS 3.5 Recognise or describe patterns in data.
- WS 3.6 Comment on the extent to which data are consistent with a hypothesis.

Maths focus Plot two variables from data

Resources needed Computers running Excel and concept-mapping software; Worksheets 6.38.1 and 6.38.2

Digital resources Presentation 6.38.1 'Air pollution'; Presentation 6.38.2 'One of these is not like the others'

Key vocabulary acid rain, global dimming, particulates, toxicity

Teaching and learning

Engage

- Display Presentation 6.38.1 'Air pollution' and ask groups to **summarise** what they know about the way these gases affect our health. [O1]

Challenge and develop

- Use a video clip to explore the link between invisible air pollution and health problems [Google search string:] Air Pollution – The 'Invisible Hazard'. [O1]
- Use 'Toxic gases' and 'Acid rain' in Section 6.38 in the Student Book to review the main hazards associated with carbon monoxide and the acidic gases SO_2 and NO_x. [O1, O2]

Explain

Low demand

- Students complete Worksheet 6.38.1 to **summarise** the sources of CO, SO_2 and NO_x and the hazards associated with them. [O1, O2]

Standard and high demand

- Pairs **create** concept maps to **summarise** the components of air pollution, their sources and the hazards associated with each pollutant. [O1, O2, O3]

Consolidate and apply

- Use 'Particulates' in Section 6.38 in the Student Book to review the main hazards associated with them. [O3]

Low and standard demand

- Students complete Worksheet 6.38.2 to explore the correlation between pollutant emissions and deaths from asthma. [O1, O2, O3]

High demand

- Pairs use the pollution data spreadsheet 6.38 to explore correlations between pollutant emissions and deaths from asthma. [O1, O2, O3]

Extend

Ask students able to progress further to complete questions 5 and 6 in Section 6.38 of the Student Book. [O3]

Plenary suggestion

Use Presentation 6.38.2 'One of these is not like the others' to show three things associated with air pollution and ask groups to **decide** why one of them is not like the others. Encourage students to find as many possible answers as they can.

Answers to Worksheet 6.38.1

1. a) and b) Cars; c) Power stations; 2. a) Poisons people; b) Kills aquatic life and plants; erodes stonework and causes smog; c) Kills aquatic life and plants; erodes stonework; 3. They contain sulfur impurities; 4. Carbon monoxide and oxides of nitrogen

Answers to Worksheet 6.38.2

Students should plot graphs like this:

1. Emissions of both pollutants have fallen since 2001; there has been an overall drop in the number of deaths from asthma; the fall has been largest for acidic gases; 2. The graph supports this hypothesis

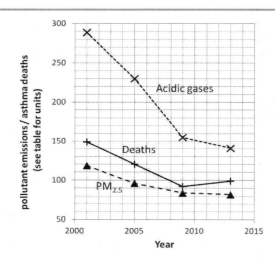

Lesson 39: Potable water

Lesson overview

OCR Specification reference

OCR C6.3

Learning objectives

- Distinguish between potable water and pure water.
- Describe the differences in treatment of ground water and salty water.
- Explain what is needed to provide potable water for all.

Learning outcomes

- Know that pure water contains water only and potable water is water that is fit to drink and contains low levels of dissolved salts and microbes. [O1]
- Describe how and why ground water is filtered and sterilised and salty water is desalinated to produce potable water. [O2]
- Explain why we need to conserve water. [O3]

Skills development

- WS 1.4 Explain everyday and technological applications of science, evaluate associated personal, social, economic and environmental implications, and make decisions based on the evaluation of evidence and arguments.
- WS 3.5 Interpret observations and other data (presented in verbal, diagrammatic, graphical, symbolic or numerical form), including identifying patterns and trends, making inferences and drawing conclusions.
- WS 4.4 Use prefixes and powers of ten for orders of magnitude (e.g. tera, giga, mega, kilo, centi, milli, micro and nano).

Maths focus Recognise and use expressions in standard form

Resources needed Worksheets 6.39.1, 6.39.2 and 6.39.3; Technician's notes 6.39

Digital resources Class computer access

Key vocabulary potable, sedimentation, desalination, reverse osmosis

Teaching and learning

Engage

- Show students three labelled containers of water samples – distilled water, tap water and seawater. Ask students to work in pairs and, for each sample, list the types of atoms, molecules, ions and substances they contain. **Discuss** their ideas as a class and **compile** a class list. [O1]

- Ask students which samples are fit to drink and introduce the terms 'pure water' and 'potable water'. Establish that pure water contains water only and potable water is water that is fit to drink. **Discuss** whether or not potable water is always pure water. [O1]

Challenge and develop

- Low demand. Establish that potable water contains low levels of dissolved substances and microbes. If appropriate, you can show/remind students about salt deposits in kettles and water pipes. [O1]

- Standard demand. Tell students that only 1% of the Earth's water is suitable for drinking and that this has to be treated before it can be consumed. Students can read 'Producing potable water' in Section 6.39 of the Student Book. They can answer question 1 on Worksheet 6.39.1. This covers water treatment stages used by Thames Water. This is also explained on the Thames Water website [Google search string:] Thames drinking water treatment. [O2]

- Tell students that chlorine is not the only disinfectant used to sterilise water – ozone gas and ultraviolet light are also used. Students can answer question 2 on Worksheet 6.39.1. This asks them to interpret data on the effectiveness of ultraviolet light in killing microbes. [O2]

Explain

- Explain that some areas of the world do not have sufficient suitable groundwater to provide the required potable water and have to use sea or salty water to provide potable water.

- Students will need computer access. They can work in pairs. In each pair, one student **researches** 'distillation' and the other 'reverse osmosis' as methods of desalinating salty water. This is the subject of question 1 on Worksheet 6.39.2. Students can use the questions as a guide to their research. They **report back** to each other and **share** notes. Check that they **understand** that desalination consumes a lot of energy. [O2]

- High demand: Explain that many of the current water systems around the world were built when water was considered to be an unlimited resource. Students can read 'Potable water for all' in Section 6.39 in the Student Book and answer questions 5 and 6. They can answer question 2 on Worksheet 6.39.2. Note that some of these questions are given marks to encourage students to give an appropriate number of points in their answers. [O3]

Consolidate and apply

- Refer back to students' lists of atoms, molecules, etc., in distilled, tap and seawater from the 'Engage' part of the lesson. Ask them to make any necessary amendments. [O1]

- Discuss the graph on Worksheet 6.39.3. This shows the supply of potable water from the Colorado river for the area and also the demand. Figures are from 1920 projected to 2060. [O2, O3].

- Either discuss the answers to the questions on Worksheet 6.39.3 as a class or ask students to answer them. [O1, O2, O3]

Extend

Ask students able to progress further to:

- **Investigate** the labels on mineral water bottles to find out if mineral water is pure water, has been disinfected or has additives.

- **Investigate** the benefits and disadvantages of using a) chlorine, b) ozone and c) ultraviolet light to disinfect water. [O2]

Plenary suggestions

Ask students to draw a large triangle with a smaller inverted triangle inside. In the three outer triangles, they write something they have seen, something they have discussed and something they have found out. In the centre triangle, they write the 'big idea' they have learned.

Ask students to write down three 'big ideas' they have learned during the lesson. Ask them to **share** their ideas in groups and **produce** a master list of ideas. They can find out which other groups **agreed** with their lists.

Answers to Worksheet 6.39.1

1. a) Large pieces of biomass/leaves, twigs, etc.; b) Coarse sand beds remove larger suspended particle; fine sand beds remove smaller suspended particles; c) To disinfect the water; d) 1257 dm^3; 2. a) To control variables; b)(i) Dysentery bacteria, *E. coli*, influenza virus; (ii) As (i) plus *Salmonella* bacteria; c) *E. coli* means raw sewage has entered the water supply; d)(i) 0.000000254 m; (ii) 0.0000254 cm

Answers to Worksheet 6.39.2

1. Student response; 2. a) Less rainfall means less ground water; b) Population is expected to increase;(c) Mend leaks, install water meters; d)(i) Consumers are more conscious of water use; use less to save money/incentive to use less; (ii) Measure water consumption of metered and unmetered households; keep variables like type of house, number of appliances, people, constant; e)(i) Drains into the rivers; and back to the sea; (ii) Filtered, disinfected and treated to remove any metal contaminated from the roadside

Answers to Worksheet 6.39.3

1. Filtration; disinfection; 2. Increases; 3. Increasing population; increase in domestic appliances requiring water; increase in industry/irrigation; 4. Increased temperature; decreased rainfall; 5. 1997/1998; 6. Conserve use; new sources; 7. a) Seawater; b) Desalination

Lesson 40: Waste water treatment

Lesson overview

OCR Specification reference

OCR C6.3

Learning objectives

- Explain how waste water is treated.
- Describe how sewage is treated.
- Compare the ease of treating waste water, groundwater and salt water.

Learning outcomes

- Explain the treatment required for waste water from urban lifestyles and industry before the water can be returned to the environment. [O1]
- Describe the stages of screening, sedimentation, anaerobic digestion and aerobic treatment of sewage. [O2]
- Comment on the relative ease of obtaining potable water from waste water, ground water and salt water. [O3]

Skills development

- WS 1.4 Explain everyday and technological applications of science, evaluate associated personal, social, economic and environmental implications and make decisions based on the evaluation of evidence and arguments.
- WS 3.3 Carry out and represent mathematical and statistical analysis.
- WS 3.5 Interpret observations and other data (represented in verbal, diagrammatic, graphical, symbolic or numerical form), including identifying trends, making inferences and drawing conclusions.

Maths focus Use ratios, fractions and percentages

Resources needed Worksheets 6.40.1, 6.40.2 and 6.40.3.

Digital resources Thames Water website [Google search string:] The sewage treatment process.

Key vocabulary sewage, anaerobic, aerobic, sedimentation.

Teaching and learning

Engage

- Remind students about the water cycle they have learned about before. Tell them that there is a fixed amount of water on this planet and that it does not change, but the water is continually recycled. Ask them to **suggest** how the treatment of groundwater and desalination to produce potable water. [O1]

Challenge and develop

Low demand

- Introduce students to Worksheet 6.40.1 and explain the diagram if necessary. This is a diagram of the natural water cycle with the urban water cycle showing the treatment of groundwater and desalination to provide potable water, and the treatment of waste water added. Students can **annotate** the diagram as in the questions. [O1]

Standard demand

- Emphasise that waste water from homes and factories is ultimately returned back to the environment via rivers, groundwater or the sea, but this needs to be treated first. Ask them to work in pairs and list the substances that must be removed from sewage from homes, agricultural waste water and industrial waste water. (You may wish to introduce the term 'organic matter' to describe sewage content.) They can **share** their lists with the rest of the class. They can use Section 6.40 in the Student Book to check their lists. [O2]

Explain

- Introduce Worksheet 6.40.2, which is based on Thames Water's treatment of sewage. You may find their website useful as an additional resource. Explain the diagram if necessary and allow students to complete the worksheet. They will need computer access and their Student Book. Check their answers. [O2]

High demand

- Emphasise that students have learned about three different methods of producing potable water – from groundwater, from salt water and from waste water – and that each of these needs treating in different ways to make potable water. Ask them to **decide** which is the easiest method to use, and then what factors may require other methods to be used. Students can complete Worksheet 6.40.3. This is a case study exploring water issues in California and the use of desalination and waste water treatment to supplement groundwater sources. [O3]

Consolidate and apply

- Students can read 'The water footprint' in Section 6.40 of the Student Book. They can answer questions 5 and 6. [O3]

- Students can work in pairs and write the words 'Potable water' in the centre of a sheet of paper. They can use arrows and labels to **construct** a mind map of the different sources and methods used to produce potable water. [O2]

Extend

Ask students able to progress further to:

- **Investigate** the source of the school's water supply and where and how it is treated to make the water potable. Most water boards have websites with this information. [O2]

- **Find out** how eco-homes save water and recycle grey water for use in appliances like washing machines and toilets.

Plenary suggestions

Each student has two strips of paper. They write a question on one strip and the answer on the other. They form groups of five or six. The strips are shuffled and redistributed so that each student has one question and one answer. A student asks their question and the one with the correct answer responds. They ask the next question.

Answers to Worksheet 6.40.1

1. Student response; 2. Student response

Answers to Worksheet 6.40.2

1. Large objects, e.g. leaves; 2. Sludge is separated from the effluent; 3. a) Without oxygen; b) Sludge is broken down; 4. a) A fuel made from biomass; b) Sewage → screening → sedimentation → anaerobic digestion of sludge → methane gas (used as fuel) and solid 'cakes' (used as fuel); 5. a) With oxygen; b) So harmful bacteria do not enter water systems; 6. To check bacteria/microbe levels

Answers to Worksheet 6.40.3

1. Climate change/increased temperatures/lower rainfall; 2. Energy is needed to distill water in desalination; and to treat waste water; if fossil fuels, are burnt to produce energy; CO_2 is produced which is a greenhouse gas; 3. Thermal methods involve heating; which consume more energy than methods used in waste water treatment; 4. 81%; 5. a) 0.154 kWh; b) 1.167 kWh; c) 0.646 kWh

Lesson 41: Practical: Analysis and purification of water samples from different sources, including pH, dissolved solids and distillation

Lesson overview

OCR Specification reference

OCR C6.3

Learning objectives

- Describe how safety is managed, apparatus is used and accurate measurements are made.
- Recognise when sampling techniques need to be used and made representative.
- Carry out a procedure to produce potable water from salt solution.
- Evaluate methods and suggest possible improvements and further investigations.

Learning outcomes

- Identify the functions and correct use of apparatus and understand how a risk assessment produces a safe procedure. [O1]
- Successfully carry out tests for pH, sodium ions and chloride ions, and a simple distillation on selected samples of salt solution. [O2]
- Suggest improvements to the procedure and apparatus used, and further investigations that could be carried out to produce potable water from salt solution. [O3]

Skills development

- WS 2.3 Apply a knowledge of a range of techniques, instruments, apparatus and materials to select those appropriate to the experiment.
- WS 2.4 Carry out experiments appropriately having due regard for the correct manipulation of apparatus, the accuracy of measurements and health and safety considerations.
- WS 2.5 Recognise when to apply knowledge of sampling techniques to ensure any samples collected are representative.
- WS 2.6 Make and record observations and measurements using a range of apparatus and methods.
- WS 2.7 Evaluate methods and suggest possible improvements and further investigations.
- AT 2 Safe use of appropriate heating devises and techniques, including the use of a Bunsen burner and a water bath or an electric heater.
- AT 3 Use of appropriate apparatus and techniques for the measurement of pH in different situations.
- AT 4 Safe use of a range of equipment to purify and/or separate chemical mixtures, including evaporation and distillation.

Resources needed Practical sheets 6.41.1 and 6.41.2; Worksheet 6.41; Technician's notes 6.41

Key vocabulary distillation, evaporation, condensation, purity, boiling point

Teaching and learning

Engage

- Students can read 'Planning your sampling' in Section 6.41 of the Student Book and answer questions 1, 2 and 3. [O2]
- Set the scenario of needing to site a distillation plant in the UK in the future to produce potable water from seawater and selecting a suitable site. Students can work in pairs to **produce** a list of requirements for such a site. These can be **shared** with the class. Requirements such as no sewage outlets or coastal industries, and availability of cheap electricity are relevant. Ask students to **decide** where and when they would take samples and how many to check the suitability of the seawater. [O2]

Challenge and develop

- Introduce the practical. This can be organised in a variety of ways according to ability, but they all need to understand the apparatus they are using and the safety procedures. Either explain these or students can complete Worksheet 6.41 and produce a risk assessment. They can use Section 6.41 of the Student Book to focus their ideas on apparatus hazards. Ensure that they appreciate the danger of broken glass in the hand when inserting the delivery tube into the conical flask. [O1]

Low demand

- Students can carry out the work on Practical sheet 6.41.1. They are testing a salt solution and their distilled water for sodium ions and chloride ions, measuring its pH and carrying out a simple distillation. [O2]

Standard demand

- Students can be provided with different samples of sea water from different locations, some suitably contaminated, and carry out the appropriate tests for sodium and chloride ions, pH and distillation to find the best site to build a desalination plant. They can use Practical sheet 6.41.1. [O2]

High demand

- Students can follow Practical sheet 6.41.2 to plan their own procedures to test and distil different samples of water. You will need to check their plans and apparatus. [O2]

Explain

- Students can complete the 'Results and evaluation' section on the practical sheets. The level of demand will follow on from the procedure used. [O3]

Consolidate and apply

- Ask students to work in pairs and list the most important safety rules for these practicals. These can be **shared** with the class. [O1]
- Ask students to **share** their results and evaluations with the rest of the class and check the uniformity of their results. [O3]

Extend

Ask students able to progress further to:

- **Find out** how flame tests can be used to detect the presence of other metal ions in the salt water samples. [O2]
- **Find out** how a Liebig condenser can be used to improve the procedure. [O3]

Plenary suggestions

Students can work in small groups and produce a freeze frame of what not to do during this practical. Other students can **guess** what the frame represents and **suggest** the safe practice.

Students can write a three-mark question on some aspect of this lesson and a mark scheme. They can try out their questions on each other.

Answers to Worksheet 6.41

Salt solution – low hazard, wipe up spillages immediately; *pH meter /universal indicator* –; *Dilute hydrochloric acid* – irritant, report spillages to the teacher; *Dilute nitric acid* – corrosive, report spillages to the teacher; *Silver nitrate solution* –low hazard, wash off skin immediately and report spillage; *Glassware* – danger of cuts from broken glass, keep away from the edge of the bench, report breakages; *Delivery tube* – danger of breaking if forced into conical flask, insert by holding the rubber bung; *Bunsen burner* – danger of burns, keep on yellow flame when not in use, keep away from clothing and hair

Answers to Practical sheet 6.41.1

1. *Salt solution*: pH 7, yellow flame, white precipitate; *Distilled water*: pH 7, no change, no change; 2. Sodium ions and chloride ions are present; 3. The flame test and chloride ion test should give negative results; 4. No; 5. Students' suggestions; e.g. use a Liebig condenser; 6. Students' suggestions; e.g. testing for microbes/other ions; 7. Depends on samples used

Answers to Practical sheet 6.41.2

1. *Salt solution*: pH 7, yellow flame, white precipitate; *Distilled water*: pH 7, no change, no change; 2. Tests for sodium ions and chloride ions are negative; 3. No; 4. Student's suggestions, e.g. use a Liebig condenser; 5. Student's suggestions, e.g. testing for microbes, other ions; 6. Depends on samples used

Lesson 42: Maths skills: Use ratios, fractions and percentages

Lesson overview

OCR Specification reference

OCR C6.3

Learning objectives

- Consider ways of comparing the amounts of gases in the atmosphere.
- Review what balanced symbol equations show.
- Compare the yields in chemical reactions.

Learning outcomes

- Use fractions and percentages to describe the compositions of mixtures. [O1]
- Use ratios to determine the mass of products expected. [O2]
- Calculate percentage yields in chemical reactions. [O3]

Skills development

- WS 4.2 Recognise the importance of scientific quantities.

Maths focus Use ratios, fractions and percentages

Resources needed Equipment as listed in the Technician's notes; Worksheet 6.42.1 and 6.42.2; Technician's notes 6.42

Digital resources Presentation 6.42 'Changing ratios'

Key vocabulary fractions, percentages, approximately, segment

Teaching and learning

Engage

- Ask pairs to sort cards cut from Worksheet 6.42.1 into three sets of five. Introduce parts per million (ppm) as a way of comparing the amounts of gases that form a very small percentage of the atmosphere. [O1]

Challenge and develop

Low demand

- Students use a video clip to demonstrate how to **interconvert** fractions, decimals and percentages [Google search string:] Learning Fractions, Decimals, and Percents | abcteach. [O1]

Standard and high demand

- Ask students to read 'The composition of air' in Section 6.42 of the Student Book and answer questions 1 and 2. [O1]

Explain

- Students explain how to **convert** numbers from fractions into percentages and vice versa. [O1]

Consolidate and apply

- Use 'Ratios of substances' in Section 6.42 in the Student Book to introduce the use of ratios to compare the numbers of molecules that take part in chemical reactions. [O2]

© HarperCollins*Publishers* Limited 2016

Low and standard demand

- **Use** Presentation 6.42 'Changing ratios' to remind students that the amount of oxygen available determines the products formed when hydrocarbons burn. Students complete worksheet 6.42.2. [O2]

High demand

- Students complete questions 3 and 4 in Section 6.42 of the Student Book. [O2]

- Standard and high demand: Students read 'Calculating the percentage yield' in Section 6.42 of the Student Book and answer questions 5 and 6. [O3]

Extend

Ask students able to progress further to use their results to **calculate** the percentage yield of carbon dioxide. They can be given the theoretical yield for this reaction, which is 0.36 grams of carbon dioxide per gram of copper carbonate. Their yields are likely to be less that 100% and they could be challenged to **suggest** why. [O3]

Plenary suggestion

Challenge groups to **suggest** pairs of reactants that combine in each of these ratios: 1 : 1, 1 : 2 and 1 : 3.

Answers to Worksheet 6.42.1

The three sets are ABGIM; CDFKL; EHJNO

Answers to Worksheet 6.42.2

1. a) 1 : 2, carbon + water; b) 1 : 3, carbon + carbon monoxide + water; c) 1 : 4, carbon dioxide + carbon monoxide + water; d) 1 : 5, carbon dioxide + water. 2. As the amount of oxygen in the mixture increases; the carbon (soot) produced is replaced by carbon monoxide; and then by carbon dioxide

When and how to use these pages: Check your progress, Worked example and End of chapter test

Check your progress

Check your progress is a summary of what students should know and be able to do when they have completed the chapter. Check your progress is organised in three columns to show how ideas and skills progress in sophistication. Students aiming for top grades need to have mastered all the skills and ideas articulated in the final column (shaded pink in the Student book).

Check your progress can be used for individual or class revision using any combination of the suggestions below:

- Ask students to construct a mind map linking the points in Check your progress
- Work through Check your progress as a class and note the points that need further discussion
- Ask the students to tick the boxes on the Check your progress worksheet (Teacher Pack CD). Any points they have not been confident to tick they should revisit in the Student Book.
- Ask students to do further research on the different points listed in Check your progress
- Students work in pairs and ask each other what points they think they can do and why they think they can do those, and not others

Worked example

The worked example talks students through a series of exam-style questions. Sample student answers are provided, which are annotated to show how they could be improved.

- Give students the Worked example worksheet (Teacher Pack CD). The annotation boxes on this are blank. Ask students to discuss and write their own improvements before reviewing the annotated Worked example in the Student Book. This can be done as an individual, group or class activity.

End of chapter test

The End of chapter test gives students the opportunity to practice answering the different types of questions that they will encounter in their final exams. You can use the Marking grid provided in this Teacher Pack or on the CD Rom to analyse results. This shows the Assessment Objective for each question, so you can review trends and see individual student and class performance in answering questions for the different Assessment Objectives and to highlight areas for improvement.

- Questions could be used as a test once you have completed the chapter
- Questions could be worked through as part of a revision lesson
- Ask Students to mark each other's work and then talk through the mark scheme provided
- As a class, make a list of questions that most students did not get right. Work through these as a class.

Marking Grid for Single Chemistry for End of Chapter 6 Test 3816

| Student Name | Getting Started [Foundation Tier] | | | | | | | Going Further [Foundation and Higher Tiers] | | | | | More Challenging [Higher Tier] | | | | | Most demanding [Higher Tier] | | | Total marks | Percentage |
	Q. 1 (AO1) 1 mark	Q. 2 (AO1) 1 mark	Q. 3 (AO1) 1 mark	Q. 4 (AO1) 1 mark	Q. 5 (AO1) 1 mark	Q. 6 (AO2) 2 marks	Q. 7 (AO2) 2 marks	Q. 8 (AO1) 1 mark	Q. 9 (AO1) 1 mark	Q. 10 (AO3) 2 marks	Q. 11 (AO2) 4 marks	Q. 12 (AO2) 4 marks	Q. 13 (AO1) 1 mark	Q. 14 (AO1) 1 mark	Q. 15 (AO2) 2 marks	Q. 16 (AO1) 2 marks	Q. 17 (AO3) 2 marks	Q. 18 (AO2) 2 marks	Q. 19 (AO3) 2 marks	Q. 20 (A2) 4 marks		

Check your progress

You should be able to:

state examples of natural products that are supplemented or replaced by agricultural and synthetic products	→ distinguish between finite and renewable resources from given information	→ extract and interpret information about resources from charts, graphs and tables
describe the process of phytomining	→ describe the process of bioleaching	→ evaluate alternative biological methods of metal extraction
describe the components of a Life Cycle Assessment (LCA)	→ interpret LCAs of materials or products from information	→ carry out a simple comparative LCA for shopping bags
describe the composition of common alloys	→ interpret the composition of other alloys from data	→ evaluate the uses of other alloys
compare quantitatively the physical properties of materials	→ compare properties of glass and clay ceramics, polymers, composites and metals	→ explain how the properties of materials are related to their uses and select appropriate materials
apply the principles of dynamic equilibrium to the Haber process	→ explain the trade-off between rate of production and position of equilibrium	→ explain how the commercially used conditions for the Haber process are related to the availability and cost of raw materials and energy supplies, control of equilibrium position and rate
describe why crude oil is a finite resource	→ identify the hydrocarbons in the series of alkanes	→ explain the structures and formulae of alkanes
describe the difference between an alkane and an alkene	→ draw displayed structural formulae of the first four members of the alkenes	→ explain why alkenes are called unsaturated molecules
explain how a voltage can be produced by metals in an electrolyte	→ evaluate the uses of cells	→ interpret data for the relative reactivity of different metals
discuss the scale of global climate change	→ discuss the risk of climate change	→ discuss the environmental implications of climate change
describe how emissions of carbon dioxide can be reduced	→ describe how emissions of methane can be reduced	→ give reasons why actions on reductions may be limited
explain how waste water is treated	→ describe how sewage is treated	→ compare the ease of treating waste, ground and salt water

© HarperCollins*Publishers* Limited 2016

Worked example

1 **Fertilisers are added to soil to provide three essential elements. Identify the essential elements.**

a NKNa   **b** (NPK)   **c** NPS   **d** NKS

> The answer NPK is correct

2 **Draw the apparatus needed to make a fertiliser *solution* from acid and alkali. Do not include crystallisation.**

1. Use a measuring cylinder to pour alkali into a conical flask.

2. Add acid to the alkali until it is neutral.

measuring cylinder

conical flask

burette

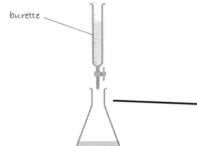

> The apparatus for a titration is correct.

3 **Fertilisers are needed to improve crop growth. Give reasons why they are needed.**

They provide more crops for a growing population.

> A better answer would include the fact that natural deposits of fertilisers are running out and not always in the place needed.

4 **Give two *different* ways that corrosion of iron can be prevented.**

The iron can be painted or covered with grease.

> This is partly correct. A better answer would include another type of prevention, e.g. sacrificial protection.

5 **Evaluate which metal would be best for making the following objects. a lighting circuit b coiled spring for holding a toy c a decorated wrist band d heating coil for a hot drinks cup**

Alloy	Electrical conductivity S/m	Hardness Mohs	MP °C	Thermal conductivity W/mK	Tensile strength N/m²
A	5.9×10^7	3	1062	380	220
B	4.6×10^6	1.5	318	35	17

a A   b A   c B   d A

> The answer is partially correct but not complete. The student identifies the best alloys for each object but has not explained why they chose them (not evaluated). The choice must be justified with a reason.

Chapter 1: Particles

Lesson 1.1 Three states of matter

1 X = solid.

Y = gas

Z = gas

2 −28 °C - W is a solid. Particles vibrate about fixed positions.

−18 °C - Melting point. The vibrational energy is enough to overcome the forces between particles. Particles start moving.

−14 °C - W is now a liquid. Distance between particles has increased. Particles have enough energy to overcome these forces of attraction and move around.

3 25 °C - Particles can move around each other.

38 °C - Particles move around faster with more energy and liquid expands. Some particles have enough energy to escape from the surface (evaporate).

42 °C - Boiling point. The particles now have enough energy to overcome the forces of attraction between them and they escape from the liquid to become gas particles.

46 °C - W is now a gas. The particles are far apart and move randomly and energetically.

4 The forces between the delocalised electrons and lattice of positive ions are strong so the energy required to overcome these forces is large.

5 The forces between molecules of ethanol are greater than the forces between propane molecules. So for ethanol, more energy is needed to break the forces and become a gas. Therefore, ethanol has a higher boiling point.

6 • If there were no forces between gas particles, they could never condense to become a liquid.

• Molecules come in all shapes and sizes which will affect the forces of attraction and the forces on collision. This will affect the temperature at which they condense and the energy released when they do so. It will also affect the way they pack together as a liquid.

• Molecules are not solid and inelastic. They are flexible and elastic. This will also affect the forces between particles and forces on collision. This will affect the temperature at which they condense and the energy released when they do so.

Lesson 1.2 Changing ideas about atoms

1 Dalton did not know about negative electrons which were discovered much later. So positive charge was not considered either.

2 In the Dalton model there were no subatomic particles, the atom was the smallest particle.

3 He had more evidence and was able to do theoretical calculations.

4 An exact amount of energy, no more, no less

5 Democritus, Dalton, JJ Thompson, Geiger and Marsden, Rutherford, N Bohr, J Chadwick

400BC 1803 1897 1909 1911 1913 1932

Atoms Atoms Electron scattering experiment nucleus orbits neutron

6 Needed different experimental evidence and the particle was not charged

Lesson 1.3 Modelling the atom

1 Positive

2 2

3 That it is very small (and in constant motion around the nucleus)

4 Nucleus is made of protons and neutrons held together with no empty space, so a small radius. The electrons are in energy levels away from the nucleus with space in between so the radius of the atom is much larger.

5 There are more electrons in the potassium atom than in the lithium atom so they will take up more space around the nucleus. (The answer to this question does not yet need to refer to successive energy levels)

6 As more positive charges are added they have greater attraction or pull on the negative charges which move towards the nucleus making the overall radius smaller

Lesson 1.4 Key concept: Sizes of particles and orders of magnitude

1 1000/0.25 = 4,000

2 5×10^5

3 $2.5 \times 10^{-11} / 1.75 \times 10^{-15} = 14,286$

4 1×10^{-12}

5 The lithium atom has one outer shell electron. It loses this electron when it becomes an ion. So it has lost a whole shell. So Li+ is smaller then Li. Also, because there is one less electron in Li+, the repulsion amongst the remaining electrons is less so they can be pulled closer to the nucleus.

6 The fluorine atom has seven outer shell electrons. It gains one electron when it becomes a negative ion. The nucleus cannot hold the 10 electrons in the F- ion as tightly as the 9 electrons in the F atom. Also, the extra electron in the ion causes more repulsion. Both these factors mean that the negative ion is larger.

7 Li to Rb = 244/152 = 1.61. Be to Sr = 215/111 = 1.94. So Be to Sr shows the greatest increase in radii.

Lesson 1.5 Relating charges and masses

1 3

2 The neon atom has 10 positively charged protons and 10 negatively charged electrons i.e. the same number of each. So the charge cancels out.

3 Li

4 20

5 9 protons, 9 electrons and with mass of 19 must be 10 neutrons

6 17 protons as it has 17 electrons and with mass 35 must be 18 neutrons

7

Mg atom	12	12	12	12	0
Mg ion	12	12	10	12	+2

8 An Mg atom has equal numbers of protons and electrons so is neutral.
An Mg ion has 12 protons and 10 electrons so two less negative charges so is charged positively

9 10

Lesson 1.6 Subatomic particles

1 Al 13 27 13 13 14
2 15 15 16
3 $^{35}_{17}Cl$
4 6 protons, 6 electrons, 7 neutrons
5 Similar as they each have 1 proton and 1 electron. Different as they have 1, 2 or 3 neutrons.
6 27.7

Lesson 1.7 Maths skills: Standard form and making estimates

1 1×10^9
2 100 000 000
3 1×10^{-18}
4 0.000 000 001
5 a 1.8×10^{13}
 b 2.4×10^8
 c 3×10^{10}
6 $6.02 \times 10^{23} \times 1 \times 10^{-10} \times 2 = 1.2046 \times 10^{14}$ m (note that the radius of the atom has been multiplied by 2 to obtain the diameter).
7 $3.0 \times 10^{26} \times 3.3 \times 10^{-22} = 9.9 \times 10^4$

End of chapter questions

1 7 (1)
2 1 mark for each correct diagram

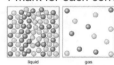

3 Students' own answers, correctly drawn (2)
4 Liquid. G changes from solid to liquid at −33 °C and from liquid to gas at 52 °C. (1)
5 b) (1)
6 d) 10×10^{-10} (1)
7 Proton: 1, +1; Electron -1; Neutron no charge (2)
8 7.5 (1)
9 a) the same number of protons (1)
10 Students' own answers, correctly drawn (2)
11 Rutherford's theory differed in that instead of 1 particle or a ball of matter, he proposed that the atom had a positively charged nucleus, but that much of it was empty space. Bohr took this nuclear model and adapted it to show how electrons orbit at specific distances from the nucleus in shells. (2)
12 Nucleus is made of protons and neutrons held together with no empty space, so a small radius. The electrons are in energy levels away from the nucleus with space in between so the radius of the atom is much larger
13 16 (1)
14 14 d and 16 (1)
15 $^{39}_{19}X$ (2)

16 a) 19; b) 18 (2)
17 The number of protons in an atom is the atomic number. The total number of protons and neutrons is the mass number. If an atom has a negative charge, it has more electrons than protons. A 2-ion would have 2 more electrons than protons. (4)
18 It is a non-metal as it gains electrons to fill its outer shell. It is less reactive than the element 2,7 as the atom has two electrons being added to its outer shell rather than just one. It is less reactive than the atom 2, 6 as the two electrons being added are further away from the nucleus and experience less pull. Electrons being added to its outer shell are further away from the nucleus and experience less pull (4)
19 Students' own drawings (4)
20 Any 5 from Democritus, Dalton, JJ Thompson, Geiger and Marsden, Rutherford, N Bohr, J Chadwick
400BC 1803 1897 1909 1911 1913 1932
It took a long time for scientists to accept new theories, work was theoretical because atoms can't be seen and experimentation had gradual results.

Chapter 2: Elements, Compounds and Mixtures

Lesson 2.1 Key concept: Pure substances

1. Dissolve the mixture in water. The salt will dissolve but the sand won't. Filter to remove the sand. Heat the salt solution to concentrate and leave to crystallise.
2. Use chromatography. The colours will separate.
3. The magnesium is added until there is excess. It is easy to filter off the excess magnesium. Also, if there was any sulfuric acid left in solution, it would be more difficult to purify the magnesium sulfate.
4. Filter so that the sediment and grit was removed. Then fractionally distill the ethanol / water mixture.
5. Sam. Impurities lower melting points. His melting point is higher, which means it is closer to the data book value.
6. Akira had the purest sample since the boiling was lower than Ben's and closer to the data book value.
7. a Yes it was. The boiling point of water is 100 °C. Impurities raise the boiling point.
   b Evaporate / distil the water / crystallise the solid.
   c It was not pure. It melted over a wide range and not sharply.

Lesson 2.2 Relative Formula Mass

1. 80
2.
	Protons	Neutrons
^{12}C	6	6
^{13}C	6	7
^{14}C	6	8

3. Some elements have more than one isotope which have the same atomic number but different mass number (different number of neutrons). The atomic masses are averaged according to the proportion of each isotope in a naturally occurring sample.
4. 120
5. 187.5
6. $MgBr_2$ +   2 $AgNO_3$ → $Mg(NO_3)_2$ +   2 AgBr
   184   2 x 170   148   2 x 188
   Reactants = 524
   Products = 524
7. Molecular mass of R = 44. Molecular formula of R = C_3H_8

Lesson 2.3 Mixtures

1. Filter paper in a filter funnel, with a flash beneath
2. Can be separated by dissolving and filtering
3. Chromatography
4. Distillation
5. Dissolve the mixture in water, filter off the sand. Evaporate off the alcohol to leave copper sulfate solution. Crystallise the copper sulfate from the solution to get pure crystals.
6. d, b, a, c

Lesson 2.4 Formulations

1. Most mass: sugar. Least mass: salt

2. If the components are not in the correct proportions, it will affect the properties of the cement e.g. setting time, strength etc. This could be dangerous since it is used in building.
3. Gold has many uses, each requiring different properties. If they are not precisely prepared, electrical circuits may not work etc.
4. They contain the same percentage of nitrogen.
   NPK 4:1:3 contains 12.5 % P and 37.5 % K.
   NPK 4:2:2 contains 25 % P and 25 % K.
5. E. It has too much active ingredient and too little filler.
   C. It has too little lubricant and too much filler.
6. Oil and solvent are harmful to the environment and to the user. Water based paints were formulated to limit the harm caused. They are as good or better than solvent based paints.

Lesson 2.5 Chromatography

1. Because there is no dark blue spot in the food.
2. You would have to obtain some different samples of known green food colours and spot them along with the unknown food colour. If they match, then the food colour has been identified.
3. 56/70 = 0.80
4. 0.68 x 90 = 61.2 mm
5. A - A single substance but not the same substance as in B or C. It doesn't contain the pure drug.
   B - Contains two substances - a mixture. One of these is the drug. The other is not in A or C. So the drug is not pure.
   C - Contains one substance, the pure drug.
6. Pure drug: 7.1/9.2 = 0.77
   Spot A: 4.2/9.2 = 0.46
   Spot B: 5.3/9.2 = 0.58

Lesson 2.6 Practical: Investigate how paper chromatography can be use in forensic science to identify an ink mixture used in a forgery

1. The ink contains dyes which would move up the paper with the solvent. This would interfere with the other spots.
2. Capillary tube.
3. The dyes in the ink spots are soluble in the solvent. If the solvent covers the ink spots, the dyes would just dissolve in the solvent.
4. It should be left as long as possible to allow good separation. However, the paper needs to be removed before the solvent front reaches the top.
5. Red, blue and green.
6. Yellow.
7. R_f = 6.7/12.5 = 0.54
8. Jo: R_f = 4.5/10 = 0.45
   Alex: R_f = 5.6/12.5 = 0.45
9. Sam: R_f = 4.2/12.4 = 0.339
   Jo: 0.339 x 10 = 3.39 = 3.4 cm.
10. a The distance for the yellow spot is too large. Jo's solvent front did not travel as far so the value should be less than Alex and Sam's.

b Repeat the experiment and remeasure the distance. Calculate the R_f and compare it to Sam and Alex's R_f value.

11 Dyes in the ink have different solubilities in different solvents. Therefore, they will be carried different distances up the chromatography paper. This will alter their R_f value.

Lesson 2.7 Maths skills: Use an appropriate number of significant figures

1 25 mm

2 21 mm

3 27 mm; pure drug 21.5 mm, A 12.5 mm, B 15.5 mm and 21.5 mm, C 21.5 mm

4 pure drug 0.79629, A 0.4629, B 0.5750 and 0.79629, C 0.79629
C is likely to be the pure drug as it produces only one spot with the same R_f value as the reference pure drug.

5 0.80.
The reason 2 significant figures are chosen is because both measurements are taken to 2 significant figures. So the calculation needs to be taken to the same number of significant figures.

6 0.46, 0.56 and 0.80.

Lesson 2.8 Comparing metals and non-metals

1 Shiny, malleable

2 Dull, brittle

3 Test it with universal indicator –it should go yellow/orange

4 Magnesium nitrate

5 If UI goes blue it is an alkaline oxide which means it was a metal oxide
If UI goes red it is an acidic oxide which means it was a non-metal oxide

6 Non-metal

7 Low melting point and density suggest that X is non–metallic. However, only metals are malleable and good conductors of electricity (although graphite is a good conductor it is not malleable and has a very high melting point). Therefore X is a metal and the oxide will be basic.

Lesson 2.9 Electron structure

1 H is ring with 1 electron. Li is two rings, first with 2 electrons, second with 1 electron

2 2,5

3 a Shell pattern 2,8,1

b Sodium / Na.

c Group 1.

4 2,7

5 Ca / Calcium. 20 neutrons

6 2,8,8

7 Group number is the same as the number of electrons in the outer shell, except for group 0 which have complete outer shells.

8 a 13

b 13

c Group 3

Lesson 2.10 Metals and non-metals

1 Non-metal

2 Metal

3 Aluminium has an electron pattern 2,8,3, so it loses 3 electrons to make a stable ion. Metals lose electrons to make ions.

4 Element number 8 has an electron pattern 2,6, so it gains 2 electrons to make a stable ion. Non-metals gain electrons to make ions.

5 It has only one space left to fill with an electron to be a stable ion, non-metals gain electrons. It can gain an electron from potassium (a metal that loses one electron to form a stable ion) to form potassium fluoride.

6 The atom with 3 electrons (Li) has a pattern 2,1. The 1 outer electron is transferred to the atom with a pattern 2,7 (F). This has one space to fill. Both are now stable ions and form a compound of formula LiF

Lesson 2.11 Chemical bonds

1 Ionic bonding is between oppositely charged metal and non-metal ions. Covalent bonding is between non-metal atoms which share electrons.

2 Metallic. It cannot be covalent because it is a good electrical conductor. It cannot be ionic since it conducts when liquid. Covalent substances have low melting points.

3 It carries charge and when it moves it conducts electricity.

4 Electrons in covalent substances cannot move. They do not have delocalised electrons.

5 Electrons are on the "outside" of the atom and can therefore interact with other atoms/molecules, etc. The nucleus is shielded from other atoms.

Lesson 2.12 Ionic bonding

1 It loses its outer shell electrons and forms a stable positive ion.

2 Sodium loses an outer electron. This is transferred to chlorine which gains an electron. Positive sodium ions (Na+) attract negative chloride ions (Cl-) and a lattice is formed.

3 2,8

4 1

5 It has gained one electron. It has 17 protons in the nucleus and 18 electrons in shells around the nucleus. Overall it has a charge of 1-.

6

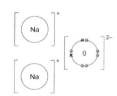

Lesson 2.13 Ionic compounds

1 (ii)

2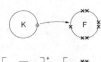

3 Potassium atoms have one outer shell electron which they can donate to sulfur. In doing so they become 1+ ions. Sulfur has 6 outer shell electrons and can accept 2 electrons to become a 2- ion. The ions are stable with the electronic structure of a noble gas. So 2 potassium atoms donate an electron each to 1 sulfur atom. The empirical formula is therefore K_2S

4 Empirical formula is $CaCl_2$.

5 a Na_3N

 b Al_2O_3.

Lesson 2.14 Properties of ionic compounds

1 There are strong forces of attraction between oppositely charged ions. Much energy is needed to overcome these forces.

2 Ionic compounds have high melting points. However, they do not conduct when solid but do when liquid. So D cannot be an ionic compound.

3 The ions are fixed in the lattice and cannot move.

4 The charge on the magnesium ion is 2+ and on the oxygen ion 2-. The higher the charge the greater the forces of attraction between the ions. Sodium and potassium ions have a 1+ charge and chloride ions a 1- charge. So more energy is needed to separate the ions in magnesium oxide.

Lesson 2.15 Properties of small molecules

1 There are weak forces of attraction between CO_2 molecules. So not much energy is required to separate the molecules.

2 Carbon dioxide is a larger molecule than carbon monoxide so the intermolecular forces are stronger.

3 Butane is a larger molecule than propane. Therefore, the intermolecular forces between butane molecules are greater. It takes more energy to separate butane molecules than propane.

4 When nitrogen boils only very weak intermolecular forces have to be broken. Not much energy is needed. The covalent bond within nitrogen is much stronger than the intermolecular forces and is not broken.

5 Pure water does not contain (enough) ions to conduct. Sea water contains ions which can move and carry charge so can conduct.

6 Pentane. The molecules can get closer to each other so the intermolecular forces are greater. Spheres cannot get as close together.

7 The intermolecular forces between chlorine molecules are greater. This is because chlorine has more electrons than fluorine. So more energy is required to separate chlorine molecules.

Lesson 2.16 Covalent bonding

1 H-F

2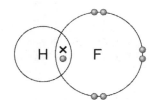

3 3 bonding pairs (and 1 non-bonding pair).

4 CH_4

5 Same as H_2O with O replaced by S.
Draw outer shell only for sulfur.

6 Oxygen has a double covalent bond which is 4 electrons (2 shared pairs). This means that each oxygen atom has a share in 8 electrons - full shell stability.

7 Nitrogen has a triple covalent bond which is 6 electrons (3 shared pairs). This means that each nitrogen atom has a share in 8 electrons - full shell stability.

8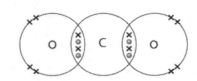

Lesson 2.17 Giant covalent structures

1 Similarity: Silicon and carbon atoms have 4 covalent bonds. Same shape (tetrahedral) around silicon and carbon atom. Difference: Only carbon atoms in diamond (element). Silicon and oxygen atoms in silicon dioxide (compound).

2 Molecules are atoms covalently bonded together into relatively small units. Silicon dioxide is a giant lattice of huge numbers of covalently bonded Si and O atoms.

3 They both have the same giant covalent structure - a network of strong covalent bonds.

4 It is extremely hard and has a high melting point. The latter allows it to withstand the high temperatures generated in the cutting tool due to friction.

5 The bonds in graphite do not act in all directions. It has a layered structure and the forces between the layers are weaker so they can slide over each other.

6 Graphite is a giant covalent substance and has a very high melting point. It can therefore withstand high temperatures in furnaces.

© HarperCollins*Publishers* Limited 2016

7 Graphite has a very high melting point and can conduct electricity. It can therefore withstand high temperature.

8 They do not have any mobile / delocalised electrons. All the electrons are locked in covalent bonds.

Lesson 2.18 Polymer structures

1 Intermolecular forces are weak. However, because polythene has long chains, the intermolecular forces are stronger so polythene is a solid.

2 Intermolecular forces are relatively weak in poly(ethene). Therefore, these forces can be broken and the chains can slide over each other.

3 The strands of polymer are connected by cross-links, so the strands cannot slip past each other.

4

5 a Condensation. 2 different molecules react together.

   b The intermolecular forces are relatively strong.

Lesson 2.19 Metallic bonding

1 Magnesium has a configuration of 2,8,2. The 2 outer electrons are "spare" and can be delocalised

2 Lithium has 1 outer shell electron (configuration 2,1) which is free to move (delocalised) throughout the lithium ion lattice. It is a sea of electrons in a positive lithium ion lattice.

3 Sodium has 1 electron in its outer shell. This is not localised to one Na atom but free to move throughout the whole structure. The outer electron is delocalised.

4 Silver has delocalised electrons which can move throughout the positive ion lattice. This movement is an electrical current. Silver oxide is an ionic compound. It does not have delocalised electrons and the ions cannot move when solid.

5 As the metal is heated, the ions in the lattice vibrate more and move apart (they occupy a greater volume). The higher the temperature the greater the expansion. However, the strong attraction between the delocalised electrons and the positive ions means that the metallic bonds do not break (until it reaches its melting point).

6 Aluminium has 3 outer shell electrons that can be delocalised whereas sodium only has 1 outer shell electron. So aluminium has more charge carriers and therefore has a greater electrical conductivity

Lesson 2.20 Properties of metals and alloys

1 Brass is composed of copper and zinc. Bronze is composed of copper and tin.

2 Steel is harder than iron (the distortion in the layers in steel means that they cannot as easily slide over each other). Steel does not rust as

easily as iron. This means that steel is a very useful construction material.

3 False. Steel is alloy containing carbon. Carbon is a non-metal.

4 The layers of atoms in copper can slide over each other. Therefore, copper can be pulled into wires.

5 Delocalised electrons can move through the structure

6 a Pure silver. The presence of copper distorts the layers and makes it more difficult for them to slide past each other.

   b 0.075 g.

7 When different metal atoms are in the main metal lattice, they distort the layers. This makes it more difficult for the layers to slide over each other. So the alloy is harder. Other properties of alloys are also different e.g. melting points are lower. The ability to change the properties of an alloy is very useful.

8 When copper atoms are added to the main metal lattice, they distort the layers. This makes it more difficult for the layers to slide over each other. So the alloy is harder.

9 The movement of delocalised electrons through the lattice is disrupted. This is due to the metal added to the copper.

Lesson 2.21 Key Concept: The outer electrons

1 Two from Ne, Ar, Kr, Xe

2 One

3 Two

4 Two

5 Both lithium and sodium have one outer electron to lose. In sodium the electron is further away from the 'pull' of the nucleus than is the outer electron of lithium. So the outer electron of sodium is more easily lost. Sodium is therefore more reactive than lithium.

6 Chlorine is more reactive than bromine or iodine. This is because the spare space in the outer shell of electrons is nearer to the nucleus so incoming electrons have a 'greater pull' from the nucleus than the electrons transferring into a bromine or iodine atom. It is however less reactive than fluorine as the incoming electrons are even nearer to the nucleus in the smaller atom of fluorine.

Lesson 2.22 The periodic table

1 Aluminium

2 Chlorine

3 2,8,6; Sulfur, Group 6, Period 3

4 Group 7, Period 3

5 F: 2,7; Cl: 2,8,7; They both have 7 electrons in their outer shells

6 Oxygen, Sulfur, Selenium, Tellurium and Polonium

Lesson 2.23 Developing the periodic table

1 That every 8 elements there was often a pattern of behaviour

2 Not all the elements had been discovered.

3 Gallium

4 The chemical behaviour of some elements that were grouped together was very different. Mendeleev swapped elements so that their chemical behaviour was similar.

5 He predicted the atomic 'weight' and density, which were very close in the discovered element, germanium. It fitted in the space in the table that he predicted. The evidence from this discovery supported his predictions.

6 a 20.

b Mn and Cu, Nb and Rh Re and Pt and Hg

7 $^{40}_{18}Ar$ and $^{39}_{19}K$. The relative atomic mass of $^{40}_{18}Ar$ suggests it should be placed after $^{39}_{19}K$.

8 Disagree – it is possible to keep on adding protons to the nucleus to create increasingly heavy elements / there may be extreme conditions in which "new" elements exist e.g. in a supernova. Agree – all stable elements have been discovered.

Lesson 2.24 Diamond

1 It shares its electron orbits with four other electron orbits forming four covalent bonds.

2 They form strong covalent bonds in all directions

3 The (Si-O) bonds in silicon dioxide are weaker than the (C-C) bonds in diamond. So less energy is required to break them.

4 Diamond has no free electrons - they are all locked in covalent bonds so cannot move. Graphite has delocalised electrons between layers which can move. When electrons move, a current flows.

5 Carbon has 4 outer shell electrons. It can accept four more electrons, one from each carbon. It then has 8 electrons in its outer shell which is full shell stability.

6 The network of strong covalent bonds makes diamond very hard and therefore it is capable of grinding other materials. Also, it has a very high melting point which means it will not be affected by the heat generated during grinding.

Lesson 2.25 Graphite

1 a 4

b 3

2 The forces between layers are weak - the layers slide over each other.

3 Diamond: Shared electron pair between carbons for each of the four covalent bonds per atom. No spare delocalised electrons. Graphite: Shared electron pair between carbons for each of the three covalent bonds per atom. Spare electron delocalised between layers.

4 The delocalised electrons can carry heat energy. So when one part of graphite is heated, the delocalised electrons transmit the energy to the other parts.

5 a Graphite can conduct electricity as a solid like metals. It has a very high melting point like many metals.

b Carbon is a non-metal. It is not malleable, ductile, shiny etc. It forms covalent bonds. Metals form a giant metallic structure which is a lattice of positive metal ions in a sea of delocalised electrons. Metals form positive ions and form ionic compounds.

6 Graphite has a high melting point so will not be affected by the high temperatures. It can conduct electricity.

7 Graphite and diamond both contain networks of strong covalent bonds between carbon atoms. They both adopt a giant covalent structure. They therefore have similar melting points since the attractive force between atoms is similar.

Lesson 2.26 Graphene and fullerenes

1 Similarities: Composed of carbon atoms only (they are forms of the element carbon). Contain hexagonal rings. Covalent bond between carbons. Differences: Graphene is a single layer of atoms and one atom thick. Fullerenes form hollow 3D shapes. Fullerenes can contain 5 or 7 membered carbon rings.

2 Diamond: Giant covalent. Network of strong covalent bonds. 4 covalent bonds per carbon

Graphite: Giant covalent. Strong covalent bonds within layers. 3 covalent bonds per carbon within layers. Weaker forces between layers.

Fullerenes: Simple molecular structures e.g. C60. They form hollow shapes. For instance Buckminsterfullerene is spherical. They contain rings of 6 carbons (and often 5 or 7 membered rings). 3 covalent bonds per carbon.

3 The network of carbon-carbon bonds is very strong. So a great deal of force is required to break the bonds.

4 Lower than diamond. Buckminsterfullerene is a molecule. The intermolecular forces between the C_{60} molecules are much weaker than the network of strong covalent bonds in diamond. So it takes much less energy to separate C_{60} molecules than to break diamonds covalent bonds.

5 Graphite: Giant covalent layered structure. Strong covalent bonds within layers. 3 covalent bonds per carbon within layers. Weaker forces between layers.

Graphene: Single layer of graphite, one atom thick. Hexagonal rings of carbon are connected to each other by strong covalent bonds. 3 covalent bonds per carbon.

6 Graphite is soft because the carbon layers inside a stick of graphite slide over each other very easily. However, graphene is like a single layer of graphite. The bonds in graphene (and in the graphite layer) are very strong and require lots of energy to break them.

Lesson 2.27 Nanoparticles, their properties and uses

1 Between 1 and 100 nanometres

2 100 times smaller

3 Free nanoparticles could get into our lungs causing health problems.

They can be used in sunscreens to prevent health problems from excess sun exposure.

4   12:1

5   Have a higher surface area to volume ratio
More surface for action as a catalyst

Lesson 2.28 Maths skills: Using ratios in mixture, empirical formulae and balanced equations

1 a   1 : 7

  b   1 : 5

  c   Caleb's

2 a   7 : 5, 3 : 1

  b   Gold alloy B

3 K_2S

4   2 : 1, 2 : 4 : 1

5 a $MgSO_4$

  b $MgCl_2$

  c $FeCl_2$

  d $Al_2(SO_4)_3$

6   4 : 5

7   1 : 1

8   3 : 2

9 $2 N_2 + O_2 \rightarrow 2 N_2O$

End of chapter questions

1   Ionic (1)

2   Carbon dioxide (1)

3   Nitrogen (N), Phosphorus (P) and Potassium (K).

4   2,8,1 (1)

5   It has one electron in its outer shell so is in Group 1. It has three electron shells so is in period 3 (2)

6   Filter the mixture to remove the sand. (1) Carry out chromatography on the filtrate. (1)

7   2 marks for all 3 correct. 1 mark for 2 correct.

   K^+ ion

   NO2   small molecule

   $-(XZ)-_n$ Polymer

8   Distillation. Heat the mixture and the ethanol will distill of at 80 $^\circ$C - before the water at 100 $^\circ$C.(1)

9 a   0.050/(0.050 + 0.79 + 0.11) x 100 = 5.3 %   (1)

  b   The proportions need to be correct so that the correct dose of aspirin is given

10   d) Giant metallic (1)

11   **A** Low boiling points suggest weak intermolecular forces between molecules.(1) Simple molecular substances do not conduct electricity because they have no delocalisable electrons. (1)

   B Giant metallic elements (and graphite) conduct electricity when solid, ionic compounds do not. (1) Metals have high melting and boiling points due to strong metallic bonds. (1)

12   honeycomb/lattice structure (2)

13   Silver has a sea of delocalised electrons that can move throughout the lattice of positive silver ions. (1)

14   The electron pair is shared between the 2 chlorine atoms (in the covalent bond). (1)

15   1 mark for the correct total number of outer shell electrons. 1 mark for the shared electron pair between the P and F atoms.

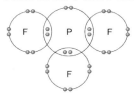

16   **Level 3**: At least 6 points. All 3 structures correctly identified. 1 point must be about conductivity. (5-6)
   Level 2: At least 4 points. 2 structures correctly identified. 1 point must be about conductivity. (3-4)
   Level 1: At least 2 points. 1 structure correctly identified. (1-2)
   Nothing written worthy of credit. (0)

Indicative content

Structure A

- Giant metallic.
- High melting and boiling point due to a strong metallic bonds.
- A lot of energy is needed to overcome the strong attraction between the sea of delocalised electrons and lattice of positive metal ions.
- Conducts electricity when solid (and liquid) so cannot be diamond or simple molecular since these do not conduct.
- Graphite does conduct when solid but has a much higher melting and boiling point.
- It cannot be giant ionic which does not conduct when solid.
- Giant metallic substances are insoluble in water due to the strong metallic bonds.

Structure B

- Giant ionic.
- High melting and boiling point due to a strong electrostatic attraction between oppositely charged ions.
- A lot of energy is needed to overcome this strong attraction.
- Conducts electricity when liquid since the ions can move and carry charge.
- It does not conduct when solid since the ions cannot move (fixed in solid lattice) and so cannot carry charge.
- Cannot be diamond or simple molecular since these do not conduct.
- Cannot be giant metallic or graphite since these conduct when solid.
- Some ionic compounds are soluble. The charged ends of the water molecules attract the charged ions.

Structure C

- Giant covalent (diamond).
- Very high melting and boiling point due to a network of strong covalent bonds.
- Doesn't conducts electricity so cannot be a giant metallic or giant ionic (or graphite).
- Cannot be simple molecular since the melting and boiling points for simple molecular

substances are low (weak intermolecular forces).

- Giant covalent substances are insoluble in water due to the difficulty of breaking the network of strong covalent bonds.

Electrical conductivity

- **A** has delocalised electrons which can move through the lattice of positively charged metal ions (giant metallic structure) when solid (or liquid).
- **B** has ions that can move when liquid and carry charge. When solid the ions cannot move since they are fixed in the lattice.
- **C** has covalent bonding. The electron pairs cannot move since they are "fixed" in covalent bonds between atoms.

17 There are only weak intermolecular forces between methane molecules. (1) The energy required to break these forces is small. (1)

18 The electron pattern shows one electron in the outer shell so the element is from group 1. It loses an electron to form an ion and is therefore a metal, and it reacts violently because the electron that is lost is a long way from the nucleus and so is readily lost (2)

19 It is a non metal as it gains electrons to fill its outer shell. It is less reactive than the element 2,7 as the atom has two electrons being added to its outer shell rather than just one. It is less reactive that the atom 2, 6 as the two electrons being added are further away from the nucleus and experience less pull. electrons being added to its outer shell are further away from the nucleus and experience less pull (4)

20 a 80.4/6.5 = 12.4. (1) In iron (giant metallic), the delocalised electrons can move and carry thermal energy. (1) In sodium chloride, the ions are fixed in the lattice and cannot transmit much thermal energy. (1)

  b The giant covalent structure of SiO_2 consists of a network of strong covalent bonds. Simple molecular sulfur has weak intermolecular forces between molecules. (1)

 © HarperCollins*Publishers* Limited 2016

Chapter 3: Chemical reactions

Lesson 3.1 Elements and compounds

1 Element, compound, compound
2 Beryllium and chlorine
3 Potassium and bromine
4 Lead and iodine
5 Sodium oxide
6 The element iron and the compound carbon dioxide.
7 Hydrogen reacts with oxygen to form water. The oxygen had to be chemically removed from substance D. So substance D is a compound - zinc oxide.

Lesson 3.2 Atoms, formulae and equations

1 C S_8 Cl_2 elements as one type of atom
   CO_2 SO_3 compounds as two types of atom
2 a carbon and oxygen, sulfur, chlorine, sulfur and oxygen, carbon, carbon and hydrogen
   b 2, 1, 1, 2, 1, 2
   c 3, 8, 2, 4, 60, 14
3 Magnesium, sulfur, oxygen and carbon, hydrogen
4 LiCl, $MgCl_2$, K_2O
5 $2Na + Cl_2 \rightarrow 2NaCl$
6 $4Al + 3O_2 \rightarrow 2Al_2O_3$
7 $N_2 + 2 O_2 \rightarrow 2 NO_2$ D=1, E=2, F=2

Lesson 3.3 Moles

1 18 g
2 $3 \times (39 + 80) = 357$ g
3 $2 \times 6.02 \times 10^{23} = 1.204 \times 10^{24}$
4 a 28 g/mol
   b 81 g/mol
   c 84 g/mol
   d 132 g/mol
5 72/18 = 4 moles
6 a 2 moles
   b 4 moles of $H_2 = 4 \times 2 = 8$ g.

Lesson 3.4 Key concept: Conservation of mass and balanced equations

1 ZY
2 4
3 Na = 2, S = 2, O = 3
4 Al = 2, S = 3, O = 12
5 a d = 2, e = 1, f = 1, g = 2.
   b d = 1, e = 5, f = 3, g = 4.

Lesson 3.5 Key concept: Amounts in chemistry

1 $24 + 32 + (4 \times 16) = 120$
2 a $40 + ((14 + (3 \times 16)) \times 2) = 164$
   b $(2 \times 27) + ((32 + (4 \times 16)) \times 3) = 342$
3 40 g/mol
4 $2 \times 6.02 \times 10^{23} = 1.024 \times 10^{24}$
5 Moles $ZnCO_3 = 6.25/125 = 0.05$. Moles ZnO = 0.05 since 1:1 ratio. So mass ZnO = $0.05 \times 81 = 4.05$ g.

6 Moles $C_3H_8 = 660/44 = 15$. Moles CO_2 produced = $15 \times 3 = 45$ (1:3 ratio). Mass CO_2 = $45 \times 44 = 1,980$ g.

Lesson 3.6 Mass changes when gases are in reactions

1 0.7 g of mass lost as carbon dioxide gas.
2 0.7 g
3 0.9 g
4

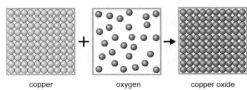

copper   oxygen   copper oxide

5 1.6 g
6 8 minutes. Graph is horizontal meaning no more mass is being lost so the reaction is finished. The acid was in excess and all the magnesium carbonate has been used up.
7

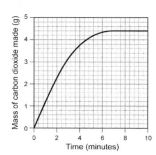

8 a Nitric acid. Zinc was left at the end of the reaction so was not the limiting reactant.
   b 3.62 g of zinc and 6.97 g of nitric acid.
9 Mass of O_2 gained = $2 \times 32 = 64$. Mass of CO_2 lost = $2 \times 44 = 88$. So $88 – 64 = 24$ g of mass lost overall.

Lesson 3.7 Using moles to balance equations

1 "Tonne" moles $MgCO_3 = 84/84 = 1$. 1:1 ratio. So mass of MgO = $1 \times 40 = 40$ tonnes.
2 Moles $Al_2O_3 = 204/102 = 2$.
   Moles Al = 108/27 = 4
3 By conservation of mass, mass of $O_2 = 204 - 108 = 96$ g.
   Moles $O_2 = 96/32 = 3$.
   Moles $Al_2O_3 = 2$.
   Moles Al = 4.
   $2 Al_2O_3 \rightarrow 4 Al + 3 O_2$

Lesson 3.8 Key Concept: Limiting reactants and molar masses

1 Rate = 50/23.5 = 2.13 cm3/s
2 a 33 cm3
   b 65 cm3
3 0.0125 g is half of 0.025 g. So half the volume of hydrogen will be collected - about 16.5 cm^3.

325   © HarperCollins*Publishers* Limited 2016

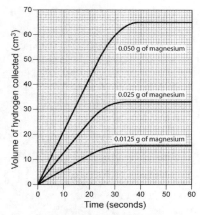

4 It is the reactant that limits the amount of product that can be formed.

5 The amount of product will quadruple.

6 3/24 = 0.125 moles Mg. Ratio Mg:$MgCl_2$ = 1:1. So mass $MgCl_2$ = 0.125 x 95 = 11.9 g

7 These are the formula masses of Mg and $MgCl_2$.

8 a 2 moles since the ratio of Mg:$MgCl_2$ = 1:1.

b 4 moles since ratio of Mg:HCl = 1:2.

Lesson 3.9 Amounts of substances in equations

1 Moles Mg = 6.0/24 = 0.25. 1:1 ratio. So 0.25 x 40 = 10 g.

2 Moles MgO = 2.0/40 = 0.05. 1:1 ratio. So 0.05 x 24 = 1.2 g.

3 Ratio C_3H_8:H_2O = 1:4. So 6:24. 24 moles H_2O.

4 Moles $ZnCO_3$ = 1.25/125 = 0.01. 1:1 ratio. So mass ZnO = 0.01 x 81 = 0.81 g.

5 Moles CuO = 7.95/79.5 = 0.1. 1:1 ratio. So mass $CuCO_3$ = 0.1 x 123.5 = 12.35 g.

Lesson 3.10 Endothermic and exothermic reactions

1 The temperature increased and heat was given out so this is an exothermic reaction.

2 Plants use the Sun's energy in the form of light to make glucose from carbon dioxide and water. The reaction takes in energy which is endothermic.

3 7.75 g.

4 Plot the graph. Volume on the x axis. Draw best fit straight lines through the points going up and going down. They should intersect at about 25 cm^3 / 31 $^\circ$C.

As the acid is added, the exothermic neutralisation produces heat and the temperature rises. The peak of temperature rise is where exactly the correct moles of acid has been added to exactly neutralise the alkali (in other words if the ratio of acid to alkali is 1:1 then the same number of moles of acid have been added to the alkali). This occurs at 24 cm^3 / 31.5 $^\circ$C. As further acid is added, the temperature decreases as the heat is just spread out in a greater mass of solution (no further reaction takes place).

Note that to accurately work out the end point of the neutralisation from the graph, draw best fit straight lines through the points going up and going down. They should intersect at about 24 cm^3 / 31.5 $^\circ$C.

5 To prevent heat loss - polystyrene is a good insulator.

6 Some suggestions:

• Change the starting concentration of the alkali and repeat. Keep the same concentration of acid and same volumes.

• Change the concentration of the acid and repeat. Keep the same concentration of alkali and the same volumes.

• Repeat the experiment with other strong and weak acids and alkalis. Keep the same volumes.

7 1. Heat loss to the surrounding. Polystyrene is not a perfect insulator and there is no lid. So heat energy is lost to the surroundings. The apparatus can be modified by adding a lid with a hole for the thermometer.

2. The temperature data suggest that the thermometers have an uncertainty of plus or minus 1 $^\circ$C. Since the temperature changes are not large, an accurate thermometer would be needed (e.g. one that reads to 0.2 $^\circ$C).

3. The end point may occur in between the volumes of acid added. So to accurately work out the maximum temperature rise, two best fit straight line should be drawn. One as the temperature rises and one as it falls. Then extrapolate the lines. The maximum temperature rise is where the lines intersect. This is the point of neutralisation.

Lesson 3.11 Reaction profiles

1 Heat energy from the lit fuse.

2

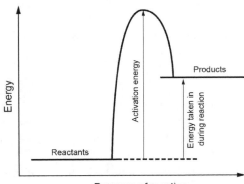

3

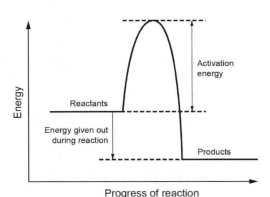

4 2 H-H and 1 O=O reactant bonds have to be broken. This requires energy (endothermic). 4 O-H product bonds (2 H_2O) are formed which releases energy (exothermic). Reaction is exothermic because less energy is required to break reactant bonds than is released forming product bonds.

Lesson 3.12 Energy change of reactions

1 The reaction is endothermic as the temperature goes down.
Reactant bonds have to be broken. This requires energy (endothermic). Product bonds are formed which releases energy. However, overall reaction is endothermic because more energy is required to break reactant bonds than is released forming product bonds.

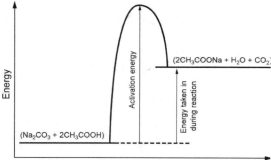

2 2076

3 Reactant bonds: 4 x C-C + 12 x C-H + 8 x O=O = 10,316
Product bonds: 10 x C=O + 12 x O-H = 13,640
Energy change = 10,316 - 13,640 = -3,324 kJ/mol

Lesson 3.13 Maths skills: Recognise and use expressions in decimal form

1 C

2 $37.2^{\circ}C$

3 0.001

4 0.00001

5 100 000 000

6 10 080 J

7 3500 J

Lesson 3.14 Oxidation and reduction in terms of electrons

1 Increasing reactivity: Copper, iron zinc, magnesium

2 Tin displaces copper (ions) but not zinc or iron(II). So tin is more reactive than copper but less reactive than zinc and iron.

3 $Mg + NiSO_4 \rightarrow MgSO_4 + Ni$

4 $Mg + Cu^{2+} \rightarrow Mg^{2+} + Cu$

5 $Mg \rightarrow Mg^{2+} + 2\ e^-$
$Fe^{2+} + 2\ e^- \rightarrow Fe$

6 Aluminium is oxidised - it has lost 3 electrons. Cr^{3+} has been reduced since it has gained 3 electrons

7 a $Mg \rightarrow Mg^{2+} + 2e^-$ Magnesium has lost electrons and has been oxidised. Ag+ + $e^- \rightarrow$ Ag   Silver has gained electrons and has been reduced.

b $Mg + 2\ Ag^+ \rightarrow Mg^{2+} + 2\ Ag$

Lesson 3.15 Electron transfer, oxidation and reduction

1 Sodium loses one electron and transfers it to chlorine which gains one electron. Both end up with the stable electron arrangement of a noble gas.

2 They do not need to lose or gain electrons from their outer shell.

3 a a) Zinc is oxidised since oxygen is added. In terms of electrons, zinc loses 2 electrons and is oxidised to Zn^{2+}. Oxygen gains electrons and is reduced to oxide ions, O^{2-}.

b Iron(II) oxide is reduced to iron since oxygen is removed. Fe^{2+} gains 2 electrons which is reduction.

4 a Silver gains one electron and is reduced (Ag^+ + $e^- \rightarrow$ Ag).

b Electrons are lost by the hydroxide ions(OH^-). This is oxidation (4 $OH^- \rightarrow O_2 + 2\ H_2O + 4e^-$).

5 Both calcium and magnesium have two outer shell electrons. However, calcium has more shells and the outer shell is further from the nucleus.

Lesson 3.16 Neutralisations of acids and salt production

1 potassium hydroxide + hydrochloric acid $\rightarrow$ potassium chloride + water

2 copper carbonate + nitric acid $\rightarrow$ copper nitrate + water + carbon dioxide

3 Zinc nitrate

4 Copper sulfate

5 $ZnCl_2$

6 $Cu(NO_3)_2$

7 $CaCO_3 + 2\ HNO_3 \rightarrow Ca(NO_3)_2 + H_2O + CO_2$

Lesson 3.17 Soluble salts

1 If the solid is in excess, all of the acid will have reacted and totally converted into the soluble salt. Also, if some acid was unreacted, it would contaminate the salt solution. The excess solid is easy to filter off.

2 It is heated to make the salt solution more concentrated. This means that the crystals form more effectively (saturated solution).

3 copper oxide + sulfuric acid → copper sulfate + water

4 potassium hydroxide + hydrochloric acid → potassium chloride + water

5 $ZnCO_3 + 2 HCl → ZnCl_2 + H_2O + CO_2$

6 $MgO + 2 HNO_3 → Mg(NO_3)_2 + H_2O$

Lesson 3.18 Reaction of metals with acids

1 Magnesium chloride

2 Iron + sulfuric acid → iron(II) sulfate + hydrogen

3 $Zn + 2 HCl → ZnCl_2 + H_2$

4 $2 Fe + 6 HCl → 2 FeCl_3 + 3 H_2$ (or half the ratios)

5 $Zn → Zn^{2+} + 2 e^-$
$2 H^+ + 2 e^- → H_2$

6 Iron is oxidised since it has lost electrons. Hydrogen ions, H+, have been reduced since they have gained electrons.

7 Balanced equation: $2 Al + 3 H_2SO_4 → Al_2(SO_4)_3 + 3 H_2$
Half equations:
$Al → Al^{3+} + 3e^-$
$2 H^+ + 2e^- → H_2$ (or $3 H^+ + 3e^- → 1½ H_2$)

Lesson 3.19 Practical: Preparing a pure, dry sample of a soluble salt from an insoluble oxide or carbonate

1 Magnesium carbonate and sulfuric acid.

2 Magnesium mass is measured with a balance. Sulfuric acid solution is measured with a measuring cylinder (though a pipette would be more accurate).

3 For magnesium powder: g. For sulfuric acid solution: cm^3 (ml).

4 Safety spectacles, gloves and a laboratory coat should be worn.

5 The sulfuric acid was the limiting reagent. There was excess magnesium carbonate.

6 Filtration.

7 The salt solution is heated to evaporate some water and concentrate the solution. The solution is then left so that the rest of the water evaporates and the salt crystallises.

8 It can be washed with a small volume of the solvent (water). It can then be dried. Also, recrystallisation can be carried out. This involves dissolving the solid salt in the minimum amount of hot water. Then leaving it to cool and crystallise again. The impurities remain in the solution. This can be repeated several times depending on the purity required.

9 Measure the melting point and compare to a data book value. When impurities are present,

the melting point will be lower than the data book value. The purer the salt, the closer to the data book value.

10 When impurities are present, the melting point will be lower than the data book value. The purer the salt, the closer to the data book value.

11 It is not possible to tell whether there is a trend in melting points e.g. 41.6 °C, 41.4 °C, ? °C. A third measurement would have confirmed whether the readings were randomly scattered about the "true" value or part of a trend. Generally, when melting points are measured, the first one is rough and the second accurate. But it is not possible to say whether this is true with the data provided.

12 $MgCO_3 + H_2SO_4 → MgSO_4 + H_2O + CO_2$

13 Moles $MgSO_4$ = 6/120 = 0.05.
Mass $MgCO_3$ = 0.05 x 84 = 4.2 g
Mass H_2SO_4 = 0.05 x 98 = 4.9 g

14 Sulfuric acid is the limiting reagent. So an excess of $MgCO_3$ is added to ensure all of the sulfuric acid is reacted. If acid was in excess, the salt solution would be acidic, making purifying the salt more difficult.

15 Washing the crystals with water may dissolve some of the crystals. Some of the salt will be left in solution after a recrystallisation. So the percentage yield will decrease as purity increases.

16 For Alex's readings, if +0.2 °C is added to 41.6 °C, it comes to 41.8 °C. So it is still 0.9 °C from the data book value. It is a similar story for Sam's readings. Alex's readings are more accurate since they are closer to the data book value.

Lesson 3.20 pH and neutralisation

1 Hydrogen ion / H^+

2 An acid produces hydrogen ions in aqueous solution. Alkalis produce hydroxide ions in aqueous solution.

3 Red and pH 1. Green and pH 7. Purple and pH 14.

4 7

5 a $3 H^+ / PO_4^{3-}$
b Phosphoric acid: high concentration of H^+ ions. Sodium hydroxide: high concentration of OH^- ions.

6 $H^+ (aq) + OH^- (aq) → H_2O (l)$

7 a 8 - 10 (alkaline). There is an excess of OH^- and a low concentration of H^+.

b Magnesium hydroxide contains OH^- ions and hydrochloric acid in the stomach H^+ ions. These react together to form water. The acid is therefore neutralised.

328

© HarperCollins*Publishers* Limited 2016

Lesson 3.21 Strong and weak acids

1   Hydrogen ions and sulfate ions.

2   The citric acid molecule is not fully ionised. This means that pH is higher than for a strong acid - the concentration of hydrogen ions is lower.

3   Students' own answers.

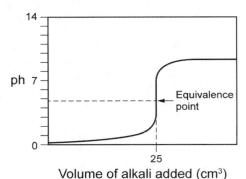

Volume of alkali added (cm^3)

4   60 g/dm^3

Lesson 3.22 Make order of magnitude calculations

1   The east side. It has regions where the pH of rainfall is less than 4.3

2   4000 - 900 = 3100 thousand tonnes

3   1990 = 6,950 thousand tonnes. 2011 = 2100 thousand tonnes. So 2100/6950 = 0.30 relative decrease.

4   36.52 g/dm^3

5   2.3 g/dm^3

6   6

7   pH 7 so 1 x 10^{-7}

Lesson 3.23 Practical: Investigate the variables that affect temperature changes in reacting solutions, such as acid plus metal, acid plus carbonates, neutralisations, displacement of metals

1   Different reactants will most likely produce different energy changes when they react. So temperature changes are likely to be different. Changing concentration will affect the temperature change. Changing concentration changes the number of "reactions" per volume and therefore the energy change per volume.

2   Type of metal; Type of ions in solution; Concentration of ions in solution; Size of pieces of metal

3   It is necessary to see how experimental results change due to one variable changing. If two are changed at once, it would not be possible to determine the contribution of each variable to the change.

4   The identity of the metal and the acid. The concentration of the acid. The volume of the acid solution. The container for the reaction. The procedure for the experiment.

5 a   7.2, 5.4, 11.1, 3.7.

b   11.1 °C. It should be discarded.

c   As the concentration of the acid increases the temperature change increases when reacting with carbonates. So their hypothesis was correct.

6   The evidence supports the original hypothesis: "As the concentration of the acid increases the temperature change increases when reacting with carbonates". This is clearly seen in the data over 2 trials. They could have gathered more evidence to support their conclusion e.g. use a wider range of concentrations.

7   If lumps of carbonate are used, there may not be the same number or size between trials. This will affect the surface area and therefore rate. Rate can affect heat loss. It cannot be assumed that this will not have an affect on the temperature change.

8   If heat is lost, the temperature increases will be smaller. This will affect the validity of the results.

Lesson 3.24 The process of electrolysis

1   In solid ionic compounds, the ions are fixed in the lattice and cannot move. When molten or aqueous, the ions can move and carry charge. Therefore, electricity can flow.

2   Copper at the cathode. Chlorine at the anode.

3   Aluminium ions, Al^{3+} are positive and attracted to the negative cathode.

4   $Cu^{2+} + 2 e^- \rightarrow Cu$

   $2 Cl^- \rightarrow Cl_2 + 2 e^-$

5 a $2 Br{-} \rightarrow Br_2 + 2e^-$

b   Mg.

c   Magnesium is above hydrogen in the reactivity series. Magnesium is too reactive to be discharged. So the hydrogen ions in water are discharged.

Lesson 3.25 Electrolysis of molten ionic compounds

1   Cathode. Positive ions are attracted to the negative cathode.

2   Anode. Negative ions migrate to the positive anode.

3   Sodium at the cathode since sodium ions are positive and the cathode is negative. Bromine at the anode since bromide ions are negative.

4   Cathode: $Cu^{2+} + 2 e^- \rightarrow Cu$

   Anode: $2 Br{-} \rightarrow Br_2 + 2 e^-$

Lesson 3.26 Electrolysis of aqueous solutions

1   Silver would be discharged in preference to hydrogen as it is lower in the reactivity series.

2   Copper would still be discharged at the cathode as it is lower in the reactivity series than sodium. Chlorine would be discharged in preference to oxygen, (provided it was present in sufficient concentration) as chlorine is lower in the reactivity series than hydroxide ions.

3 At the cathode: At the anode :
$2H^+ + 2e^- \rightarrow H_2$ $4OH^- - 4e^- \rightarrow 2H_2O + O_2$

4 At the cathode: At the anode:
$Cu^{2+} + 2e^- \rightarrow Cu$ $2Cl^- - 2e^- \rightarrow Cl_2$
The discharge of copper is a reduction reaction as electrons are gained. The discharge of chlorine is an oxidation reaction as electrons are lost.

5 $2H^+ + 2e^- \rightarrow H_2$ $4OH^- -4e^- \rightarrow 2H_2O +O_2$

Lesson 3.27 Practical: Investigating what happens when aqueous solutions are electrolysed using inert electrodes

1 The cathode (if the metal is less reactive than H^+) since positive metal ions are attracted to the negative cathode.

2 Hydrogen. Insert a lighted splint into the gas. A popping sound indicates the presence of hydrogen gas.

3 No they don't. The order of increasing reactivity for the metals is:
Silver, copper, zinc, sodium
The less reactive the metal the easier it is to deposit on the cathode.
The products formed (going down the column) were: Hydrogen, copper, hydrogen, silver.

4 The reactivity series.

5 The order of increasing reactivity for the metals is:
Silver, copper, H^+, zinc, sodium
The products formed (going down the column) were: Hydrogen, copper, hydrogen, silver.
Copper and silver are less reactive than H^+ so are discharged.
Sodium and zinc are more reactive than H^+ so H^+ is discharged as H_2.

6

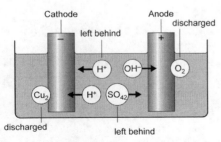

7 Zinc Sulfate: Oxygen. Copper sulfate: Oxygen. Sodium chloride: Chlorine. Silver nitrate: Oxygen.

8 The order of increasing reactivity for the metals is:
Silver, copper, H^+, zinc, sodium
So Cu^{2+} and Ag^+ more readily accept electrons (and are reduced) than H^+ ions. Sodium and zinc will not accept electrons in preference to H^+ since they are more reactive than H^+.

The overall conclusion is that if the metal is less reactive than H+, it will deposit on the cathode.

9 The overall conclusion is that if the metal is less reactive than H+, it will deposit on the cathode. This contradicts the original hypothesis. However, the evidence supports the alternative hypothesis "Does the ability to deposit on an electrode link to the reactivity of the metal ion in solution". This only applies to inert electrodes. If other electrodes were used (e.g. metal electrodes) then more experiments would have to be carried out.

End of chapter questions

1 The lighted splint supplies the activation energy. (1)

2 Which of these statements about a fuel cell is false?

3 Sports injury packs and self-heating cans

4 b endothermic (1)

5 2 g. It is easier to make a salt when the acid is the limiting reagent. (1)

6 Initially red due to the hydrochloric acid. (1) As sodium carbonate reacts, the colour will eventually change through orange to green and then blue. (1)

7 C 20 28 34 36 37 exothermic reaction
cm^3 0 15 20 24 26 rate of reaction by gas volume
$^{\circ}$C 21 18 16 15 14 endothermic

8 Magnesium oxide (1)

9 Hydrogen (1)

10 Hydrogen and Iodine (2)

11 2 marks for named reactants and balanced symbol equation. 1 mark for named reactants and word equation
Use zinc solid and aqueous sulfuric acid (could use zinc carbonate / zinc oxide).
$Zn + H_2SO_4 \rightarrow ZnSO_4 + H_2$
2 marks for all 4 steps of the method. 1 mark for 2 or 3 steps correct.
• Zinc is added to the sulfuric acid until no more of the zinc reacts and it is in excess.
• The excess zinc is filtered off leaving a solution of the zinc sulfate.
• The zinc sulfate solution is then heated a little to concentrate it.
• The concentrate is left to cool and crystallise to produce solid zinc sulfate

12 1 mark for an explanation of the reasoning.
Zinc displaced copper from solution so zinc is more reactive than copper.
X is less reactive than zinc since it did not displace it from solution.
X is more reactive than copper since it displaced it from solution.

So increasing reactivity order: Copper, X, zinc. (1)

13   a) 6.02×10^{23}

14   Produces water that is non toxic and can be used for other things (1)

15   0.25 moles (2)

16   Level 3: Correctly identifies at least three reactants and the salt product. At least one correct symbol equation. At least two correct observations. (5-6) Level 2: Correctly identifies at least two reactants and the salt product. At least one correct equation (word or symbol). At least one correct observation. (3-4) Level 1: Correctly identifies at least one reactant and the salt product. (1-2) Nothing written worthy of credit. (0)

Indicative content

Reactants and salt product

- Reactant: Calcium / Ca. Product: Calcium sulfate / $CaSO_4$.

- Reactant: Calcium oxide / CaO. Product: Calcium sulfate / $CaSO_4$.

- Reactant: Calcium hydroxide / $Ca(OH)_2$. Product: Calcium sulfate / $CaSO_4$.

- Reactant: Calcium carbonate / $CaCO_3$. Product: Calcium sulfate / $CaSO_4$.

Word Equations

- Calcium + sulfuric acid $\rightarrow$ calcium sulfate + hydrogen

- Calcium oxide + sulfuric acid $\rightarrow$ calcium sulfate + water

- Calcium hydroxide + sulfuric acid $\rightarrow$ calcium sulfate + water

- Calcium carbonate + sulfuric acid $\rightarrow$ calcium sulfate + water + carbon dioxide.

Symbol Equations

- $Ca + H_2SO_4 \rightarrow CaSO_4 + H_2$

- $CaO + H_2SO_4 \rightarrow CaSO^4 + H_2O$

- $Ca(OH)_2 + H_2SO_4 \rightarrow CaSO4 + 2\ H_2O$

- $CaCO_3 + H_2SO4 \rightarrow CaSO_4 + H_2O + CO_2$

Observations

- Fizzing when calcium added to acid.

- Calcium disappears / dissolves when added to acid.

- Fizzing when calcium carbonate added to acid.

- Calcium carbonate disappears / dissolves when added to acid.

17 a   A 20; B 100; C 74; D 2; E 100

   b C_7H_{16} (4)

18 a $SnO_2 + C \rightarrow Sn + CO_2$ (1)

   b   Oxygen is removed from the tin(IV) oxide which is reduction. Oxygen is added to the carbon which is oxidation. (1)

   c   Tin is below carbon in the reactivity series. (1) Since it is less reactive than carbon, it can be reduced by carbon. Electrolysis is used for metals that are more reactive than carbon. (1)

19   It uses only hydrogen and oxygen to produce water which is non-polluting.

20   **Level 3:** Correctly calculates energy change for methane and propane (one ecf allowed). At least three points for bond energy explanation. (5-6)

   Level 2: Correctly calculates energy change for methane or propane (one ecf allowed). At least two points for bond energy explanation. (3-4)

   Level 1: Correctly calculates reactant or product bond energy for methane or propane reactions. At least one point for bond energy explanation. (1-2) Nothing written worthy of credit. (0)

Indicative content

Calculating Energy Change
for Methane

- Reactant bond energy = $(4 \times 414) + (2 \times 498)$ = 2652

- Product bond energy = $(2 \times 806) + (4 \times 465)$ = 3472

- Overall energy change = 2652 – 3472 = –820 kJ/mol. Negative sign not essential.

Calculating Energy Change for Propane

- $C_3H_8 + 5\ O_2 \rightarrow 3\ CO_2 + 4\ H_2O$

- Reactant bond energy = $(8 \times 414) + (5 \times 498) + (2 \times 346)$ = 6494

- Product bond energy = $(6 \times 806) + (8 \times 465)$ = 8556

- Overall energy change = 6494 – 8556 = –2062 kJ/mol. Negative sign not essential.

- Propane gives out (1242 KJ/mol) more energy than methane.

Explaining Energy Change for
Methane

- Bond breaking is endothermic. Bond making is exothermic.

- C–H bonds in methane and O=O bonds in oxygen are broken. This takes in energy.

- C=O bonds in carbon dioxide and O–H bonds in water are formed. This gives out energy.

- Less energy is required to break the reactant bonds in methane than is given out when product bonds form.

- The reaction is exothermic because more energy is released in forming new bonds than is taken in breaking reactant bonds. (6)

© HarperCollins*Publishers* Limited 2016

Chapter 4: Predicting and Identifying Reactions and Products

Lesson 4.1 Exploring Group 0

1. Neon 10, Argon 18. Argon has the higher atomic mass.

2. $56^{\circ}C$ (temperature greater than $-107^{\circ}C$)

3. As the atomic mass goes up the boiling point increases.

4. The diameter increases from helium to radon as extra electron shells are added. As the diameter increases, so does the boiling point, because the larger the diameter the more energy needed to move atoms away from each other.

5. One ring with 2 electrons. The outer shell has no more space to accept electrons, and it is a stable configuration so it doesn't want to lose electrons either.

6. Ar is 2,8,8. The atom has no spaces to accept electrons so is unreactive. Argon atoms have a larger diameter and have more electrons. Therefore the forces between argon atoms is greater. More energy is needed to overcome these stronger forces so the boiling point is higher.

7 a. $Xe + 2 F_2 \rightarrow XeF_4$

   b. Noble gases are unreactive because they have a full outer shell which makes them stable.

Lesson 4.2 Exploring Group 1

1. Less dense than water.

2. Hydrogen

3. Test with universal indicator which turns blue.

4. Down the group the elements are more reactive.

5. Rubidium reacts violently with water. Reactivity increases down Group 1.

6. It has one electron in its outer shell, which if it loses it becomes a more stable ion.

7. One electron ring with two electrons, square bracket around the ring, positive charge top right hand side outside bracket.

8 a. $4 K + O_2 \rightarrow 2 K_2O$

   b. Potassium ion, K^+: 2,8,8 Oxide ion, O_2^-: 2,8

   c. KOH / potassium hydroxide.

Lesson 4.3 Exploring Group 7

1. $2Na + Br_2 \rightarrow 2NaBr$

2. LiF

3. 35   liquid   254   solid

4. The picture shows a dark solid.

5. As the molecular mass increases, the boiling point increases.

6. Chlorine and bromine, they are higher in the group and more reactive so will displace iodine from a solution of potassium iodide.

7. $Cl_2 + 2KI \rightarrow 2KCl + I_2$

8 a. Solid.

   b. $Cl_2 + 2 NaAt \rightarrow At_2 + 2 NaCl$

   c. Astatine will not react with sodium iodide. Astatine is less reactive than iodine.

Lesson 4.4 Transition metals

1 a. melting point goes up to Fe then down as the atomic number increases

   b. density increases as atomic number increases

2. Zinc is predicted have melting point of about 600 $^{\circ}C$ and a density of 9 g/dm^3

3. They remain unchanged at the end of the reaction so can be used again.

4. Iron(III) ions ($Fe3+$) were reduced to iron(II) ($Fe2+$) ions.

5. 3

6. CuO - $Cu2+$ (and O_2^-)
   Cu2O - $Cu+$ (and O_2^-)

Lesson 4.5 Reaction trends and predicting trends

1. It is lower down in the Group. Its outer electron is further away from the nucleus than the outer electron of sodium (so experiences less 'pull' from the nucleus), so is lost more easily.

2. Astatine has more electron shells than chlorine. The electron to be captured is therefore further from the nucleus so the attraction is less.

3. Neon has a stable outer shell of electrons, so neither gains or loses them so is unreactive.

4. Sodium only has to lose one electron to react whereas aluminium needs to lose three electrons.

5. Rubidium will react violently with water, acid and oxygen, as it is further down group 1 than potassium so will lose its outer electron more easily. It will form an alkaline oxide as it is a metal.

6. Strontium will react vigorously with acid, with greater vigour than calcium, as it is lower down the group, and so loses its electrons more easily, as there is less 'pull' on them by the nucleus.

7 a. $H_2 + Cl_2 \rightarrow 2 HCl$

   b. Fluorine will react more violently because it is more reactive than chlorine. Reactivity decreases down Group 7.

Lesson 4.6 Reactivity series

1. Sodium is more reactive than magnesium when placed in water. Magnesium is more reactive than zinc when reacted with acids. So sodium must be more reactive than zinc.

2. The single outer shell electron in potassium is further from the nucleus (potassium has one more shell than sodium). So there is less attraction from the nucleus and the electron is more easily lost.

3. $Zn + Cu^{2+} \rightarrow Zn^{2+} + Cu$

Lesson 4.7 Test for gases

1. Hydrogen + oxygen $\rightarrow$ water

2. Three conditions for combustion – source of ignition, fuel and oxygen. When the splint goes into oxygen this increases the combustion. The splint is the fuel, the glow is the ignition source and the increase in oxygen increases the combustion.

3. Calcium hydroxide + carbon dioxide $\rightarrow$ calcium carbonate + water

4 As more CO_2 is bubbled through the cloudy limewater the calcium carbonate reacts to form calcium hydrogen carbonate which is soluble in water and so the solution goes clear again

5 Yellow Green

6 Bromine

7 Iodine

Lesson 4.8 Metal hydroxides

1 Add sodium hydroxide solution to copper(II) sulfate solution in a test tube. A gelatinous blue precipitate of copper(II) hydroxide is formed.

2 The metal ion is iron(II) / Fe^{2+}. The precipitate is iron(II) hydroxide.

3 Calcium ions / Ca_{2+}

4 A: Copper sulfate. Copper forms a blue precipitate with NaOH.

B: Magnesium sulfate. Magnesium forms a white precipitate with NaOH.

5 $MgCl_2$ (aq) + 2 NaOH (aq) → $Mg(OH)_2$ (s) + 2 NaCl (aq)

6 Fe(III) is Fe^{3+}. To balance charge, it needs 3 OH- ions to make $Fe(OH)_3$.

7 $FeCl_3$ (aq) + 3 NaOH (aq) → $Fe(OH)_3$ (s) + 3 NaCl (aq)
Fe^{3+} (aq) + 3 OH- (aq) → $Fe(OH)_3$ (s)

Lesson 4.9 Test for anions

1 Pass the carbon dioxide into limewater. It will turn milky if carbon dioxide is present.

2 Add some acid (e.g. nitric acid) to the carbonate. If it bubbles, pass the gas into limewater. If the limewater turns milky, it is a carbonate.

3 A yellow precipitate.

4 It has dissolved salts in it e.g. sodium chloride, potassium chloride etc.

5 Barium chloride solution.

6 KI (aq) + $AgNO_3$ (aq) → AgI (s) + KNO_3 (aq)

7 $CaCl_2$ (aq) + 2 $AgNO_3$ (aq) → 2 AgCl (s) + $Ca(NO_3)_2$ (aq)

Lesson 4.10 Flame tests

1 Blue

2 It needs to be clean to avoid contamination from other metal compounds. It needs to be moistened so that the metal compound sticks to the wire.

3 Copper chloride.

4 It could be contaminated with a different ionic compound which is interfering with the test.

5 Calcium carbonate - red.
Copper(II) chloride - green
Potassium chloride - lilac.

6 Flame colours may be contaminated from previous tests. Some metals do not give colours. Different metal ions may give very similar colours.

7 Potassium chloride will give a lilac flam and calcium chloride a red flame.

8 a a Calcium carbonate / $CaCO_3$.

b $CaCO_3$ (s) + 2 HCl (aq) → $CaCl_2$ (aq) + H_2O (l) + CO_2 (g)

9 a Sodium.

b Sodium chloride.

c 2 Na (s) + Cl_2 (g) → 2 NaCl (s)

Lesson 4.11 Instrumental methods

1 Methane

2 It is more rapid, more accurate and more sensitive.

3 Criminals sometimes leave traces of their DNA at crime scenes. The DNA can be chemically cut up and analysed. A DNA fingerprint is (almost) unique to an individual. So the DNA pattern can be compared to patterns in a criminal database or compared to a sample taken from a suspect.

4 10 000 000/400 = 25 000

5 DNA fragments separated and identified by electrophoresis.

6 Pentane has a molecular mass of 72. The peak at 72 is pentane itself (unfragmented).

Lesson 4.12 Practical: Use chemical tests to identify the ions in unknown single ionic compounds

1 Bubble the gas through limewater. If it turns milky/cloudy it is carbon dioxide.

2 Dissolve each salt in water and add NaOH solution using a teat pipette. Wear safety spectacles, gloves and a lab coat. The following hydroxide precipitates will be seen:
Copper: Blue precipitate.
Iron(II): Light green precipitate.
Iron(III): Brown precipitate.

3 Lithium: Crimson. Sodium: Yellow. Potassium: Lilac.

4 White hydroxide precipitates are formed in both cases.

5 Carry out a flame test. Calcium ions will give a red flame (no colour for magnesium ions).

6 A white hydroxide precipitate forms. However, this redissolves in excess NaOH, which distinguishes aluminium ions from calcium and magnesium.

7 $CaCO_3$ / calcium carbonate.

8 $FeCl_3$ / iron(III) chloride

9 NaBr / sodium bromide.

10 K_2SO_4 / potassium sulfate

11 CuI / copper(I) iodide

End of chapter questions

1 Hydrogen (1)

2 Helium (1)

3 Yellow, sodium; Crimson, lithium; Lilac potassium (2)

4 Sodium. (1)

5 a F, G, D, E (1)

b Ease with which the metal loses electrons to form a positive ion. The less energy required for this the greater the reactivity. (1)

6 Blue precipitate (1)

7 Barium chloride, white precipitate. (1)

8 2 marks for all 3 correct. 1 mark for 2 correct.
Cream - bromide
Yellow - iodide
White – chloride

9 Sodium hydroxide, NaOH (1)

10 7 (1)

11 Group 1 metals become more reactive as you go down table because the single electron they lose from the outer shell is further away from the nucleus so there is less pull from the nucleus (2)

12 a X. Boiling points increase down the group. The closer to the top of the group the lower the atomic number. (1)

  b Reactivity decreases down the group. So X is more reactive. (1)

13 Iron(III) oxide (1)

14 1 mark for an explanation of the reasoning.
Zinc displaced copper from solution so zinc is more reactive than copper
X is less reactive than zinc since it did not displace it from solution.
X is more reactive than copper since it displaced it from solution.
So increasing reactivity order: Copper, X, zinc. (1)

15 Flame test - red colour. (1)
Add sodium hydroxide and a white precipitate forms. (1)

16 D: 17.5 °C. (1) E: 2.33 (1).

17 **Level 3**: At least 9 valid points. Answer has logical structure and the discussion of chemical and physical properties is balanced. (5-6)

  Level 2: At least 6 valid points. Answer has logical structure in part. (3-4)

  Level 1: At least 3 valid points. Answer lacks logical structure. (1-2)

  Nothing written worthy of credit. (0)

Indicative content

Similarities
- Both are chemically reactive.
- Both have the same number of shells / energy levels (3 shells).

Differences in physical properties
- Sodium is a metal, chlorine is a non–metal.
- Sodium exists as a metallic solid. Chlorine exists as a covalent gaseous molecule, Cl_2 (diatomic).
- Sodium has a higher boiling point and melting point than chlorine.
- Sodium can form alloys with other metals. Chlorine cannot form alloys.
- Sodium is a good electrical and thermal conductor. Chlorine is a poor electrical and thermal conductor.

Difference in chemical properties
- Sodium reacts with non–metals. Chlorine reacts with non–metals and metals.
- Sodium forms positive sodium ions, Na^+, when it reacts with non–metals. An ionic compound is formed.
- Cl forms negative halide ions, Cl^-, when it reacts with metals. An ionic compound is formed.
- Chlorine reacts with non–metals to form molecular covalent compounds.

- Chlorine can undergo displacement reactions with halide ions.
- Differences in the way sodium and chlorine react with water e.g. sodium fizzes.
Electronic structure

18 Increasing order of reactivity:
Cu, Fe, Z, Mg (1)
1 marks for at least 2 of the following points:
Mg displaces the ions of the other metals from their solutions so it is most reactive.
X displaces iron and copper ions from their solutions so is more reactive than Fe and Cu.
Fe displaces copper ions from its solution so is more reactive than Cu.
Cu does not displace any of the others so is least reactive.

19 The precipitate for each of the ions will be mixed up. So the colour will not be easy to identify. (1)

20 The silver nitrate test identifies the chloride ion. (1) Magnesium, calcium and aluminium ions all give white precipitates with NaOH so this test only narrows the identity down. (1) If the precipitate doesn't redissolve, then it is not an aluminium ion. A red flame colour indicates calcium ions are present. Magnesium ions do not give a colour in a flame test. (1)

21 The electron pattern shows one electron in the outer shell so the element is from group 1. It loses an electron to form an ion and is therefore a metal, and it reacts violently because the electron that is lost is a long way from the nucleus and so is readily lost (4)

22 Dissolve the ionic compound in water and divide in two. Add sodium hydroxide solution. If a gelatinous light green precipitate is seen, iron(II) ions are present. (1) Add silver nitrate solution. If a cream precipitate is seen, bromide ions are present. (1)

23 a carbon dioxide

  b $CaCO_3 + 2HCl – CaCl_2 + CO_2 + H_2O$
  $CaCO_3 – CaO + CO_2$

  c Lithium produces a crimson flame very closely matched to the red flame calcium. Another test would identify Ca or Li. (4)

© HarperCollins*Publishers* Limited 2016

Chapter 5: Monitoring and Controlling Chemical Reactions

Lesson 5.1 Concentration of solutions

1 a) 200 g/dm^3.
    b) 20 g/dm^3.
    c) 160 g/dm^3
    So b, c, a.

2 a) 32 g/dm^3 b) 12.8 g/dm^3 c) 12.8 g/dm^3

3 4.2 x (250/1000) = 1.05 g

4 5.4 x (35/100) = 1.89 g

5 1/2 = 0.5 dm^3

6 0.18/0.6 = 0.3 mol/dm^3

7 Moles = (500/1000) x 3 = 1.5

8 4.90/98 = 0.05 moles H_2SO_4. Concentration = 0.05 x (1000/200) = 0.25 mol/dm^3.

9 a 0.0250 x (1000/125) = 0.200 mol/dm^3.

  b 0.200 x 63 = 12.6 g/dm^3.

10 a 18.25/36.5 = 0.5 mol/dm^3.
  b 0.500 x 6.02 x 10^{23} = 3.01 x 10^{23}.

Lesson 5.2 Using concentrations of solutions

1 To the bottom of the curve (meniscus). 42.5 cm^3

2 Pipette delivers a fixed volume. The burette delivers a variable volume.

3 Because it is used to give a rough idea of the end point. It is not meant to be accurate.

4 26.8 / 26.9 / 26.7. Average titre = 26.8 cm^3

5 Moles hydrochloric acid = (23.8/1000) x 0.11 = 0.002618. Ratio HCl:NaOH = 1:1. Concentration NaOH = 0.002618 x (1000/25) = 0.105 ml/dm^3

Lesson 5.3 Amounts of substance in volumes of gases

1 44 g

2 (8.8/44) x 24 = 4.8 dm^3

3 Moles NO_2 = 46/46 = 1.
    Moles Kr = 84/84 = 1.
    1 mole (molecular/formula mass) of any gas occupies 24 dm^3 at rtp.

4 1.5 x 24 = 36 dm^3

5 Moles N_2 = 7/28 = 0.25. Volume = 0.25 x 24 = 6 dm^3

6 Ratio C_3H_8:H_2O = 1:4. Therefore 1.5:6 dm^3. So 6 dm^3 of H_2O is produced.

7 C_5H_{12} + 8 O_2 → 5 CO_2 + 6 H_2O
    Ratio C_5H_{12}:CO_2 = 1:5. So for 2 dm^3 of C_5H_{12}, 10 dm^3 of CO_2 is formed.

8 Moles N_2 = 42/28 = 1.5. So 4.5 moles H_2 needed (1:3 ratio). Volume = 4.5 x 24 = 108 dm^3.

Lesson 5.4 Key concept: Percentage yield

1 Loss in filtration; loss due to evaporation; loss in transferring liquids

2 30%

3 60%

4 63g

5 a 85%
  b Loss in filtration

6 a 35.5g
  b 18.2g

Lesson 5.5 Key concept: Amounts in chemistry

1 24 + 32 + (4 x 16) = 120

2 a 40 + ((14 + (3x16)) x 2) = 164
  b (2 x 27) + ((32 + (4 x16)) x 3) = 342

3 40 g/mol

4 2 x 6.02 x 10^{23} = 1.024 x 10^{24}

5 Mass NaOH = 0.5 x 40 = 20 g

6 Moles H_2 = 4/2 = 2. 1 mole of any gas occupies 24 dm^3 at rtp. So 2 x 24 = 48 dm^3

7 Moles $ZnCO_3$ = 6.25/125 = 0.05. Moles ZnO = 0.05 since 1:1 ratio. So mass ZnO = 0.05 x 81 = 4.05 g.

8 Moles C_3H_8 = 660/44 = 15. Moles CO_2 produced = 15 x 3 = 45 (1:3 ratio). Mass CO_2 = 45 x 44 = 1,980 g.

Lesson 5.6 Atom economy

1 100 % (only one product)

2 b) Greater than 0 but less than 100 %

3 Method 1: (568/676) x 100 = 84 %
    Method 2: (568/624) x 100 = 91 %

4 If there were 2 products and only 1 was the desired product, it could have an atom economy of 64 %. If both products were desired, then it has a 100 % atom economy.

5 The production of SO_2 is a by-product of burning coal. The atom economy is 100 % since there is only one product. This would suggest that there is no waste. However, SO_2 is a pollutant, forming acid rain in the presence of water.

6 Atom economy of the Haber process is 100 % since ammonia is the only product. For the ammonium chloride reaction, the atom economy is 19 %. The Haber process is more sustainable if atom economy is considered. However, there are other factors. For instance, if all the products were desirable for the ammonium chloride process then its atom economy would also be 100 %. Also, the following need to be taken into account: amount of energy consumed to make ammonia, whether the raw materials are finite and non-renewable and whether the process is polluting.

Lesson 5.7 Maths skills: Change the subject of an equation

1 24.5 + 10 = X + 4.5.
    X = (24.5 + 10) - 4.5 = 30 g.

2 49 + X = 91
    X = 91 - 49 = 42 g.

3 Moles MgO = 62.5/40 = 1.5625
    Ratio 1:1
    Mass $MgSO_4$ = 1.5625 x 120 = 187.5 g

4 Moles $MgCO_3$ = 25.2/84 = 0.3
    Ratio 1:1
    Mass $MgSO_4$ = 0.3 x 120 = 36 g

5 Moles alkali = 0.12 x (25/1000) = 0.003
    Concentration of acid = 0.003 x (1000/27.2) = 0.110 mol/dm^3

6   Moles alkali = 0.15 x (25/1000) = 0.00375
Volume of acid = 0.00375/0.13 = 0.0288 dm^3 = 28.8 cm^3

Lesson 5.8 Measuring rates

1   1.6 - 1.7 g.

2   During the first minute. The concentration of hydrochloric acid is greatest at the start of the reaction.

3   18.4 cm^3

4   58 seconds (curve becomes horizontal).

5   Keep the total volume of sodium thiosulfate and hydrochloric acid constant. Dilute the sodium thiosulfate. Run the experiment again and record the time taken for the cross to disappear. Repeat for different concentrations. Analyse the time data for trends.

6   1. Temperature affects rate and is not controlled (room temperature may vary and the reaction itself is exothermic). Place the conical flask in a temperature controlled water bath.
2. The end point when the cross disappears relies on eye sight judgment. Use colourimeter (measures change in intensity of light as it passes through the precipitate) or data logger. 3. Starting the stopwatch and mixing simultaneously is difficult. So at least two operators needed.

Lesson 5.9 Calculating rates

1 a   Slow rate. Long time taken (minutes to hours).

   b   Fast rate. Very short time taken.

   c   Very slow rate. Extremely long time taken.

2   30/50 = 0.6 cm^3/s

3   30/20 = 1.5 cm^3/s

4   Check students' sketches.
Gradient = 52/40 = 1.3 cm^3/s

5   The steeper / larger gradient for the blue line indicates a greater rate of reaction than for the red line i.e. greater volume collected per unit time.

6 a   20 seconds. Red = 52/40 = 1.3 cm^3/s. Blue = 28.5/10 = 2.85 cm^3/s.

   b   40 seconds. Red = 52/40 = 1.3 cm^3/s. Blue = (70 - 24) / (30 - 0) = 1.53 cm^3/s.

Lesson 5.10 Factors affecting rates

1   34 seconds

2   As the temperature increases, the time taken decreases and therefore the rate increases.

3   Time taken would decrease and rate would increase.

4   It is not valid. The aim was to see how concentration affected the rate. The concentration and temperature change in the 2 experiments so variables have not been controlled.

5   1.5 g

6   As the size of the chips decrease, the time taken to produce a given volume of CO_2 decreases and therefore rate increases. In other words increasing surface area increases rate.

7   Powder: 2.2/0.8 = 2.75 g/min.
Small chips: 2.2/4.6 = 0.48 g/min.

Single chip: 2.2/7.6 = 0.29 g/min.

8   The same mass of calcium carbonate and concentration of hydrochloric acid was used each time.

9   • Increase temperature. Rate would increase.
• Increase concentration of HCl. Rate would increase.
• Increase surface area of Mg (use Mg powder for instance). Rate would increase.

Lesson 5.11 Collision theory

1   Increasing pressure means there are more particles per volume. Therefore there are more collisions and more successful collisions. The rate increases.

2   • HCl molecules move faster as the temperature increases.
• This means that the proportion of HCl molecules that have an energy equal to or greater than the activation energy increases.
• So the number of successful collisions with magnesium increases and the rate of reaction increases.
• Also, as temperature increases, the increased kinetic energy means that there are more collisions with magnesium.

3   Increasing concentration of HCl means there are more molecules per volume. Therefore there are more collisions and more successful collisions. The rate increases.

4   1 x 10^6/5 x10^{10} = 2 x 10^{-5} s

Lesson 5.12 Catalysts

1   • Magnesium chloride: Not a catalyst. No decrease in time to collect 100 cm^3 of gas.
• Copper chloride: Not a catalyst. Although time was reduced (greater rate), it did not remain unchanged at the end (it was a different colour). In fact Zinc displaces copper ions to form copper. It is the copper that acts as a catalyst.
• Copper: It is a catalyst. There was a decrease in time to collect 100 cm^3 of gas. Copper remained unchanged at the end of the reaction.

2   Zn (s) + H_2SO_4 (aq) $\xrightarrow{\text{Cu}}$ $ZnSO_4$ (aq) + H_2 (g)

3   Vanadium pentoxide provides an alternative reaction pathway with lower activation energy. So a greater number of molecules have an energy equal to or greater than the activation energy. The number of successful collisions increases and so does the rate.
2 SO_2 + O_2 ⇌ 2 SO_3

4 a   2 H_2O_2 (aq)→ 2 H_2O (l) + O_2 (g)

   b   Catalase. Provides and alternative reaction pathway with lower activation energy.

   c   75 kJ/mol. Any value greater than 58 kJ/mol since catalysts lower activation energy.

   d   Hydrogen peroxide is toxic to cells. Catalase increases the rate of decomposition so it prevents the cells being damaged.

336 © HarperCollins*Publishers* Limited 2016

Lesson 5.13 Factors increasing the rate

1 21 seconds (when the line becomes horizontal).

2 Metal and acid e.g. magnesium and hydrochloric acid.

3 The limiting reactant has been used up so there are no more collisions between the reactant particles. No more product is produced and rate drops to zero.

4 $60/10 = 6$ cm^3/s

5 Table 5.1: 13 cm^3. As the concentration of hydrochloric acid increases, a greater volume of hydrogen is collected in 10 seconds. Rate increases.

Table 5.2: 13 cm^3. As the temperature increases, a greater volume of hydrogen is collected in 10 seconds. Rate increases.

6 65 cm^3. The same concentration of HCl was used and it was in excess. The same mass of Mg was used.

7
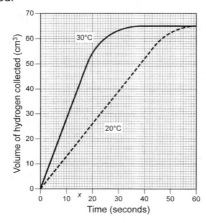

8 a When twice the mass (moles) of magnesium was used, twice the volume (moles) of hydrogen were produced.

  b The gradient is greater for the reaction with higher mass. So the rate is greater for the higher mass.

9 The final volumes for the 2 masses are 33 and 66 cm^3. 0.0495 g is halfway between the 2 masses. Therefore the final volume will be 49.5 cm^3. The curve will have a gradient in between the red and the blue curve.

10 a $CaCO_3$ (s) + 2 HNO_3 (aq) → $Ca(NO_3)_2$ (aq) + H_2O (l) + CO_2 (g)

  b The rate of reaction has increased (approximately doubled) between experiments 1 and 2 (0.37 cm^3/s and 0.71 cm^3/s). The total volume has doubled which suggests that the concentration of nitric acid has doubled. Calcium carbonate must have been in excess.

Lesson 5.14 Practical: Investigate how changes in concentration affect the rates of reactions by a method involving the production of a gas and a method involving a colour change

1 The higher the temperature, the greater the number of successful collisions and the greater the rate.

2 Temperature.

3 Concentration (and volume) of acid.

The identity of the acid e.g. hydrochloric acid.

Mass of Mg and its particle size (e.g. ribbon or powder).

4 The volume of gaseous hydrogen produced per time is being used to follow rate. A gas syringe is a convenient and accurate way of measuring gas volume.

5 The bung must be replaced quickly to avoid hydrogen gas escaping.

6 The effect that increasing temperature has on the rate of reaction between calcium carbonate and hydrochloric acid.

7 45 cm^3 at 30 ^{o}C is anomalous. This point should be discarded.

8 Students plot and sketch 3 lines, check accuracy of the sketches.

9 Increasing temperature increases the rate of reaction between calcium carbonate and hydrochloric acid.

10 1.3 cm^3/s

11 The evidence supports their conclusion. Increasing temperature increases the rate of reaction between calcium carbonate and hydrochloric acid. The experiment was carried out twice at each temperature with similar results so the experiment is repeatable. This gives validity to their conclusion.

12 They repeated their experiment twice at each of the temperatures and obtained similar results.

13 They used the same apparatus each time and controlled the variables. They only changed the temperature - only one variable was changed. So their results were valid.

14 They would need to place the conical flask in a thermostatted water bath so that the temperature was constant throughout the run.

They could place the magnesium in a small tube in the conical flask. The open end of the tube would be above the acid level. To start the reaction, the conical flask could be tipped so that the acid made its way into the tube. This would avoid losing hydrogen through replacing the bung at the start.

15 Carry out the experiment with different metals and different acids.

Extend to carbonates and metal oxides.

16 The particle model can be applied to all chemical reactions so their conclusion is valid for chemical reactions in general.

• Reactant molecules move faster as the temperature increases.

- This means that the proportion of reactant molecules that have an energy equal to or greater than the activation energy increases.
- So the number of successful collisions between reactant molecules increases and the rate of reaction increases.
- Also, as temperature increases, the increased kinetic energy means that there are more collisions with magnesium.

Lesson 5.15 Reversible reactions and energy changes

1 a Irreversible.

b Reversible.

c Reversible.

2 It is reversible. Boil / evaporate the water and solid sodium chloride appears.

3 A reaction that takes in energy from the surroundings. It is often accompanied by a temperature decrease.

4 An increase in pressure causes the position of equilibrium to shift to the side with the smallest number of molecules. In the case of the Haber process, this is to the product side (4 to 2 molecules). As the pressure is increased, there is a greater increase in the number of successful collisions between reactant molecules compared to product molecules.

5 No effect. It doesn't alter the relative energies of reactants and products. It just lowers the activation energy.

Lesson 5.16 Equilibrium

1 As soon as the product is formed, it is lost. Therefore, more product is formed and that is lost. And so on. So an equilibrium cannot be reached unless all the reactants and products are together.

2 The rate at which salt dissolves equals the rate at which the ions form solid salt again.

3 The position of equilibrium lies to the left - the reactant side.

4 The position of equilibrium will shift left to the reactant side. This is to minimise the increase in product concentration.

5 Adding water: A (white) precipitate of BiOCl would form since the position of equilibrium will shift left to counteract the increased water concentration. Adding hydrochloric acid: The (white) precipitate of BiOCl would disappear since the position of equilibrium will shift right to counteract the increased hydrochloric acid concentration.

Lesson 5.17 Changing concentration and equilibrium

1 SO_3 would need to be removed to make more product. The position of equilibrium shifts right to minimise the reduction in SO_3 concentration.

2 Remove / liquefy the methanol. The position of equilibrium would shift right to try to increase the methanol concentration again. Alternatively, increase the concentration of CO / H_2. The

position of equilibrium would shift right to try to decrease the CO / H_2 concentration again.

3 Sodium chloride is a source of chloride ions, Cl-. So the position of equilibrium would shift left to try to reduce the Cl- concentration again. More white $PbCl_2$ precipitate would be seen to form.

Lesson 5.18 Changing temperature and equilibrium

1 a A: 54 %. B: 46 %

b A: 58 %. B: 42 %

2 As the temperature is increased, heat is added. The position equilibrium moves in the direction that takes in (and minimises) the added heat and lower the temperature again. Endothermic reactions take in heat from the surroundings. So the products would be favoured.

3 As temperature increases heat is added. The position of equilibrium will move in the direction which removes the added heat - the endothermic direction. The graph shows that as temperature increases, the position of equilibrium shifts left since the % conversion decreases. So left is endothermic and the reaction is exothermic in the forward direction.

4 a As temperature increases, the position of equilibrium shifts right since the % conversion increases

b Endothermic in the forward direction. As temperature is increased more heat is added. The equilibrium will shift in the direction that removes the heat - the endothermic direction. This is to the right since % conversion has increased.

c As pressure increases, the position of equilibrium shifts to the left to the side with fewest molecules. With fewer molecules the pressure is less. So the increase in pressure is relieved. This is reflected in a decrease in % conversion with increasing pressure.

Lesson 5.19 Changing pressure and equilibrium

1 49 % reactants and 51 % products.

2 Increasing pressure results in a decrease in percentage of products (decrease in yield). This suggests that there are more product molecules than reactant molecules.

3 a Increasing pressure shifts the position of equilibrium to the right to the side with fewest molecules (5 to 2).

b Decreasing pressure shifts the position of equilibrium to the right to the side with the larger number of molecules (2 to 3).

c Decreasing pressure shifts the position of equilibrium to the right to the side with the larger number of molecules (2 to 3).

4 In both reactions there is a compromise between rate and yield. Low temperature increases yield because the position of equilibrium shifts right in the exothermic direction. But the rate is too slow at low temperatures. So rate is increased to give reasonable rates at the expense of yield.

High pressure in the Haber process increases rate and also moves the position of equilibrium to the right to the side with fewest molecules i.e. increases yield. The same would be true of the Contact process. However, at low pressures, the yield of SO_3 is high anyway. The extra cost of having high pressure equipment is not worth the small increase in yield or rate.

Lesson 5.20 Maths skills: Use the slope of a tangent as a measure of rate of change

1 a Fastest at the very start (time = between 0-10 seconds).

b In the middle of the curved section where the gradient is decreasing and heading to 0 (horizontal).

c The time at the start of the horizontal section..

2 a At the start of the reaction. The concentration of the acid is the highest. There are a greater number of molecules per volume so more successful collisions.

b When the plot is horizontal. All of the limiting reactant will have reacted so there will not be any more collisions.

3 a 59.5 - 21 = 38.5 cm^3

b 38.5/20 = 1.93 cm^3/s.

c 21/10 = 2.1 cm^3/s.

d The gradient between 10 - 30 seconds is smaller. The reaction rate decreases with time so the gradient decreases with time.

4 a 42/15 = 2.8 cm^3/s.

b 52/40 = 1.3 cm^3/s.
The gradient for the red line is less. So the rate of reaction for the red line is less. It may be that the concentration of acid is lower or the temperature is lower for the red reaction.

5 Note that tangents are difficult to judge by eye.
22 seconds: (70 - 29)/ 32 = 1.28 cm^3/s
28 seconds: (70 - 47)/ 42 = 0.55 cm^3/s
The rate decreases as the reaction progresses. The rate at 28 seconds is less than the rate at 22 seconds. This is because the concentration of acid is decreasing resulting in fewer successful collisions.

6 a a 2.7 cm^3/ min. Rate decreases as the reaction proceeds. The smaller tangent value means slower rate. Therefore the smaller tangent value must have been taken closer to the end.

b Using 8.5 cm^3/ min: 8.5/60 = 0.14 cm^3/s

End of chapter questions

1 d) they increase the rate of reaction (1)

2 increases then decreases to zero. (1)

3 Increase temperature / increase pressure (gas) / increase concentration (solutions) / increase surface area (solid) / add a catalyst. Any 2 for 2 marks.

4 The carbon dioxide escapes / leaves the bottle / it is not a closed system. (1)

5 Rate increases. Greater concentration means more molecules per volume and therefore more collisions. (1)

6 One of the products is a gas so leaves the flask. (2)

7 78.6% (2)

8 The forward and reverse reactions occur at the same rate, and the concentrations of reactants and products do not change. (1)

9 At higher pressure there is a higher concentration of reactant molecules per volume. So the collision frequency is greater. (1)

10 1. Place on a mass balance. Monitor the change in mass with time as the reaction proceeds. (1) 2. Collect the gas in a gas syringe / over water with measuring cylinder or burette filled with water. Measure the volume of gas produced with time. (1)

11a (iii), (i). (ii)

b (i) 98.6%, (ii) 1.5%, (iii) 100% (4)

12 55 kJ/mol was for the catalysed reaction. Catalysts lower the activation energy (1) by providing an alternative reaction pathway. (1)

13 It is not a closed system - the gaseous products can escape. (1)

14 Dust of these substances is finely divided so has a large surface area. Reactions with oxygen are much quicker with a larger surface area. (1)

15 The silver nitrate test identifies the chloride ion. (1) Magnesium, calcium and aluminium ions all give white precipitates with NaOH so this test only narrows the identity down. (1) If the precipitate doesn't redissolve, then it is not an aluminium ion. A red flame colour indicates calcium ions are present. Magnesium ions do not give a colour in a flame test.

16 18 dm^3 (2)

17 a Temperature change. (1)

b Concentration of hydrochloric acid. (1)

c 23 °C / 40 cm^3. 33 °C / 76 cm^3 (1)

d As temperature increases, so does the reaction rate. / For a 10 °C rise in temperature, the rate doubles. (1)

18 As temperature decreases, the number of reactant molecules with energy equal to or greater than the activation energy decreases. (1) So the number of successful collisions decreases. Rate decreases. (1)

19 The higher the pressure the greater the yield from the data.
The lower the temperature the greater the yield from the data. (2)

20 a 1700/17 = 100 moles NH_3. Ratio: N_2:NH_3 is 1:2. So 50 moles of N_2 needed. Mass N_2 = 50 x 28 = 1,400 g.

b Moles NH_3 = 6/24 = 0.25. So 0.25 x 17 = 4.25 g.

c 3.012 x 10^{21}

Chapter 6: Global Challenges

Lesson 6.1 Extraction of metals

1 Oxygen is removed from zinc oxide so it is reduced. Carbon has had oxygen added so it is oxidised.

2 Carbon has been oxidised since oxygen has been added.

3 Iron(III) oxide has been reduced since oxygen has been removed. Carbon monoxide has been oxidised since oxygen has been added.

4 a $Cu_2O + C \rightarrow 2\,Cu + CO$

b Copper(II) oxide is being reduced (oxygen removed). to copper Carbon is being oxidised (oxygen added) to carbon monoxide.

5 $2\,ZnS + 3\,O_2 \rightarrow 2\,ZnO + 2\,SO_2$

6 Fe_2O_3 is composed of Fe^{3+} ions. These are reduced (gain 3 electrons) to Fe.

Lesson 6.2 Using electrolysis to extract metals

1 Sodium is a very reactive metal and above carbon in the reactivity series. Therefore reduction by carbon will not work.

2 The negative oxide ions, O^{2-}, are attracted to the positive anode and form oxygen (oxidation since electrons are lost). Al^{3+} ions are attracted to the negative cathode. At the cathode they are reduced (they gain electrons). Molten aluminium is formed.

3 The electrolysis cell has to be heated to high temperatures. Large electrical currents are needed. Both of these require a large amount of energy. This is very costly.

4 O^{2-} ions are attracted to the positive anode. They transfer their electrons to the anode and oxygen forms. The electrons flow around the circuit where they are transferred to Al^{3+} ions at the negative cathode. Aluminium forms.

5 Cathode: $Ca^{2+} + 2\,e^- \rightarrow Ca$
Anode: $2\,Cl^- \rightarrow Cl_2 + 2\,e^-$

6 Half equations: $K^+ + e^- \rightarrow K$ and $2\,F^- \rightarrow F_2 + 2e^-$.
Overall equation: $2\,KF \rightarrow 2\,K + F_2$

Lesson 6.3 Alternative methods of metal extraction

1 • Copper ores are becoming scarce so new ways of extracting copper from low grade ores is needed.
• Demand for copper is high.
• Traditional mining is damaging to the environment.

2 Some plant roots selectively absorb compounds containing metals e.g. copper compounds. These plants are then harvested and burned. Their ash contains the metal compound.

3 Bacteria are used to produce a leachate solution containing copper compounds. Then the copper compound can be reacted with scrap iron. Iron is more reactive than copper and displaces it from the copper compound.

4 The fact that it is slow makes bioleaching less economical. The sulfuric acid may cause environmental problems e.g. leak into the ground or surface water and make it acidic.

5 Bioleaching costs are high and the process is quite slow. So it is often not economic for a company to use this method on a large scale in remote areas. Mining has adverse environmental consequences. However, it is capable of producing large quantities of the metal rapidly. So it is more economic.

6 $1.9 \times 10^7 \times 0.2 \times 0.9 = 3.42 \times 10^6$ tonnes of copper.

Lesson 6.4 The Haber process

1 Nitrogen and hydrogen.

2 Nitrogen and hydrogen are purified. They are reacted together at high temperature and pressure to form ammonia. The remaining nitrogen and hydrogen is recycled back to the start.

3 Cryogenic (fractional) distillation of air. The air is liquefied by cooling to low temperatures. Then it is fractionally distilled. This is carried out several times to remove oxygen, since oxygen damages the catalyst.

4 Reacting methane (from fossil fuels) and steam without any sulfur present produces hydrogen.

5 Water, H_2O. Vast quantities available.

6 a As pressure is increased, the position of equilibrium moves to the right (to the side with fewest molecules). So the percentage yield of ammonia increases with increasing pressure, whichever temperature is selected.

b Since the reaction is exothermic, a lower temperature moves the position of equilibrium to the right. So the lower the temperature the greater the percentage yield of ammonia.

7 The graph shows that the percentage yield does not greatly increase above 200 atmospheres. Also, the higher the pressure the more expensive it is to build a plant due to safety considerations. Higher pressure also requires more energy to sustain.

Lesson 6.5 Production and use of NPK fertilisers

1 Pour the ammonia solution into the conical flask using a measuring cylinder. Add the sulfuric acid solution to the burette. Add an indicator (e.g. universal indicator) to the flask. Run the sulfuric acid into the conical flask until the indicator changes to the desired colour (green for universal indicator). Methyl orange is a more effective indicator.

2 a Potassium phosphate.

b Ammonium sulfate.

3 B calcium phosphate + nitric acid → calcium nitrate + phosphoric acid
C phosphoric acid + ammonia → ammonium phosphate
Ammonia from the Haber process is used to produce nitric acid. Nitric acid is used in reaction B. Ammonia is used in reaction C. Phosphoric acid from reaction B is reacted with ammonia in reaction C..

4 Vast quantities of artificial fertiliser are made every year. Phosphate rock is central to the process.

Phosphate rock is a finite resource and will eventually run out.

5 $CaCO_3 + 2 HNO_3 \rightarrow Ca(NO_3)_2 + H_2O + CO_2$

6 $Ca_3(PO_4)_2 + 6 HNO_3 \rightarrow 3 Ca(NO_3)_2 + 2 H_3PO_4$

Lesson 6.6 Life cycle assessment and recycling

1 Paper is renewable (trees can be grown in tens of years). Plastic comes from non-renewable crude oil. So Paper as a raw material is more sustainable in this case.

2 Both paper and plastic can be recycled. Paper is biodegradable - it breaks down quite quickly in landfill sites. Plastic can take hundreds of years to break down. So paper is easiest to dispose of.

3 It is the assessment of the environmental impact at every stage including the recycling of the product into a new product.

4 a Qualitative and subjective.

  b Quantitative and objective.

5 The students do not have any quantitative data to make objective statements about CO_2 produced. So their LCA will have to be based on subjective opinion. Their opposing viewpoints may lead them to different conclusions.

6 Manufacturing: They use totally different processes. Use and operation: Plastic bags potentially last longer since paper bags tend to rip. However much plastic bag use is once only. Disposal: Both paper and plastic bags can be re-used. However, plastic bags are stronger and can be reused more often. Both paper and plastic can be recycled. Transport: Paper bags are bulkier and heavier than the equivalent plastic bags. Therefore transport costs for paper bags will be higher. Emissions and waste: The carbon footprint for a plastic bag is less than for a paper bag. Both cause environmental problems. Waste paper is biodegradable but plastic is often non-biodegradable and can last for hundreds of years in landfill sites.

7 a This data is an average. As such, it is not necessarily directly applicable to all situations.

  b Cost. The manufacturer has to take into account cost and this may influence which materials and methods are used. Cost also influences the consumer. The consumer may buy a cheaper product even though the LCA suggests it might have a greater environmental impact.

Lesson 6.7 Ways of reducing the waste of resources

1 • Quarrying may result in extra noise and heavy traffic.

• There may be air pollution e.g. dust.

• The quarry has a visual impact and may disfigure the local environment.

• There may be damage to plant and animal life

2 Metal and glass are produced from a limited supply of raw materials. Extracting these raw materials has a negative environmental impact e.g. in the energy consumed. If metal and glass is recycled (or reused) it conserves the Earth's

resources and has less of an environmental impact.

3 Reusing a product requires less energy and resources than recycling. For instance, if a plastic bag is recycled, it must be broken down and remade into another product.

4 Crude is a finite, non-renewable resource. Crude oil fractions are used as fuel, for making plastics and chemicals and in many other processes. So the use of crude oil needs to be limited.

5 Use cotton for clothes and fabrics rather than polymers such as polyester and nylon.

Make furniture out of wood rather than plastic.

Use paper cups rather than plastic cups.

6 Reduce: Cars consume energy, cause pollution and use valuable resources. These could be reduced by car sharing, walking, cycling, taking public transport or not buying a car in the first place.

Reuse: Parts of cars can be reused e.g. for spare parts for cars of the same model.

Recycle: The metal and some of the plastics in cars can be recycled.

7 It will save resources if they are manufactured near to the market. Cars are large and heavy. It requires much energy to transport cars long distances (fuel for ships, car transporters etc.).

8 Recycle any waste metal from the process back into the production line or use it for other purposes.

Make sure that the assembly runs on the minimum amount of energy or use renewable energy.

9 a Reduction in use of energy since 80 % less bags are being manufactured. Reduction in use of crude oil since plastic is derived from substances that are derived from crude oil. Reduction in waste in landfill sites. Reduction in energy needed to recycle the bags.

  b 1.52×10^9

Lesson 6.8 Alloys as useful materials

1 a X since it has the highest carbon content.

  b Y since it has the lowest carbon content.

2 X 50 units (the higher the carbon content the stronger the steel)

Y 27 units (the lower the carbon content the softer the steel)

Z 35 units

3 Copper: 5309 g. Tin: 280 kg. Phosphorus: 11.2 g.

4 By varying the percentages, the properties of the alloy are varied e.g. hardness, melting point. This means that the alloy can be tailored to a specific need.

5 The pewters containing 4 % and 15 % lead. Lead is toxic.

6 There will be less distortion of the metal layers since there is a smaller percentage of other metals. The layers can therefore more easily slide over each other.

7 a Titanium is very reactive and would react with the oxygen in air.

b 280

Lesson 6.9 Corrosion and its prevention

1 The desert is dry with low humidity whereas the UK has lots of rain. Without water, rusting cannot occur.

2 Dissolved oxygen is present in shallow water. Rusting requires both water and oxygen.

3 It prevents water (and oxygen) from coming into contact with the iron.

4 If the paint is scratched or comes off, then oxygen and water will come into contact with the iron again and it will rust.

5 Magnesium is more reactive than zinc so offers better protection to the iron.

6 Paint. Prevents water (and oxygen) from coming into contact with iron.

Galvanising. Prevents water (and oxygen) from coming into contact with iron. Zinc is more reactive than iron so it also provides sacrificial protection. When the surface is scratched, the zinc reacts rather than the iron.

7 a $Mg \rightarrow Mg^{2+} + 2e^-$

b The electrons from the magnesium are accepted by the Fe^{3+} ions which are reduced to iron atoms again.

c It dissolves as Mg^{2+} is formed.

Lesson 6.10 Ceramics, polymers and composites

1 Soda-lime composition: sand, sodium carbonate, limestone.
Borosilicate glass: sand, boron trioxide.
So both contain sand but differ in the number and identity of the other components.

2 It contains glass fibres - the reinforcement - and a (polyester) resin which is the matrix (binder).

3 The long polymer chains are linked by cross-links. The appearance is like a net.

4 Drinks bottles need to be moulded so the plastic has to be soft when heated up. Also, drinks bottles need to be be flexible. Thermosetting plastic tends to be hard and brittle.

5 Polymer: It is too flexible (lower number equals greater flexibility) and its low density means that it would not withstand impacts very well (although it would be light).
Ceramic: This is too rigid and inflexible and would likely shatter in an accident.

6 On balance, the polymer would be the best choice with glass fibre coming second. The polymer has high tensile strength, is very flexible and has a relatively low density.
Glass fibre has the highest tensile strength but is not as flexible and has 3 times the density.

Lesson 6.11 Maths skills: Translate information between graphical and numerical form

1 13 %

2 5 x 3.6 = 18°

3 About 55 %

4 About 200 atm

5

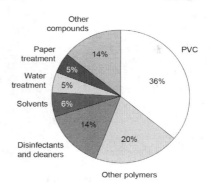

6

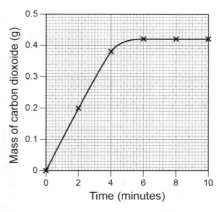

Lesson 6.12 function groups and homologous series

1 CH_4

2 4

3 1 carbon and 2 hydrogens are added each time ($-CH_2-$).

4 C_8H_{18}

5 n = number of carbons = 18. So $C_{18}H_{18} + 2 = C_{18}H_{38}$

6 $CH_3CH_2CH_2CH_2CH_2CH_2CH_3$; $(CH_3)_2CHCH_2CH(CH_3)_2$ or other suitable isomers

7

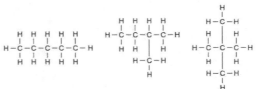

Pentane   2 methyl butane   2,2 dimethyl propane

Lesson 6.13 Structure and formulae of alkenes

1   Hexene. C_6H_{12}.

2   Alkanes have all single C-C bonds. Alkenes have a carbon-carbon double bond, C=C.

3   Because there are double the number of hydrogens compared to carbons. This is show by the general formula for an alkene with n = 16, C_nH_{2n}.

4   a) and d) are alkenes and will turn bromine water colourless.
     b) and c) are alkanes.

5    $CH_2CHCH_2CH_2CH_2CH_2CH_2CH_3$ /
     $CH_3CHCHCH_2CH_2CH_2CH_2CH_3$ /
     $CH_3CH_2CHCHCH_2CH_2CH_2CH_3$

6   C_2H_4, C_3H_6, C_4H_8 C_5H_{10}
     $C_2H_4 + H_2 \rightarrow C_2H_6$
     $C_3H_6 + H_2 \rightarrow C_3H_8$
     $C_4H_8 + H_2 \rightarrow C_4H_{10}$
     $C_5H_{10} + H_2 \rightarrow C_5H_{12}$

7 a C_nH_{2n-2}.
   b $C_{10}H_{18}$.

6.14 Reactions of alkenes

1    $C_2H_4 + O_2 \rightarrow 2 C + 2 H_2O$

2    $C_4H_8 + 4 O_2 \rightarrow 4 CO + 4 H_2O$

3   They are more prone to incomplete combustion. So they would produce more carbon monoxide which is toxic. They also burn with more smoky flames and this would not be suitable for domestic use. Alkenes are also more reactive than alkanes and are more useful as starting material to make other organic molecules.

4   C_5H_{10}. It is an alkene and the others are alkanes. Alkenes tend to incompletely combust.

5

6    $CH_3CHClCHClCH_2CH_3$

7   Butane, C_4H_{10}.

8

9

10

Propan-1-ol   Propan-2-ol

6.15 Alcohols

1    $CH_3CH_2CH_2CH_2CH_2CH_2OH$

2   CO_2 and H_2O (assuming complete combustion).

3   Conditions need to be free of oxygen (anaerobic). Oxygen will be present if the flask is open.

4   Bubbles since carbon dioxide gas is produced.

5    $C_4H_9OH + 6 O_2 \rightarrow 4 CO_2 + 5 H_2O$

6.16 Carboxylic acids

1   Fizzing / bubbling. Carbon dioxide is produced.

2   Potassium ethanoate.

3   Propanoic acid is only partially ionised whereas hydochloric acid is completely ionised. So with propanoic acid, there will be a lower concentration of hydrogen ions compared to hydrochloric acid.

4   Hydrogen ion, H+, and hexanoate ion, $CH_3CH_2CH_2CH_2CH_2COO-$.

5

6   Alcohol - butanol. Carboxylic acid - propanoic acid.

7   To make ethanol: React ethene with steam in the presence of an acid catalyst to produce ethanol. To make ethanoic acid: Use an oxidising agent (potassium dichromate) to oxidise ethanol to ethanoic acid. Making the ester: React the ethanol and ethanoic acid together (in the presence of an acid catalyst) to produce ethyl ethanoate.

6.17 Addition polymerisation

1   -O-O-O-

2   It has a C=C bond. Monomers add across the C=C double bond, which breaks.

3

4

5

6

7 a

$$\left[\begin{array}{c} \quad \overset{\displaystyle F}{\underset{\displaystyle F}{\overset{|}{\underset{|}{C}}}} - \overset{\displaystyle F}{\underset{\displaystyle F}{\overset{|}{\underset{|}{C}}}} \quad \end{array}\right]_n$$

b

$$H_2N(CH_2)_6NH_2 + HOC(CH_2)_4COH$$
with two $C=O$ (O double-bonded) groups above

Lesson 6.18 Condensation polymerisation

1 Addition polymers are formed from a single alkene monomer and have the structure -[X]-[X]-[X]-. The polymer has the same empirical formula as the monomer.
Condensation polymers are usually formed from 2 different monomers and have the structure -[X][Y]-[X][Y]-[X][Y]-. A small molecule such as water is lost when the monomers react.

2 Ethanediol has 2 alcohol groups. Hexanedioic acid has 2 carboxylic acid groups. The alcohol reacts with the carboxylic acid to form the ester. Since both groups in each monomer can react, a chain can be formed linked by ester groups. A molecule of water is given off every time an ester group is formed.

3 -[-CH$_2$-CH$_2$-OOC-COO-CH$_2$-CH$_2$-OOC-COO-]- $_n$

4

$$\left[\begin{array}{c} -OCH_2\,CH_2 - O - \overset{O}{\underset{O}{\overset{\|}{C}}} - \bigcirc - \overset{O}{\overset{\|}{C}} - \end{array}\right]_n$$

5 Yes it can. It has two different functional groups in the same molecule. The carboxylic acid and alcohol react to form a polyester. A molecule of water is given off so it is a condensation polymer.

6 a It is a condensation polymer. A small molecule - water - is given off when the amine and carboxylic acid react to form the amide. -COOH + -NH$_2$ → -CONH- + H$_2$O.

b

$$H_2N(CH_2)_6NH_2 + HOC(CH_2)_4COH$$
with two $C=O$ (O double-bonded) groups above

Lesson 6.19 Amino acids

1 Octylamine. C$_8$H$_{17}$NH$_2$

2 7

3 They all have an amine group, a carboxylic acid group and a hydrogen atom attached to the same carbon. They all differ in the R group attached to the carbon. The general structure is H$_2$NCHRCOOH.

4 -(-HNCH(CH$_3$)CO-)- $_n$

Lesson 6.20 DNA and other naturally occurring polymers

1 Check accuracy of Students' drawings.

2

C

3 Enzymes are biological catalysts. They increase the rate of biological reactions (which occur at the active site in the enzyme) by providing an alternative reaction route with lower activation energy.

4 They are naturally occurring condensation polymers (polyamides). The amino acids are the monomers and when they form a polymer chain (protein) water is given off.

Lesson 6.21 Crude oil, hydrocarbons and alkanes

1 Coal / natural gas.

2 Fossil fuels are non-renewable and polluting. Alternative energy sources are "greener".

3 Fuels are used to produce energy. Petrochemicals are substances produced from crude oil fractions that are used for a variety of purposes such as solvents etc.

4 a Viscosity increases as the number of carbon atoms increases.

b Flammability increases as the number of carbon atoms decreases.

5 350–450

Lesson 6.22 Fractional distillation and petrochemicals

1 It contains a large number of different hydrocarbons with different properties.

2 C$_6$H$_{14}$ and C$_7$H$_{16}$. They have similar number of carbons.

3 Y must have more carbons since it has a higher boiling point (condenses at a higher temperature). So the intermolecular forces of attraction between molecules of Y are greater. These greater forces mean the Y molecules can condense at higher temperatures where there is more energy.

4 Larger hydrocarbon molecules have larger molecular masses, so the forces of attraction between them are stronger than between small molecules.

5 a Melting points increase as the number of carbon atoms increase.

b The line zig-zags for each carbon atom. This is because the different structures of atoms take different amounts of energy to melt.

Lesson 6.23 Properties of hydrocarbons

1 Diesel oil. It has lower viscosity than kerosene.

2 Kerosene should be cracked to make more petrol

Lesson 6.24 Cracking and alkenes

1 More petrol is needed than can be supplied. So diesel and kerosene (less needed than can be supplied) should be cracked.

2 High temperature and a catalyst or very high temperature and steam.

3 Pentene since it is an alkene. The bromine adds across the C=C bond.

4 D = 1, E = 1, F = 1, G = 2.

5 $C_{10}H_{22} \rightarrow C_8H_{18} + C_2H_4$. Other equations are possible.

Lesson 6.25 Cells and batteries

1 Zinc is more reactive than copper. If the metals are placed into an electrolyte, zinc in the zinc electrode loses electrons (and is oxidised) and becomes zinc ions. The electrons travel around the circuit to the copper electrode. This is electrical current. The copper ions in solution accept the electrons from the copper electrode. Copper is deposited on the copper electrode.

2 They can be reused many times which saves on valuable resources.

They create less pollution since fewer of them are used compared to disposable batteries.

They are cheaper in the long run than disposable batteries because they can be used many times.

3 Rechargeable. Although it has a lower voltage, this is likely to be sufficient for a small toy. The mass is smaller than for B and C which means that it is more suitable for a small toy (a small toy car will need less power if it is lighter). It can be recharged many times which is more convenient than disposable batteries. C could not be used in a toy since it contains acid, which is hazardous and lead which is toxic (as is the voltage). The voltage for B is too high for a toy.

4 Silver. The most negative voltage is between magnesium and silver which suggests that they differ the most widely in reactivity. Magnesium is more reactive than nickel. Nickel is more reactive than silver.

5 Ni is more reactive than silver. So nickel is oxidised.

$Ni \rightarrow Ni^{2+} + 2e^-$

$Ag^+ + e^- \rightarrow Ag$

Lesson 6.26 Fuel cells

1 $2 H_2 + O_2 \rightarrow 2 H_2O_2$

2 The energy released is converted into electricity. The hydrogen and oxygen gases are reacting at different electrodes.

3 Not many filling stations sell hydrogen. / Hydrogen is difficult to store. It has to be compressed and liquefied and stored in rugged fuel tanks. / Hydrogen is explosive and there is a hazard

associated with this. / A large proportion of hydrogen is produced from fossil fuels which are non-renewable. This requires lots of energy.

4 Hydrogen loses electrons at the negative electrode, which is oxidation. At the positive electrode, hydrogen ions are reduced since they gain electrons.

Lesson 6.27 Key concept: Intermolecular forces

1 Ethene, ethanoic acid, glycine.

2 Decane has a longer chain than octane. There are greater intermolecular forces between decane chains. So more energy is required to separate decane chains and it has a higher boiling point.

3 Propene is a small molecule (alkene) with weak intermolecular forces. Poly(propene) consists of long chains and therefore there are greater intermolecular forces compared to propene. Therefore it takes more energy to separate poly(propene) chains so it has a higher melting point.

4 It is a cross-linked polymer with chains held together with covalent bonds. This makes the polymer more rigid. Covalent bonds are much stronger than the intermolecular forces between chains. The covalent bonds prevent the chains from sliding over each other.

5 The intermolecular forces between ethane molecules is greater than between ethene molecules. It takes more energy to separate ethane molecules and it therefore has a higher boiling point.

Lesson 6.28 Maths skills: Visualise and represent 3D models

1

2 a) Propanol

b) Decane

3

Poly (tetrafluoroetFene) (PTFE)

4

5   4 bases. Adenine, guanine, cytosine and thymine. A pairs with T and C pairs with G.

6   DNA is a double helix. Its function is related to its shape. A 2D model does not show the relationship in space between the nucleotides or the importance of the helical structure.

Lesson 6.29 Proportions of gases in the atmosphere

1   Plants release oxygen during photosynthesis.

2 a   Very small proportion of water vapour.

  b   Large proportion of water vapour (very humid).

  c   Fairly humid (lots of rainfall and presence of oceans and seas).

3   There are no life forms (algae, plants, cyanobacteria etc.) on Mars that produce oxygen.

4   Breathing is the process of gases traveling in and out of the lungs. Oxygen is taken in and carbon dioxide breathed out. Respiration is the release of energy in cells from the breakdown of glucose. Oxygen is used for this process.

5   Glucose molecules contain 6 carbons. Each of these is oxidised to carbon dioxide.

6   Plants photosynthesise. In doing so they produce glucose and oxygen and remove carbon dioxide. Animals and plants respire, consuming glucose and oxygen and producing carbon dioxide. So glucose has a key role in maintaining the balance of carbon dioxide and oxygen in the atmosphere.

Lesson 6.30 The Earth's early atmosphere

1   Volcanic activity released gases - mainly carbon dioxide and some nitrogen. There was very little oxygen. Small proportions of methane and ammonia may also have been present. Water vapour was present.

2   There were no plants or bacteria that could produce oxygen (no photosynthesis).

3   There is a lack of direct evidence. Therefore different theories can fit the evidence available.

4   It is known that there was intense volcanic activity during the first billion years of Earth. It is assumed that the same composition of gases is given out by volcanoes today and those of early Earth.

5   The volcanic model of early Earth assumes that the composition of gases emitted by volcanoes is the same as it is now. This cannot be proved since there is no direct evidence. The number of stomata is known to be related to level of carbon dioxide in plants of today. So counting stomata on fossil leaves should give a better indication of the levels of carbon dioxide in the early atmosphere than volcanoes. However, it is still an indirect measurement (a proxy) and assumptions have to be made that cannot be easily tested.

6   High carbon dioxide levels: fewer stomata (each stomata allows more gas in). Low carbon dioxide: more stomata.

Lesson 6.31 How oxygen increased

1   2.7 billion years ago.

2   Anaerobic organisms live (produce energy) without oxygen. Aerobic organisms need oxygen to produce energy.

3   Glucose, which provides the plants with an energy source.

4   Plants (and algae / cyanobacteria) produce oxygen during photosynthesis. As the number of plants and number of new species grew, more oxygen was produced and more carbon dioxide was used up.

5   Algae can photosynthesise and started producing oxygen 2.7 billion years ago. Therefore algae, along with plants and cyanobacteria, helped to increase the percentage of oxygen in the early atmosphere.

6   The oxygen produced by plants and other organisms first reacted with iron in the Earth's oceans (and Earth's crust) producing iron(III) oxide. So little oxygen ended up in the Earth's atmosphere initially. This is the time lag where organisms that photosynthesised existed but there was no oxygen build up. Eventually, there was an excess of oxygen which started to build up in the atmosphere.

Lesson 6.32 Key concept: Greenhouse gases

1   Carbon dioxide / methane.

2   Greenhouse gases effectively trap hit (like an insulating blanket). They keep the temperature at a level where plant and animal life can flourish. Without these gases, the average temperature would be considerably less and detrimental to life.

3   Short wavelength radiation from the Sun enters the Earth's atmosphere. Long wavelength radiation is radiated back from the Earth. See Figure 2.16.

4   Venus is closer to the Sun than Earth. This in itself would make the average temperature higher than on Earth. It also has an atmosphere with a large proportion of carbon dioxide. Since carbon dioxide is a greenhouse gas that traps heat energy, this would also raise the average temperature. So Venus has a much higher average temperature than Earth.

5   It is reflected back into space by clouds, dust, bright surfaces like snow etc.

6   It would be significantly cooler (by about 33 $^{\circ}$C). This would make it difficult for life to exist since water would mostly freeze. Also, without carbon dioxide, plants (as well as algae and cyanobacteria) would not exist. So animals would not exist either.

Lesson 6.33 Human activities

1   Deforestation / combustion of fossil fuels.

2   Trees (and other photosynthesising plants in the forest) use carbon dioxide from the atmosphere during photosynthesis. So if there are less trees, the carbon dioxide levels are likely to rise.

3   So that other experts in the field can check the validity of the data and reproducibility of the results. Much of the evidence is open to interpretation. When the evidence is peer reviewed, the validity of any conclusion can be assessed.

4   Decrease in crop yields / increase in desertification / flooding / sea level rising / glaciers melting / changing weather patterns etc.

5   Use the climate models to test the effect of different methane concentrations on the average Earth temperature. Look for evidence of increase in methane concentration in the past and relate to temperature changes e.g. ice cores and volcanic activity.

6   The data available is not clear cut and comes from indirect sources. There are different interpretations of the same data. The data is also very complex. It cannot be reduced to a 2 minute news item since this is not enough time to discuss a very complex subject.

Lesson 6.34 Global climate change

1   Burning fossil fuels. Deforestation, increased agriculture and decomposition of rubbish.

2   Decrease in crop yields / increase in desertification / flooding / sea level rising / coastal erosion / glaciers melting / changing weather patterns such as severe storms etc.

3   Weather patterns are changes in the weather that follow cycles. Climate change is a long term change in global climate that may have a big effect on weather patterns.

4   They are a reserve of fresh water. Without this fresh water, animal, plant and human life would be in jeopardy. Also, if they melt, sea levels will rise and land will be flooded.

5   Temperature stress: Global warming is increasing Earth's average temperature. This is causing problems for people in the Artic where glaciers are melting. Some areas will become much drier (desertification) as weather patterns change.

·   Water stress: Global warming will cause glaciers, which are a source of fresh water, to melt. Along with changes in rainfall patterns, this will limit the availability of fresh water.

·   Food production: As temperatures rise and weather patterns change (such as rainfall patterns), the capacity to grow crops may decrease.

·   Distribution of wildlife: As weather patterns and plant distribution change, animals will inevitably be affected. Species may die out or change their migration patterns.

Lesson 6.35 Carbon footprint and its reduction

1   Energy is used to produce the game (offices for staff, use of computers etc.), to advertise it, to download it and/or to manufacture it. Energy is required to play it in the form of electricity for a PC, laptop, tablet or game console. All of these cause greenhouse gases to be emitted (assuming the power doesn't come from renewable energy such as solar power).

2   The majority of experts believe that global warming is a reality and the best response is to reduce it. To limit and reduce the impact of global warming, carbon footprints have to be reduced. This will limit the emission of greenhouse gases.

3   It takes energy to make solar panels so greenhouse gas emissions inevitably occur. However, once in operation, they convert sunlight into heat/electrical energy without any carbon dioxide emissions.

4   Manufacturing solar panels is likely to produce greenhouse gas emissions so they have a carbon footprint. However, over the long lifetime that solar panels are used, they convert sunlight into heat/electrical energy without any carbon dioxide emissions. So when operating, they have no carbon footprint. There may be a carbon footprint when they come to the end of their life and are disposed of.

5   Cavity wall insulation: Slows heat loss through the walls so less energy used.

Double glazing: Slows heat loss through glass so less energy used.

Hot water tank with jacket on: Slows energy loss from hot water so less energy used.

Loft insulation: Slows heat loss through the roof so less energy used.

Low energy light bulbs: These consume less power than conventional light bulbs so are energy efficient.

Solar panels: These generate energy/electricity without consuming non-renewable fossil fuels and without producing greenhouse gases. They are more efficient at producing energy than conventional fossil fuel power generation.

Thermostat set low: This reduces the energy consumed. It uses the energy more efficiently than it would otherwise be if the thermostat was higher.

Appliances switched off: This prevents power consumption when on standby and therefore uses the energy more efficiently.

6   During their lifetime, trees and plants take in carbon dioxide during photosynthesis. Also, they trap the carbon dioxide by storing it as carbon in organic molecules. So they reduce the carbon footprint by reducing atmospheric carbon dioxide.

Lesson 6.36 Limitations on carbon footprint reduction

1   An increase in population means that additional methane is generated because of:

·   More livestock farming to feed an increasing population.

·   More deforestation (for wood, grazing land etc.).

·   More rice grow to feed an increasing population.

·   More rubbish generated.

2
- Drive less.
- Turn down thermostat for heating.
- Install solar panels.
- Walk/cycle more or use public transport.
- Plant a tree.
- Use local food.
- Recycle / reuse.

3 To improve public information about global warming and carbon footprints:
- Present the information and evidence more clearly.
- Limit the use of technical language.
- Highlight the fact that these issues are complex.
- Make the point that "do nothing" is not an option.
- Raise awareness of global warming and carbon footprint reduction.
- Make these issues part of the national curriculum.
- Avoid the use of sound bites and overly simplified science.
- Encourage people to find out more about these issues.

4 The information may be biased because:
- There are companies who benefit from selling fossil fuels. They may promote arguments that fossil fuels have a limited effect on global warming.
- There is disagreement about how to manage the economic impact.
- Some companies may benefit from carbon trading.
- Politicians may not fully understand the complexities of the science behind the arguments.
- There is disagreement about the science behind global warming.

5 The sports stadium may reduce emissions by generating "green" electricity. For instance by using solar and wind turbines. However, if the CO_2 emissions of thousands of spectators driving to and from the stadium, eating cooked food etc. is taken into account, the stadium has a very large carbon footprint. The stadium might say that it is the spectators that are responsible for their greenhouse gas emissions. Therefore, there would be a disagreement over the ownership of the carbon footprint.

Lesson 6.37 Atmospheric pollution from fuels

1 Sulfur dioxide / Nitrogen oxides / Carbon monoxide / Unburned hydrocarbons / Soot.

2 Insufficient oxygen to allow the fuel to react completely with oxygen. Instead of carbon dioxide, carbon monoxide (and carbon) are produced.

3 Carbon monoxide: Converted to carbon dioxide in the catalytic converter in a vehicle.
Sulfur dioxide: Reacts with limestone (calcium carbonate) in power stations / removed from coal and petroleum products by chemical or other means.

Oxides of nitrogen: Converted to nitrogen in the catalytic converter in a vehicle.
Unburnt hydrocarbons: Converted to carbon dioxide and water in the catalytic converter in a vehicle.
Soot: Trapped by filter in diesel cars.

4 Insufficient oxygen leads to incomplete oxidation of carbon in fuel.

5 $N_2 + O_2 \rightarrow 2 NO$

6 Soot, tobacco smoke and smog.

7 Particles which are less than 10 micrometers can penetrate the lungs. Particles which are less than 2.5 micrometers can penetrate the small sacs in the lungs called alveoli. Particulate contaminants such as soot are produced during incomplete combustion. Diesel cars produce a lot of particulate contaminants. Soot particles have an average diameter well below 2.5 micrometers and can potentially cause damage to the lungs. Pollution from cars causes smog. The particles in smog are Many of the particles in smog are below 2.5 micrometers and can also potentially damage lungs.

8 Use a filter to trap particulates in the exhaust. Burn the particulates after they leave the engine.

6.38 Properties and effects of atmospheric pollutants

1 The blood can carry less oxygen and therefore the heart (and lungs) has to work harder to get oxygen to the cells.

2 If appliances such as boilers are not maintained they can give out significant quantities of carbon monoxide. Carbon monoxide is colourless, odourless and toxic and is dangerous to humans and animals.

3 Fossil fuels often contain sulfur. When fossil fuels are burnt sulfur dioxide is formed. Also, when fossil fuels are burnt, oxides of nitrogen are formed. Both these dissolve in water to form an acidic solution - acid rain.

4 Acid rain causes respiratory problems in humans (particularly when photochemical smog forms) and plants. Acid rain also reacts with the materials in buildings (e.g. limestone) and damages them.

5 Average size of setting dust particles = 10 micrometers.
Average size of soot particles = 0.07 micrometers.
So 10/0.07 = 143

6 a Global dimming could lead to a reduction in the average global temperature since it reduces the amount of sunlight that reaches the Earth's surface. This is the opposite effect to global warming. However, global dimming could have masked some of the effect of global warming.

b As particulate levels drop, more sunlight will reach the Earth's surface. This will reduce global dimming and may lead to a greater increase in global warming.

Lesson 6.39 Potable water

1 • Rain collects in the ground (ground water), reservoirs, lakes and rivers.
• It is then treated in 3 stages.
• Sedimentation of particles. The particles drop to the bottom of the treatment tank.
• The water is filtered through sand so that the very fine particles are removed.
• Microbes such as bacteria are killed by sterilisation e.g. using chlorine, ozone or ultraviolet light.

2 So that the water is free of microbes when it is piped to the customer. If it was sterilised earlier, microbes might reappear later in the water treatment.

3 No. The water is contaminated with bacteria and needs to be sterilised.

4 The UK has lots of rain and therefore is self-sufficient when it comes to fresh water. Spain is a hot country with a limited supply of fresh water.

5 To distill water it has to be heated. Generally, the energy required to heat the water will come from the combustion of fossil fuels (although renewable energy could be used e.g. solar power).

6 • Collect rain water and use in the garden rather than use tap water.
• Buy toilets that use less water per flush.
• Wash fruit and vegetables in a bowl, not under a running tap.
• Use a washing machine and dishwasher only when they have a full load.
• Have a short shower rather than a bath.

7 • Recycle / reuse water.
• More efficient irrigation equipment
• Water crops only when there has not been rain.
• Design / modify the industrial process so that it uses less water.

8 a 1675/190 = 8.8 times the consumption in Sub-Saharan Africa.

b Lack of water availability, dry climate and lack of infrastructure to distribute water.

Lesson 6.40 Waste water treatment

1 Water from the oceans evaporates into the atmosphere. Then it condenses to form clouds. Clouds move and rain falls into rivers, lakes, and aquifers (an aquifer is an underground layer of water bearing rock).

2 There is a low population density in rural areas. It is not economical to pipe sewage to a treatment plant. So each house has a septic tank to deal with waste water.

3 Particles of sediment sink to the bottom of the water. However not all particles sink to the bottom. Some remain suspended.

4 Large population densities in urban areas produce large quantities of waste water and sewage. Waste water may contain pollutants that need to be removed. This cannot be dealt with "locally" due to shortage of space. So the sewerage system

has been designed to treat waste water at convenient locations.

5 Australia is one of the driest continents on Earth. Australia has a large water footprint. Most of it is used for agriculture. Therefore it will be difficult for Australia to reduce its water footprint.

6 Groundwater: Relatively easy. Can be easily extracted and does not need much treatment.
Waste water: Difficult. Contains a variety of solid waste and pollutants. Removal of these is a difficult process requiring several stages.
Saltwater: It is not difficult to distill salt water for instance. However it is expensive in terms of energy and difficult to meet demand for fresh water using this method.

Lesson 6.41 Practical: Analysis and purification of water samples from different sources, including pH, dissolved solids and distillation

1 A (0.5/1.0) x 30 = 15
B (0.2/1.0) x 30 = 6
C (0.3/1.0) x 30 = 9

2 Use a pH meter (universal indicator solution / paper could be used but would not be as accurate).

3 Is there any pollution from the dairy farm e.g. waste from cows, bacteria, antibiotics etc.
Are there any lead compounds in the water? Lead is toxic.

4 So that the water boils and the gas (steam) can be condensed into the beaker.

5 To condense the water after it has boiled in the flask.

6 The contents of the flask are heated. The water (solvent) evaporates. It then cools and condenses in the condenser. The dissolved salt does not evaporate.

7 To make sure that only water, which boils at 100 ^{o}C, is collected.

8 The forces between the oppositely charged ions in the salt (which attract) are much greater than the intermolecular force between water molecules. So at 100 ^{o}C, the water molecules have sufficient energy to break free of the liquid and become a gas but the salt doesn't.

9 The cooling is more effective / the condenser fills more completely / less air bubbles when the water flows from the bottom / greater flow of water.

10 All 3 samples were purified by distillation. The 3rd sample is the purest after distillation (closer to 100 ^{o}C) and therefore contained the least amount of salt.

11 Take more samples overall.
Take more samples in sections B and C since they are most likely to be polluted.
Sample the water at different times during the day (and night) since pollution may vary.

12 Repeat the distillation with the water that was distilled. Keep repeating until thd boiling point is (close to) 100 ^{o}C.

A more accurate thermometer would have allowed purity of the distillate to be judged more accurately.

A calibrated pH meter would give the most accurate pH readings.

13 Analyse for the presence of:
- Bacteria
- Ions such as Ca^{2+}
- Chlorine
- Suspended particles
- Organic material
- Medicines

14 a Before: 102 °C. After: 101.3 °C.

b Average in this case. It improves accuracy by reducing the effect of random error.

c Separate values. The samples indicate the pH at different locations which is of more interest - the pH may vary considerably depending on conditions. The average value would not give information about specific locations.

Lesson 6.42 Maths skills: Use ratios, fractions and percentages

1 a 95 % CO_2 / 5 % other gases.

b CO_2:other gases = 19:1.

2 75 % methane / 25 % ammonia.

3 Mg:HCl = 1:2

4 Fe:O_2 = 3:2

5 6:5

6 $N_2 + 2 O_2 \rightarrow 2 NO2$ (or N2O4 as product)

7 66 %

8 25.6 g

9 (0.58/5.37) x 100 = 10.8 %. Rounded to a whole number: 11 %

10 27.0 / 0.31 = 87.1 g

End of chapter questions

1 c) nitrogen and hydrogen are not recycled if they do not react (1)

2 Managed forest (1)

3 Bronze. (1)

4 Chlorination. (1)

5 C_3H_6 (1)

6 Addition. (1) -[X]-[X]-[X]- (1)

7 Wood is a renewable resource and is sustainable if forests are managed. (1) Crude oil is non-renewable and finite and not sustainable. (1)

8 Reuse / recycle the glass. (1) This reduces the environmental impact / reduces the use of finite resources / reduces energy consumption. (1)

9 Composites are made of a binder and reinforcement. Carbon fibre is the reinforcement and polymer resin the binder. (1)

10 a There is a demand for shorter chain hydrocarbons and alkenes. In cracking, longer chain hydrocarbons, which are not in as much demand, are cracked to produce shorter chain hydrocarbons and alkenes. (1)

b $C_7H_{16} \rightarrow C_5H_{12} + C_2H_4$ (1) Other equations are possible.

11 3 marks for method.
- Set up two boiling tubes.
- Put an identical iron nail in each.
- Add water to one tube.
- Add salt water to the other tube.
- Leave both tubes open to the air.
- Leave for a week

The nail in the salt water will rust more quickly. (1) (The presence of ions increases conductivity / increases the rate of oxidation of iron.)

12 A Metals are good electrical conductors (2) since they have delocalised electrons that can move / glass and ceramic have no free electrons that can move so are poor conductors. (2)

13 Using reverse osmosis / distillation (1)

14 To kill harmful bacteria / microbes. (1)

15 It is coated with a layer of zinc. The zinc prevents water and oxygen coming into contact with the iron. (1) It is also more reactive than iron so when the surface is scratched, it reacts instead of the iron. (1)

16 Nucleotides are joined together in a long chain. (1) Each nucleotide consists of a sugar, phosphate and a base. The four bases are A, T, G and C. (1)

17 The longer the lifetime, the longer it can act as a greenhouse gas and the greater its contribution to global warming. The greater the global warming potential, the greater its contribution to global warming. (1) Although methane has a shorter lifetime in the atmosphere, it has a greater global warming potential and can therefore make a major contribution to raising global temperatures.

18 The production of ammonia from nitrogen and hydrogen is exothermic in the forward direction. Lowering the temperature increases yield by moving the equilibrium to the right. (1) However, the rate is too slow at low temperatures. So a higher temperature is used which increased the rate at the expense of yield. (1)

19 The production of ammonia from nitrogen and hydrogen is exothermic in the forward direction. Lowering the temperature increases yield by moving the equilibrium to the right. (1) However, the rate is too slow at low temperatures. So a higher temperature is used which increased the rate at the expense of yield. (1)

20 Level 3: At least 3 correct identities, 2 points from other reactions of E and any correct balanced equation. (5-6)

Level 2: At least 2 correct identities and 2 points from other reactions of E. (3-4)

Level 1: At least 1 correct identity and 1 point from other reactions of E. (1-2)

Nothing written worthy of credit. (0)

Indicative content

Identity
- D: Hexane / C_6H_{14}.
- E. Ethene / C_2H_4.
- F. Butane / C_4H_{10}.

- G. Ethanol / C_2H_5OH

Equations
- $C_6H_{14} \rightarrow C_4H_{10} + C_2H_4$
- $C_2H_4 + H_2O \rightarrow C_2H_5OH$

Other reactions of E
- *E* is an alkene.
- Reaction with hydrogen to form ethane / an alkane.
- Conditions for reaction with hydrogen: nickel catalyst and high temperatures.
- $C_2H_4 + H_2 \rightarrow C_2H_6$
- Reaction with halogens such as bromine. Decolourisation of orange Br_2 with alkenes.
- $C_2H_4 + Br_2 \rightarrow C_2H_4Br_2$
- Addition polymerisation.
- Repeat unit or diagram of polymer chain shown.

Lesson number	Lesson title	Lesson objectives	OCR specification reference	Lesson resources (on CD ROM)	Collins Connect resources
		Chapter 1: Particles			
		C1.1 The particle model			
1.1	Three states of matter	• Use data to predict the states of substances. • Explain the changes of state. • Use state symbols in chemical equations.	C1.1a–c	Practical sheet; Worksheets 1 and 2; Technician's notes	Quick starter Homework worksheet Homework quiz Slideshow Video
		C1.1 Atomic structure			
1.2	Changing ideas about atoms	• Learn how models of the atom changed as scientists gathered more data. • Consider the data Rutherford and Marsden collected. • Link their data to our model of the atom.	C1.2a	Worksheets 1, 2 and 3; Technician's notes; Presentation	Quick starter Homework worksheet Homework quiz Video
1.3	Modelling the atom	• Explore the structure of atoms. • Consider the sizes of atoms. • Explore the way atomic radius changes with position in the periodic table.	C1.2b	Worksheet; Technician's notes; Presentation; Graph plotter	Quick starter Homework worksheet Homework quiz

 © HarperCollins*Publishers* Limited 2016

Programme of study matching chart

Lesson number	Lesson title	Lesson objectives	OCR specification reference	Lesson resources (on CD ROM)	Collins Connect resources
1.4	Key concept: Sizes of particles and orders of magnitude	• Identify the scale and measurements of length. • Explain the conversion of small lengths to metres. • Explain the relative sizes of electrons, nuclei and atoms.	C1.2c	Practical sheet; Worksheets 1 and 2; Technician's notes	Quick starter Homework worksheet Homework quiz Slideshow Video
1.5	Relating charges and masses	• Compare protons, neutrons and electrons. • Find out why atoms are neutral. • Relate the number of charged particles in atoms to their position in the periodic table.	C1.2d	Worksheet; Technician's notes; Presentation	Quick starter Homework worksheet Homework quiz
1.6	Subatomic particles	• Find out what the periodic table tells us about each element's atoms. • Learn what isotopes are. • Use symbols to represent isotopes.	C1.2e	Worksheets 1 and 2; Presentation	Quick starter Homework worksheet Homework quiz Slideshow
1.7	Maths skills: Standard form and making estimates	• Consider the sizes of particles. • Use numbers in standard form to compare sizes. • Use numbers in standard form in calculations.	CM1.2	Worksheet; Technician's notes; Presentation	Quick starter Homework worksheet Homework quiz Video

 © HarperCollinsPublishers Limited 2016

Chapter 2: Elements, compounds and mixtures

C2.1 Purity and separating mixtures

2.1	Key concept: Pure substances	• Describe, explain and exemplify processes of separation. • Suggest separation and purification techniques for mixtures. • Distinguish pure and impure substances using melting point and boiling point data.	C2.1a, b	Practical sheets 1 and 2; Worksheets 1 and 2; Technician's notes 1 and 2	Quick starter Homework worksheet Homework quiz
2.2	Relative formula mass	• Review the differences between the isotopes of an element. • Distinguish between the mass of an atom and the relative atomic mass of an element. • Use relative atomic masses to calculate relative formula masses.	C2.1c	Worksheets 1, 2 and 3; Technician's notes; Presentation	Quick starter Homework worksheet Homework quiz
2.3	Mixtures	• Recognise that all substances are chemicals. • Understand that mixtures can be separated into their components. • Suggest suitable separation and purification techniques for mixtures.	C2.1d	Worksheets 1 and 2; Practical sheet; Technician's notes	Quick starter Homework worksheet Homework quiz
2.4	Formulations	• Identify formulations given appropriate information. • Explain the particular purpose of each chemical in a mixture. • Explain how quantities are carefully measured for formulation.	C2.1e	Worksheets 1 and 2	Quick starter Homework worksheet Homework quiz

	Topic	Learning objectives	Spec ref	Resources	Additional resources
2.5	Chromatography	• Explain how to set up chromatography paper. • Distinguish pure from impure substances. • Interpret chromatograms and calculate R_f values.	C2.1f–k	Practical sheet; Worksheet; Technician's notes	Quick starter Homework worksheet Homework quiz Video
2.6	Practical: Investigate how paper chromatography can be used in forensic science to identify an ink mixture used in a forgery	• Describe the safe and correct manipulation of chromatography apparatus and how accurate measurements are achieved. • Make and record measurements used in paper chromatography. • Calculate R_f values.	PAG	Practical sheets 1 and 2; Technician's notes	Quick starter Homework worksheet Homework qu Quick starter Homework worksheet Homework quiz iz
2.7	Maths skills: Use an appropriate number of significant figures	• Measure distances on chromatograms. • Calculate R_f values. • Record R_f values to an appropriate number of significant figures.	CM2.1	Presentation	Quick starter Homework worksheet Homework quiz
		C2.2 Bonding			
2.8	Comparing metals and non-metals	• Review the physical properties of metals and non-metals. • Compare the oxides of metals and of non-metals. • Make predictions about unknown metals and non-metals.	C2.2a	Worksheet; Practical sheet; Technician's notes; Presentations 1 and 2	Quick starter Homework worksheet Homework quiz

2.9	Electron structure	• Find out how electrons are arranged in atoms. • Use diagrams and symbols to show which energy levels they occupy. • Relate each element's electron configuration to its position in the periodic table.	C2.2c	Worksheets 1, 2 and 3; Technician's notes; Presentation	Quick starter Homework worksheet Homework quiz
2.10	Metals and non-metals	• Explore the links between electron configurations of elements and their properties. • Find out what happens to the outer electrons when metals react. • Draw diagrams to show how ions form.	C2.2b	Worksheets 1 and 2; Technician's notes	Quick starter Homework worksheet Homework quiz Video
2.11	Chemical bonds	• Describe the three main types of bonding. • Explain how electrons are used in the three main types of bonding. • Explain how bonding and properties are linked.	C2.2d–h	Worksheets 1 and 2	Quick starter Homework worksheet Homework quiz
2.12	Ionic bonding	• Represent an ionic bond with a diagram. • Draw dot-and-cross diagrams for ionic compounds. • Work out the charge on the ions of metals from the group number of the element.		Practical sheet; Worksheet; Technician's notes	Quick starter Homework worksheet Homework quiz Video

356

© HarperCollins*Publishers* Limited 2016

Programme of study matching chart

2.13	Ionic compounds	<ul><li>Identify ionic compounds from structures.</li><li>Explain the limitations of diagrams and models.</li><li>Work out the empirical formula of an ionic compound.</li></ul>	Practical sheet; Worksheets 1, 2 and 3; Technician's notes	Quick starter Homework worksheet Homework quiz
2.14	Properties of ionic compounds	<ul><li>Describe the properties of ionic compounds.</li><li>Relate their melting points to forces between ions.</li><li>Explain when ionic compounds can conduct electricity.</li></ul>	Practical sheets 1 and 2; Worksheet; Technician's notes 1 and 2	Quick starter Homework worksheet Homework quiz Video
2.15	Properties of small molecules	<ul><li>Identify small molecules from formulae.</li><li>Explain the strength of covalent bonds.</li><li>Relate the intermolecular forces to the bulk properties of a substance.</li></ul>	Worksheets 1 and 2	Quick starter Homework worksheet Homework quiz Video
2.16	Covalent bonding	<ul><li>Identify single bonds in molecules and structures.</li><li>Draw dot-and-cross diagrams for small molecules.</li><li>Deduce molecular formulae from models and diagrams.</li></ul>	Worksheets 1 and 2	Quick starter Homework worksheet Homework quiz Video
2.17	Giant covalent structures	<ul><li>Recognise giant covalent structures from diagrams.</li><li>Explain the properties of giant covalent structures.</li><li>Recognise the differences in different forms of carbon.</li></ul>	Practical sheet; Worksheets 1 and 2; Technician's notes	Quick starter Homework worksheet Homework quiz

                    © HarperCollins*Publishers* Limited 2016

2.18	Polymer structures	• Recognise polymers from their unit formulae. • Explain why some polymers can stretch. • Explain why some plastics do not soften on heating.		Practical sheet; Worksheet; Technician's notes	Quick starter Homework worksheet Homework quiz
2.19	Metallic bonding	• Describe that metals form giant structures. • Explain how metal ions are held together. • Explain the delocalisation of electrons.		Practical sheet; Worksheets 1 and 2; Technician's notes	Quick starter Homework worksheet Homework quiz Video
2.20	Properties of metals and alloys	• Identify metal elements and their properties, and metal alloys. • Describe the purpose of a tin–lead alloy. • Explain why alloys have different properties to those of elements.		Practical sheet; Worksheets 1 and 2; Technician's notes	Quick starter Homework worksheet Homework quiz Slideshow 1 Slideshow 2
2.21	Key concept: The outer electrons	• Review the patterns in the periodic table. • Compare the trends in Group 1 and 7. • Relate these trends to the number of outer electrons and the sizes of atoms.		Worksheet; Technician's notes; Presentation	Quick starter Homework worksheet Homework quiz

2.22	The periodic table	• Explain how the electronic structure of atoms follows a pattern. • Recognise that the number of electrons in an element's atoms outer shell corresponds to the element's group number. • Use the periodic table to make predictions.	C2.2i	Worksheets 1, 2 and 3; Presentation	Quick starter Homework worksheet Homework quiz Video
2.23	Developing the periodic table	• Find out how the periodic table has changed over the years. • Explore Mendeleev's role in its development. • Consider the accuracy of Mendeleev's predictions.		Worksheets 1 and 2; Technician's notes; Presentation	Quick starter Homework worksheet Homework quiz
C2.3 Properties of materials					
2.24	Diamond	• Identify why diamonds are so hard. • Explain how the properties relate to the bonding in diamond. • Explain why diamond differs from graphite.	C2.3a–f	Worksheet	Quick starter Homework worksheet Homework quiz Slideshow Video
2.25	Graphite	• Describe the structure and bonding of graphite. • Explain the properties of graphite. • Explain the similarity to metals.		Worksheets 1 and 2	Quick starter Homework worksheet Homework quiz Video

 © HarperCollins*Publishers* Limited 2016

2.26	Fullerenes and graphene	• Describe the structure of graphene. • Explain the structure and uses of the fullerenes. • Explain the structure of nanotubes.		Worksheets 1 and 2	Quick starter Homework worksheet Homework quiz Video
2.27	Nanoparticles, their properties and uses	• Relate the sizes of nanoparticles to atoms and molecules. • Explain that there may be risks associated with nanoparticles. • Evaluate the use of nanoparticles for a specific purpose.	C2.3g–j	Worksheets 1, 2 and 3; Technician's notes	Quick starter Homework worksheet Homework quiz
2.28	Maths skills: Using ratios in mixture, empirical formulae and balanced equations	• Consider ways of comparing the amounts of gases in the atmosphere. • Review what balanced symbol equations show. • Compare the yields in chemical reactions.	CM1	Worksheets 1and 2 Technician's notes Presentation	Quick starter Homework worksheet Homework quiz Slideshow

          © HarperCollins*Publishers* Limited 2016

Chapter 3: Chemical reactions

C3.1 Introducing chemical reactions

3.1	Elements and compounds	• Identify symbols of elements from the periodic table. • Recognise the properties of elements and compounds. • Identify the elements in a compound.	C3.1a	Worksheets 1 and 2; Practical sheet; Technician notes	Quick starter Homework worksheet Homework quiz Slideshow
3.2	Atoms, formulae and equations	• Learn the symbols of the first 20 elements in the periodic table. • Use symbols to describe elements and compounds. • Use formulae to write equations.	C3.1b, c	Worksheets 1, 2 and 3; Practical sheet; Technician's notes; Presentation	Quick starter Homework worksheet Homework quiz Slideshow Video
3.3	Moles	• Describe the measurements of amounts of substances in moles. • Calculate the amount of moles in a given mass of a substance. • Calculate the mass of a given number of moles of a substance.	C3.1g–i	Worksheets 1 and 2; Technician's notes	Quick starter Homework worksheet Homework quiz
3.4	Key concept: Conservation of mass and balanced equations	• Explore ideas about the conservation of mass. • Consider what the numbers in equations stand for. • Write balanced symbol equations.	Key Concept C3.1d–f	Worksheet; Technician's notes; Presentation	Quick starter Homework worksheet Homework quiz Slideshow Video

 © HarperCollins*Publishers* Limited 2016

		Key Concept	Worksheets 1 and 2	Quick starter Homework worksheet Homework quiz Slideshow
3.5	Key concept: Amounts in chemistry	• Use atomic masses to calculate formula masses. • Explain how formula mass relates to the number of moles. • Explain how the number of moles relates to other quantities.		
3.6	Mass changes when gases are in reactions	C3.1j	Worksheet; Practical sheet; Technician's notes; Presentations 1 and 2; Graph plotter	Quick starter Homework worksheet Homework quiz Slideshow
		• Find out how mass can be gained or lost during a reaction. • Find the mass of carbon dioxide released per gram of copper carbonate decomposed. • Assess the accuracy of our measurements.		
3.7	Using moles to balance equations	C3.1k	Worksheet	Quick starter Homework worksheet Homework quiz Slideshow
		• Convert masses in grams to amounts in moles. • Balance an equation given the masses of reactants and products. • Change the subject of a mathematical equation.		
3.8	Key concept: Limiting reactants and molar masses	Key Concept	Technician's notes; Practical sheet; Worksheets 1 and 2; Presentation	Quick starter Homework worksheet Homework quiz Slideshow Video
		• Recognise when one reactant is in excess. • Consider how this affects the amount of product made. • Explore ways of increasing the amount of product.		

Programme of study matching chart

3.9	Amounts of substances in equations	• Calculate the masses of substances in a balanced symbol equation. • Calculate the masses of reactants and products from balanced symbol equations. • Calculate the mass of a given reactant or product.	C3.1l	Worksheet	Quick starter Homework worksheet Homework quiz Slideshow
C3.2 Energetics					
3.10	Endothermic and exothermic reactions	• Explore the temperature changes produced by chemical reactions. • Consider how reactions are used to heat or cool their surroundings. • Investigate how these temperature changes can be controlled.	C3.2a	Technician's notes; Practical sheet; Worksheet; Graph plotters 1 and 2	Quick starter Homework worksheet Homework quiz Slideshow
3.11	Reaction profiles	• Use diagrams to show the energy changes during reactions. • Show the difference between exothermic and endothermic reactions using energy profiles. • Find out why many reactions start only when energy or a catalyst is added.	C3.2b, c	Technician's notes; Worksheets 1 and 2; Presentation	Quick starter Homework worksheet Homework quiz Video
3.12	Energy change of reactions	• Identify the bonds broken and formed during a chemical reaction. • Consider why some reactions are exothermic and others are endothermic. • Use bond energies to calculate overall energy changes.	C3.2d	Technician's notes; Worksheets 1 and 2; Presentation	Quick starter Homework worksheet Homework quiz

 © HarperCollins*Publishers* Limited 2016

Programme of study matching chart

3.13	Maths skill: Recognise and use expressions in decimal form	• Read scales in integers and using decimals. • Calculate the energy change during a reaction. • Calculate energy transferred for comparison.	CM3.2	Practical sheet; Worksheet; Technician's notes; Presentation	Quick starter Homework worksheet Homework quiz

C3.3 Types of chemical reaction

3.14	Oxidation and reduction in terms of electrons	• Observe some reactions between metal atoms and metal ions. • Learn to write ionic equations and half equations. • Classify half equations as oxidation or reduction.	C3.3a, b	Technician's notes; Practical sheet; Worksheet; Presentations 1 and 2	Quick starter Homework worksheet Homework quiz
3.15	Key concept: Electron transfer, oxidation and reduction	• Review ion formation. • Classify half equations as oxidation or reduction. • Review patterns in reactivity.	Key Concept C3.3c	Worksheet; Presentation	Quick starter Homework worksheet Homework quiz Video
3.16	Neutralisation of acids and salt production	• React an acid and an alkali to make a salt. • Predict the formulae of salts. • Write balanced symbol equations.	C3.3d	Technician's notes; Practical sheet; Worksheets 1 and 2	Quick starter Homework worksheet Homework quiz Video
3.17	Soluble salts	• React an acid and a metal to make a salt. • Predict the formulae of salts. • Write balanced symbol equations and half equations.		Technician's notes; Practical sheet; Worksheets 1 and 2; Presentation and 2	Quick starter Homework worksheet Homework quiz Video

3.18	Reaction of metals with acids	• React an acid and a metal to make a salt. • Predict the formulae of salts. • Write balanced symbol equations and half equations.		Technician's notes; Practical sheet; Worksheets 1 and 2; Presentations 1 and 2	Quick starter Homework worksheet Homework quiz
3.19	Practical: Preparing a pure, dry sample of a soluble salt from an insoluble oxide or carbonate	• React a carbonate with an acid to make a salt. • Describe each step in the procedure. • Determine the purity of the product.	PAG	Technician's notes; Practical sheet; Presentations 1 and 2	Quick starter Homework worksheet Homework quiz
3.20	pH and neutralisation	• Estimate the pH of solutions. • Identify weak and strong acids and alkalis. • Investigate pH changes when a strong acid neutralises a strong alkali.	C3.3g–k	Technician's notes 1, 2 and 3; Practical sheet; Worksheet; Presentation	Quick starter Homework worksheet Homework quiz
3.21	Strong and weak acids	• Explore the factors that affect the pH of an acid. • Find out how the pH changes when an acid is diluted. • Find out how the concentrations of solutions are measured.		Worksheet	Quick starter Homework worksheet Homework quiz Video
3.22	Maths skill: Make order of magnitude calculations	• Explore the factors that affect the acidity of rain. • Find out how acid concentrations are compared. • Explore the link between hydrogen ion concentration and pH.	CM3.3	Technician's notes; Practical sheet; Presentation	Quick starter Homework worksheet Homework quiz

		PAG		
3.23	Practical: Investigate the variables that affect temperature changes in reacting solutions, such as, acid plus metals, acid plus carbonates, neutralisations, displacement of metals	• Devise a hypothesis. • Devise an investigation to test your hypothesis. • Decide whether the evidence supports your hypothesis.		Technician's notes; Practical sheet; Presentation	Quick starter Homework worksheet Homework quiz
			C3.4 Electrolysis		
3.24	The process of electrolysis	• Explore what happens when a current passes through a solution of ions. • Find out what an electrolyte is and what happens when it conducts electricity. • Find out how electricity decomposes compounds.	C3.4a	Technician's notes; Practical sheet; Worksheet; Presentation	Quick starter Homework worksheet Homework quiz Slideshow Video
3.25	Electrolysis of molten ionic compounds	• Look in detail at the electrolysis of lead bromide. • Communicate the science behind the extraction of elements from molten salts. • Write balanced half equations for electrolysis reactions.	C3.4b	Worksheet; Presentation	Quick starter Homework worksheet Homework quiz

366 © HarperCollins*Publishers* Limited 2016

3.26	Electrolysis of aqueous solutions	• Investigate the products formed when copper sulfate is electrolysed. • Predict what products other solutions will give. • Write half equations for reactions at electrodes.	C3.4c–e	Technician's notes; Worksheet; Practical sheet; Presentation	Quick starter Homework worksheet Homework quiz
3.27	Practical: Investigating what happens when aqueous solutions are electrolysed using inert electrodes	• Devise a hypothesis. • Devise an investigation to test your hypothesis. • Decide whether the evidence supports your hypothesis.	PAG	Technician's notes; Practical sheet; Presentations 1 and 2	Quick starter Homework worksheet Homework quiz

© HarperCollins*Publishers* Limited 2016

Chapter 4: Predicting and identifying reactions and products

C4.1 Predicting chemical reactions

4.1	Exploring Group 0	• Explore the properties of noble gases. • Find out how the mass of their atoms affects their boiling points. • Relate their chemical properties to their electronic structures.	C4.1a, b	Worksheet; Graph plotter 1; Presentations 1 and 2	Quick starter Homework worksheet Homework quiz Slideshow
4.2	Exploring Group 1	• Explore the properties of Group 1 metals. • Compare their reactivity. • Relate their reactivity to their electronic structures.		Worksheets 1, 2 and 3; Technician's notes; Presentation	Quick starter Homework worksheet Homework quiz
4.3	Exploring Group 7	• Explain why Group 7 non-metals are known as 'halogens'. • Compare their reactivity. • Relate their reactivity to their electronic structures.		Worksheets 1, 2 and 3; Technician's notes 1 and 2; Presentation	Quick starter Homework worksheet Homework quiz Slideshow
4.4	Transition metals	• Compare the properties of transition metals with those of Group 1 metals. • Explore the uses of transition metals. • Find out why they can form compounds with different colours.	C4.1c	Worksheet; Technician's notes; Presentation	Quick starter Homework worksheet Homework quiz Slideshow

Programme of study matching chart

4.5	Reaction trends and predicting reactions	• Review the patterns in the periodic table. • Compare the trends in Group 1 and Group 7. • Relate these trends to the way atoms form ions.	C4.1d	Worksheet; Presentation	Quick starter Homework worksheet Homework quiz
4.6	Reactivity series	• Compare the reactivity of metals. • Observe some reactions between metal atoms and metal ions. • Consider why some metals are more reactive than others.	C4.1e, f	Practical sheet; Worksheet; Presentations 1 and 2	Quick starter Homework worksheet Homework quiz Slideshow
C4.2 Identifying the products of chemical reactions					
4.7	Tests for gases	• Recall the tests for four common gases. • Identify the four common gases using these tests. • Explain why limewater can be used to detect carbon dioxide.	C4.2a	Practical sheet; Technician's notes; Presentation	Quick starter Homework worksheet Homework quiz
4.8	Metal hydroxides	• Recognise the precipitate colour of metal hydroxides. • Explain how to use sodium hydroxide to test for metal ions. • Write balanced equations for producing insoluble metal hydroxides.	C4.2b–d	Practical sheet; Worksheet; Technician's notes	Quick starter Homework worksheet Homework quiz Slideshow

© HarperCollins*Publishers* Limited 2016

4.9	Tests for anions	• Identify the tests for carbonates. • Explain the tests for halides and sulfates. • Identify anions and cations from the results of tests.		Practical sheets 1 and 2; Worksheet; Technician's notes 1 and 2	Quick starter Homework worksheet Homework quiz
4.10	Flame tests	• Carry out flame-test procedures. • Identify the colours of flames of ions. • Identify species from the results of the tests.	C4.2e	Practical sheet; Worksheet; Technician's notes	Quick starter Homework worksheet Homework quiz
4.11	Instrumental methods	• Identify advantages of instrumental methods compared with the chemical tests. • Describe some instrumental techniques. • Explain the data provided by instrumental techniques.	C4.2f, g	Worksheets 1 and 2	Quick starter Homework worksheet Homework quiz Slideshow Video
4.12	Practical: Use chemical tests to identify the ions in unknown single ionic compounds	• Describe how to carry out experiments safely using the correct manipulation of apparatus for the qualitative analysis of ions. • Make and record observations using flame tests and precipitation methods. • Identify unknown ions in chemical compounds.	PAG	Practical sheets 1 and 2; Technician's notes	Quick starter Homework worksheet Homework quiz

Chapter 5: Monitoring and controlling chemical reactions

C5.1 Monitoring chemical reactions

5.1	Concentration of solutions	• Relate mass, volume and concentration. • Calculate the mass of solute in solution. • Relate concentration in mol/dm^3 to mass and volume.	C5.1a–c	Practical sheet; Worksheet; Technician's notes	Quick starter Homework worksheet Homework quiz
5.2	Using concentrations of solutions	• Describe how to carry out titrations. • Calculate concentrations in titrations in mol/dm^3 and in g/dm^3. • Explain how the concentration of a solution in mol/dm^3 is related to the mass of the solute and the volume of the solution.		Practical sheet; Worksheet; Technician's notes	Quick starter Homework worksheet Homework quiz Slideshow
5.3	Practical: Finding the reacting volumes of solutions of acid and alkali by titration	• Use an acid to neutralise a known volume of alkali. • Use a burette to determine the volume of an acid needed. • Use the results to determine the concentration of an alkali.		Technician's notes; Practical sheet; Presentation and 2	Quick starter Homework worksheet Homework quiz
5.4	Amounts of substance in volumes of gases	• Explain that the same amount of any gas occupies the same volume at room temperature and pressure (rtp). • Calculate the volume of a gas at rtp from its mass and relative formula mass. • Calculate the volumes of gases from a balanced equation and a given volume of a reactant or product.	C5.1d–f	Practical sheet 3.12; Worksheet 3.12; Technician's notes 3.12	Quick starter Homework worksheet Homework quiz Video

			Key Concept		
5.5	Key Concept: Percentage yield	• Calculate the percentage yield from the actual yield. • Identify the balanced equation needed for calculating yields. • Calculate theoretical product amounts from reactant amounts.	C5.1g, h	Practical sheet; Worksheet; Technician's notes	Quick starter Homework worksheet Homework quiz Video
5.6	Atom economy	• Identify the balanced equation of a reaction. • Calculate the atom economy of a reaction to form a product. • Explain why a particular reaction pathway is chosen.	C5.1i–k	Worksheets 1 and 2.	Quick starter Homework worksheet Homework quiz
5.7	Maths skills: Change the subject of an equation	• Use equations to demonstrate conservation. • Rearrange the subject of an equation. • Carry out multi-step calculations.	CM5.1	Worksheet	Quick starter Homework worksheet Homework quiz Video
	C5.2 Controlling reactions				
5.8	Measuring rates	• Measure the volume of gas given off during a reaction. • Use the results to measure the reaction rate. • Explore how the rate changes during the reaction.	C5.2a	Technician's notes; Practical sheet; Worksheet; Presentation 1 and 2; Graph plotter 1	Quick starter Homework worksheet Homework quiz
5.9	Calculating rates	• Find out how to calculate rates of reaction. • Use graphs to compare reaction rates. • Use tangents to measure rates that change.	C5.2b	Technician's notes; Practical sheet; Worksheet; Presentations 1 and 2	Quick starter Homework worksheet Homework quiz

 © HarperCollins*Publishers* Limited 2016

5.10	Factors affecting rates	• Measure the time taken to produce a specific amount of product. • See how a reactant's temperature or concentration can affect this time. • Investigate the effect of breaking up a solid reactant into smaller pieces.	C5.2c	Technician's notes; Practical sheet; Worksheets 1 and 2; Graph plotter 1; Presentation	Quick starter Homework worksheet Homework quiz Video
5.11	Collision theory	• Find out about the collision theory. • Use collision theory to make predictions about reaction rates. • Relate activation energies to collision theory.	C5.2d	Worksheets 1 and 2; Presentations 1 and 2	Quick starter Homework worksheet Homework quiz
5.12	Catalysts	• Investigate catalysts. • Find out how catalysts work. • Learn how they affect activation energy.	C5.2f–i	Technician's notes; Practical sheet; Worksheet	Quick starter Homework worksheet Homework quiz Videos
5.13	Factors increasing the rate	• Interpret graphs. • Consider what determines the reaction rate. • Explore the effect of changing the amounts of reactants used.	C5.2e	Technician's notes; Practical sheet; Worksheet; Presentation	Quick starter Homework worksheet Homework quiz Slideshow

Programme of study matching chart

5.14	Practical: Investigate how changes in concentration affect the rates of reactions by a method involving the production of a gas and a method involving a colour change	• Devise a hypothesis. • Devise an investigation to test a hypothesis. • Decide whether the evidence supports a hypothesis.	PAG	Technician's notes; Practical sheets 1 and 2; Presentations 1 and 2; Graph plotter 1	Quick starter Homework worksheet Homework quiz

C5.3 Reversible reactions

5.15	Reversible reactions and energy changes	• Investigate reversible reactions. • Explore the energy changes in a reversible reaction. • Find out how reaction conditions affect reversible reactions.	C5.3a	Technician's notes; Practical sheet; Worksheet; Presentation	Quick starter Homework worksheet Homework quiz
5.16	Equilibrium	• Recognise reactions that can reach equilibrium. • Find out what happens to the reactants and products at equilibrium. • Use Le Chatelier's principle to make predictions.	C5.3b	Technician's notes; Worksheet; Presentation	Quick starter Homework worksheet Homework quiz
5.17	Changing concentration and equilibrium	• Distinguish between reactants and products. • Explore how changing their concentrations affects reversible reactions. • Use Le Chatelier's principle to make predictions about changing concentrations.	C5.3c	Technician's notes; Worksheet; Presentation	Quick starter Homework worksheet Homework quiz
5.18	Changing temperature and equilibrium	• Distinguish between exothermic and endothermic forward reactions. • Explore how changing the temperature affects reversible reactions. • Use Le Chatelier's principle to make predictions about changing temperatures.		Technician's notes; Worksheets 1 and 2; Presentation	Quick starter Homework worksheet Homework quiz

5.19	Changing pressure and equilibrium	<ul><li>Recognise the number of product and reactant molecules in a reaction.</li><li>Explore how changing the pressure affects reversible reactions.</li><li>Use Le Chatelier's principle to make predictions about changing pressures.</li></ul>		Worksheet; Presentation	Quick starter Homework worksheet Homework quiz
5.20	Maths skills: Use the slope of a tangent as a measure of rate of change	<ul><li>Practice drawing graphs.</li><li>Use graphs to compare reaction rates.</li><li>Use tangents to measure rates that change.</li></ul>	CM5.3	Worksheets 1 and 2; Presentations 1 and 2	Quick starter Homework worksheet Homework quiz Video

 © HarperCollins*Publishers* Limited 2016

Chapter 6: Global challenges

C6.1 Improving processes and products

6.1	Extraction of metals	• Find out where metals come from. • Extract iron from its oxide using carbon. • Consider how other metals are extracted from their ores.	C6.1a, b	Technician's notes; Practical sheet; Worksheet; Presentation	Quick starter Homework worksheet Homework quiz Slideshow
6.2	Using electrolysis to extract metals	• Review the connection between the reactivity series and the ways metals are extracted. • Consider how aluminium is extracted from aluminium oxide. • Learn the oxidation and reduction reactions involved.		Worksheet; Presentations 1 and 2	Quick starter Homework worksheet Homework quiz Slideshow
6.3	Alternative methods of metal extraction	• Describe the process of phytomining. • Describe the process of bioleaching. • Evaluate alternative biological methods of metal extraction.	C6.1c	Practical sheet; Worksheets 1 and 2; Technician's notes.	Quick starter Homework worksheet Homework quiz
6.4	The Haber process	• Apply principles of dynamic equilibrium to the Haber process. • Use graphs to explain the trade off with rate and equilibrium. • Explain how commercially used conditions relate to cost.	C6.1d–g	Practical sheet; Worksheets 1 and 2; Technician's notes	Quick starter Homework worksheet Homework quiz

6.5	Production and use of NPK fertilisers	• Describe how to make a fertiliser in the laboratory. • Explain how fertilisers are produced industrially. • Compare the industrial production with laboratory preparation.	C6.1h	Practical sheet; Worksheet; Technician's notes	Quick starter Homework worksheet Homework quiz
6.6	Life cycle assessment and recycling	• Describe the components of a life cycle assessment (LCA). • Interpret LCAs of materials or products from information. • Carry out a simple comparative LCA for shopping bags.	C6.1i–l	Worksheets 1 and 2	Quick starter Homework worksheet Homework quiz
6.7	Ways of reducing the use of resources	• Describe ways of recycling and reusing materials. • Explain why recycling, reusing and reducing are needed. • Evaluate ways of reducing the use of limited resources.		Worksheets 1 and 2.	Quick starter Homework worksheet Homework quiz Video
6.8	Alloys as useful materials	• Describe the composition of common alloys. • Interpret the composition of other alloys from data. • Evaluate the uses of other alloys.	C6.1m	Worksheets 1, 2 and 3	Quick starter Homework worksheet Homework quiz
6.9	Corrosion and its prevention	• Show that air and water are needed for rusting. • Describe experiments and interpret results on rusting. • Explain methods for preventing corrosion.	C6.1n, o	Practical sheets 1 and 2; Worksheets 1 and 2; Technician's notes 1 and 2	Quick starter Homework worksheet Homework quiz Video

6.10	Ceramics, polymers and composites	• Compare quantitatively properties of materials. • Compare glass, ceramics, polymers, composites and metals. • Select materials by relating their properties to uses.	C6.1p	Worksheets 1, 2 and 3	Quick starter Homework worksheet Homework quiz Video
6.11	Maths skill: Translate information between graphical and numerical form	• Represent information from pie charts numerically. • Represent information from graphs numerically. • Represent numeric information graphically.	CM6.1	Worksheets 1 and 2	Quick starter Homework worksheet Homework quiz Video
C6.2 Organic chemistry					
6.12	Function groups and homologous series	• Identify the first four hydrocarbons in the alkane series. • Name the first four compounds in homologous series. • Identify the functional group of a series.	C6.2a, C6.2b	Worksheet	Quick starter Homework worksheet Homework quiz
6.13	Structure and formulae of alkenes	• Describe the difference between an alkane and an alkene. • Draw the displayed structural formulae for the first four members of the alkenes. • Explain why alkenes are called unsaturated molecules.		Worksheets 1 and 2	Quick starter Homework worksheet Homework quiz

Programme of study matching chart

6.14	Reaction of alkenes	<ul><li>Describe the addition reactions of alkenes.</li><li>Draw the full displayed structural formulae of the products alkenes make.</li><li>Explain how alkenes react with hydrogen, water and the halogens.</li></ul>		Practical sheet; Worksheets 1 and 2; Technician's notes	Quick starter Homework worksheet Homework quiz
6.15	Alcohols	<ul><li>Recognise alcohols from their name or from given formulae.</li><li>Describe the conditions used for the fermentation of sugar using yeast.</li><li>Write balanced chemical equations for the combustion of alcohols.</li></ul>		Practical sheets 1 and 2; Worksheets 1 and 2; Technician's notes	Quick starter Homework worksheet Homework quiz
6.16	Carboxylic acids	<ul><li>Describe the reactions of carboxylic acids.</li><li>Recognise carboxylic acids from their formulae.</li><li>Explain the reaction of ethanoic acid with an alcohol.</li></ul>		Practical sheet; Worksheets 1 and 2; Technician's notes	Quick starter Homework worksheet Homework quiz Video
6.17	Addition polymerisation	<ul><li>Recognise addition polymers and monomers from diagrams.</li><li>Draw diagrams of the formation of a polymer from an alkene.</li><li>Relate the repeating unit of the polymer to the monomer.</li></ul>	C6.2a–g	Worksheets 1 and 2	Quick starter Homework worksheet Homework quiz
6.18	Condensation polymerisation	<ul><li>Explain the basic principles of condensation polymerisation.</li><li>Explain the role of functional groups in producing a condensation polymer.</li><li>Explain the structure of the repeating units in a condensation polymer.</li></ul>		Practical sheet; Worksheets 1 and 2; Technician's notes	Quick starter Homework worksheet Homework quiz

					Resources
6.19	Amino acids	• Describe the functional group of an amine. • Identify the two functional groups of an amino acid. • Explain how different amino acids build proteins.		Worksheet	Quick starter Homework worksheet Homework quiz
6.20	DNA and other naturally occurring polymers	• Describe the components of natural polymers. • Explain the structure of proteins and carbohydrates. • Explain how a molecule of DNA is constructed.	C6.2h, i	Worksheets 1 and 2	Quick starter Homework worksheet Homework quiz
6.21	Crude oil, hydrocarbons and alkanes	• Describe why crude oil is a finite resource. • Identify the hydrocarbons in the series of alkanes. • Explain the structure and formulae of the alkanes.	C6.2l–o	Worksheets 1 and 2	Quick starter Homework worksheet Homework quiz
6.22	Fractional distillation and petrochemicals	• Describe how crude oil is used to provide modern materials. • Explain how crude oil is separated by fractional distillation. • Explain why the boiling points of the fractions are different.	C6.2j, k	Practical sheet; Worksheets 1 and 2; Technician's notes	Quick starter Homework worksheet Homework quiz Video
6.23	Properties of hydrocarbons	• Describe how different hydrocarbon fuels have different properties. • Identify the properties that influence the use of fuels. • Explain how the properties are related to the size of the molecules.		Worksheets 1 and 2	Quick starter Homework worksheet Homework quiz Video

6.24	Cracking and alkenes	• Describe the usefulness of cracking. • Balance chemical equations as examples of cracking. • Explain why modern life depends on the uses of hydrocarbons.		Practical sheet; Worksheet Technician's notes	Quick starter Homework worksheet Homework quiz Slideshow 1 Slideshow 2
6.25	Cells and batteries	• Make simple cells and measure their voltages. • Consider the importance of cells and batteries. • Find out how larger voltages can be produced.	C6.2p	Technician's notes; Practical sheet; Worksheets 1 and 2	Quick starter Homework worksheet Homework quiz
6.26	Fuel cells	• Find out how fuel cells work. • Compare and contrast the uses of hydrogen fuel cells, batteries and rechargeable cells. • Learn what reactions take place inside hydrogen fuel cells.	C6.2q	Technician's notes; Worksheet; Presentations 1 and 2	Quick starter Homework worksheet Homework quiz Video
6.27	Key concept: Intermolecular forces	• Identify the bonds within a molecule and the forces between molecules. • Explain changes of state. • Explain how polymer structure determines its ability to stretch.	Key Concept	Worksheets 1 and 2	Quick starter Homework worksheet Homework quiz Video
6.28	Maths skills: Visualise and represent 3D models	• Use 3D models to represent hydrocarbons. • Use 3D models to represent polymers. • Use 3D models to represent large biological molecules.	CM6.2	Worksheets 1 and 2	Quick starter Homework worksheet Homework quiz

382 © HarperCollins*Publishers* Limited 2016

Programme of study matching chart

C6.3 Interpreting and interacting with Earth systems

6.29	Proportions of gases in the atmosphere	• Review the composition of the atmosphere. • Measure the percentage of oxygen in the atmosphere. • Consider why it stays the same.	C6.3a	Worksheets 1 and 2; Technician's notes	Quick starter Homework worksheet Homework quiz
6.30	The Earth's early atmosphere	• Explore the origins of the Earth's atmosphere. • Consider the evidence that ideas about the early atmosphere are based on. • Consider the strength of the evidence these ideas are based on.		Worksheet	Quick starter Homework worksheet Homework quiz Video
6.31	How oxygen increased	• Explore the processes that changed the oxygen concentration in the atmosphere. • Consider the role of algae. • Consider why oxygen levels in the atmosphere didn't rise when oxygen was first produced.	C6.3b	Worksheet; Presentation	Quick starter Homework worksheet Homework quiz Slideshow
6.32	Greenhouse gases	• Review the greenhouse effect. • Explain how greenhouse gases trap heat. • Consider the consequences of adding greenhouse gases to the atmosphere.	Key concept C6.3c–f	Worksheet; Presentation	Quick starter Homework worksheet Homework quiz
6.33	Human activities	• Consider the factors that affect the quality of scientific reports. • Consider the reliability of computer models. • Find out what peer review involves.		Worksheet; Presentations 1 and 2	Quick starter Homework worksheet Homework quiz Video

6.34	Global climate change	• Explore the consequences of climate change. • Consider the risks to human health. • Judge the seriousness of these consequences.	Worksheet; Presentation	Quick starter Homework worksheet Homework quiz Slideshow
6.35	Carbon footprint and its reduction	• Find out what a carbon footprint is. • Consider factors that contribute to our carbon footprints. • Explore ways of reducing our carbon footprints.	Worksheet; Presentation	Quick starter Homework worksheet Homework quiz Video
6.36	Limitations on carbon footprint reduction	• Review the uncertainties about carbon emissions. • Consider factors which limit our ability to reduce our carbon footprints. • Decide which factors are most important.	Worksheets 1 and 2; Presentation	Quick starter Homework worksheet Homework quiz
6.37	Atmospheric pollutants from fuels	• Explore the products formed when fuels burn. • Distinguish between complete and incomplete combustion. • Write equations for complete and incomplete combustion.	Worksheets 1, 2, 3 and 4; Technician's notes; Presentation	Quick starter Homework worksheet Homework quiz Slideshow
6.38	Properties and effects of atmospheric pollutants	• Review the hazards associated with air pollutants. • Investigate correlations between pollutant emissions and deaths from asthma. • Consider whether these support the hypothesis that air pollution makes asthma worse.	Worksheets 1 and 2; Presentations 1 and 2	Quick starter Homework worksheet Homework quiz

6.39	Potable water	• Distinguish between potable water and pure water. • Describe the differences in treatment of ground water and salty water. • Explain what is needed to provide potable water for all.	C6.3g–j	Worksheets 1, 2 and 3; Technician's notes	Quick starter Homework worksheet Homework quiz Video
6.40	Waste water treatment	• Explain how waste water is treated. • Describe how sewage is treated. • Compare the ease of treating waste, ground and salt water.		Worksheets 1, 2 and 3	Quick starter Homework worksheet Homework quiz Video
6.41	Practical: Analysis and purification of water samples from different sources, including pH, dissolved solids and distillation	• Describe how safety is managed, apparatus is used and accurate measurements are made. • Recognise when sampling techniques need to be used and made representative. • Carry out a procedure to produce potable water from salt solution. • Evaluate methods and suggest possible improvements and further investigations.	PAG	Practical sheets 1 and 2; Worksheet; Technician's notes	Quick starter Homework worksheet Homework quiz
6.42	Maths skills: Use ratios, fractions and percentages	• Consider ways of comparing the amounts of gases in the atmosphere. • Review what balanced symbol equations show. • Compare the yields in chemical reactions.	CM6.3	Worksheets 1 and 2; Technician's notes; Presentation	Quick starter Homework worksheet Homework quiz Video

Using Collins OCR Gateway GCSE (9–1) Chemistry Student Book and Teacher Pack to teach OCR GCSE (9–1) Combined Science

The following table gives an overview of the Student Book pages and Lesson Plans that can be used to teach Combined Science.

Student Book and Lesson Plan Reference	Chapter	Notes on how to use the resources for Combined Science
Chapter 1	**Particles**	
Student Book only	Chapter Introduction	Combined Science students need to know and understand all content.
1.1	Three states of matter	Combined Science students need to know and understand all content.
1.2	Changing ideas about atoms	Combined Science students need to know and understand all content.
1.3	Modelling the atom	Combined Science students need to know and understand all content.
1.4	Key concept: Sizes of particles and orders of magnitude	Combined Science students need to know and understand all content.
1.5	Relating charges and masses	Combined Science students need to know and understand all content.
1.6	Subatomic particles	Combined Science students need to know and understand all content.
1.7	Maths skills: Standard form and making estimates	Combined Science students need to know and understand all content.
	Check your progress	Combined Science students need to know and understand all content.
	Worked example	Combined Science students need to know and understand all content.
	End of chapter questions	Combined Science students need to know and understand all content.

Chapter 2	Elements, compounds and mixtures	
Student Book only	Chapter Introduction	Combined Science students need to know and understand all content.
2.1	Key concept: Pure substances	Combined Science students need to know and understand all content.
2.2	Relative formula mass	Combined Science students need to know and understand all content.
2.3	Mixtures	Combined Science students need to know and understand all content.
2.4	Formulations	Combined Science students need to know and understand all content.
2.5	Chromatography	Combined Science students need to know and understand all content.
2.6	Practical: Investigate how paper chromatography can be used in forensic science to identify an ink mixture used in a forgery	Combined Science students need to know and understand all content.
2.7	Maths skills: Use an appropriate number of significant figures	Combined Science students need to know and understand all content.
2.8	Comparing metals and non-metals	Combined Science students need to know and understand all content.
2.9	Electron structure	Combined Science students need to know and understand all content.
2.10	Metals and non-metals	Combined Science students need to know and understand all content.
2.11	Chemical bonds	Combined Science students need to know and understand all content.
2.12	Ionic bonding	Combined Science students need to know and understand all content.
2.13	Ionic compounds	Combined Science students need to know and understand all content.
2.14	Properties of ionic compounds	Combined Science students need to know and understand all content.
2.15	Properties of small molecules	Combined Science students need to know and understand all content.
2.16	Covalent bonding	Combined Science students need to know and understand all content.
2.17	Giant covalent structures	Combined Science students need to know and understand all content.
2.18	Polymer structures	Combined Science students need to know and understand all content.
2.19	Metallic bonding	Combined Science students need to know and understand all content.
2.20	Properties of metals and alloys	Combined Science students need to know and understand all content.
2.21	Key concept: The outer electrons	Combined Science students need to know and understand all content.

2.22	The periodic table	Combined Science students need to know and understand all content.
2.23	Developing the periodic table	Combined Science students need to know and understand all content.
2.24	Diamond	Combined Science students need to know and understand all content.
2.25	Graphite	Combined Science students need to know and understand all content.
2.26	Graphene and fullerenes	Combined Science students need to know and understand all content.
2.27	Nanoparticles, their properties and uses	Combined Science students do not need to know any of the content on this page.
2.28	Maths skills: Using ratios in mixture, empirical formulae and balanced equations	Combined Science students need to know and understand all content.
	Check your progress	Combined Science students do not need to know the content in the following box: [purple] compare 'nano' dimensions...
	Worked example	Combined Science students need to know and understand all content.
	End of chapter questions	Combined Science students need to know and understand all content.

Chapter 3	Chemical reactions	
Student Book only	Chapter Introduction	Combined Science students need to know and understand all content.
3.1	Elements and compounds	Combined Science students need to know and understand all content.
3.2	Atoms, formulae and equations	Combined Science students need to know and understand all content.
3.3	Moles	Combined Science students need to know and understand all content.
3.4	Key concept: Conservation of mass and balanced equations	Combined Science students need to know and understand all content.
3.5	Key concept: Amounts in chemistry	Combined Science students do not need to know any of the content on this page.
3.6	Mass changes when gases are in reactions	Combined Science students need to know and understand all content.
3.7	Using moles to balance equations	Combined Science students need to know and understand all content.
3.8	Key concept: Limiting reactants and molar masses	Combined Science students need to know and understand all content.
3.9	Amounts of substances in equations	Combined Science students need to know and understand all content.
3.10	Endothermic and exothermic reactions	Combined Science students need to know and understand all content.
3.11	Reaction profiles	Combined Science students need to know and understand all content.
3.12	Energy change of reactions	Combined Science students need to know and understand all content.
3.13	Maths skills: Recognise and use expressions in decimal form	Combined Science students need to know and understand all content.
3.14	Oxidation and reduction in terms of electrons	Combined Science students need to know and understand all content.
3.15	Key concept: Electron transfer, oxidation and reduction	Combined Science students need to know and understand all content.
3.16	Neutralisation of acids and salt production	Combined Science students need to know and understand all content.
3.17	Soluble salts	Combined Science students need to know and understand all content.
3.18	Reaction of metals with acids	Combined Science students need to know and understand all content.

3.19	Practical: Preparing a pure, dry sample of a soluble salt from an insoluble oxide or carbonate	Combined Science students need to know and understand all content.
3.20	pH and neutralisation	Combined Science students need to know and understand all content.
3.21	Strong and weak acids	Combined Science students need to know and understand all content.
3.22	Maths skills: Make order of magnitude calculations	Combined Science students need to know and understand all content.
3.23	Practical: Investigate the variables that affect temperature changes in reacting solutions, such as acid plus metals, acid plus carbonates, neutralisations, displacement of metals	Combined Science students need to know and understand all content.
3.24	The process of electrolysis	Combined Science students need to know and understand all content.
3.25	Electrolysis of molten ionic compounds	Combined Science students need to know and understand all content.
3.26	Electrolysis of aqueous solutions	Combined Science students need to know and understand all content.
3.27	Practical: Investigating what happens when aqueous solutions are electrolysed using inert electrodes	Combined Science students need to know and understand all content.
	Check your progress	Combined Science students need to know and understand all content.
	Worked example	Combined Science students need to know and understand all content.
	End of chapter questions	Combined Science students need to know and understand all content.

© HarperCollins*Publishers* Limited 2016

Chapter 4	Predicting and identifying reactions and products	
Student Book only	Chapter Introduction	Combined Science students do not need to know the content in the boxes: Metals and non-metals What do transition metal compound solutions look like? Testing gases How can we analyse positive ions? Metals can react to make salts How can we analyse negative ions? Testing with instruments How can we use instrumental techniques for analysis?
4.1	Exploring Group 0	Combined Science students need to know and understand all content.
4.2	Exploring Group 1	Combined Science students need to know and understand all content.
4.3	Exploring Group 7	Combined Science students need to know and understand all content.
4.4	Transition metals	Combined Science students do not need to know any of the content on this page.
4.5	Reaction trends and predicting reactions	Combined Science students need to know and understand all content.
4.6	Reactivity series	Combined Science students need to know and understand all content.
4.7	Test for gases	Combined Science students do not need to know any of the content on this page.
4.8	Metal hydroxides	Combined Science students do not need to know any of the content on this page.
4.9	Tests for anions	Combined Science students do not need to know any of the content on this page.
4.10	Flame tests	Combined Science students do not need to know any of the content on this page.
4.11	Instrumental methods	Combined Science students do not need to know any of the content on this page.
4.12	Practical: Use chemical tests to identify the ions in unknown single ionic compounds	Combined Science students do not need to know any of the content on this page.

Check your progress	Combined Science students do not need to know the content in the following boxes: [green] explain that transition metals… [blue] explain that transition metals… [purple] explain that transition metals… [green] carry out flame test… [blue] identify the colours of flames… [purple] identify substances… [green] recognise the precipitate… [blue] explain how to use sodium… [purple] write balanced equations of… [green] identify the tests for… [blue] explain the tests for halides… [purple] identify anions and cations from… [green] identify advantages of… [blue] describe some instrumental… [purple] explain the data provided by… [green] describe flame emission… [blue] identify the advantages… [purple] interpret an instrumental result …
Worked example	Combined Science students need to know and understand all content.
End of chapter questions	Combined Science students do not need to answer the following questions: 3, 6, 7, 8, 15, 19, 20, 22, 23

Chapter 5	Monitoring and controlling chemical reactions	
Student Book only	Chapter Introduction	Combined Science students do not need to know the content in the boxes: Measuring quantities How can we measure amounts of substances? Making salts and crystals Maximising chemical yields
5.1	Concentration of solutions	Combined Science students do not need to know any of the content on this page.
5.2	Using concentrations of solutions	Combined Science students do not need to know any of the content on this page.
5.3	Practical: Finding the reacting volumes of solutions of acid and alkali by titration	Combined Science students do not need to know any of the content on this page.
5.4	Amounts of substance in volumes of gases	Combined Science students do not need to know any of the content on this page.
5.5	Key concept: Percentage yield	Combined Science students do not need to know any of the content on this page.
5.6	Atom economy	Combined Science students do not need to know any of the content on this page.

5.7	Maths skills: Change the subject of an equation	Combined Science students do not need to know any of the content on this page.
5.8	Measuring rates	Combined Science students need to know and understand all content.
5.9	Calculating rates	Combined Science students need to know and understand all content.
5.10	Factors affecting rates	Combined Science students need to know and understand all content.
5.11	Collision theory	Combined Science students need to know and understand all content.
5.12	Catalysts	Combined Science students need to know and understand all content.
5.13	Factors increasing the rate	Combined Science students need to know and understand all content.
5.14	Practical: Investigate how changes in concentration affect the rates of reactions by a method involving the production of a gas and a method involving a colour change	Combined Science students need to know and understand all content.
5.15	Reversible reactions and energy changes	Combined Science students need to know and understand all content.
5.16	Equilibrium	Combined Science students need to know and understand all content.
5.17	Changing concentration and equilibrium	Combined Science students need to know and understand all content.
5.18	Changing temperature and equilibrium	Combined Science students need to know and understand all content.
5.19	Changing pressure and equilibrium	Combined Science students need to know and understand all content.
5.20	Maths skills: Use the slope of a tangent as a measure of rate of change	Combined Science students need to know and understand all content.
	Check your progress	Combined Science students do not need to know the content in the following boxes: [green] relate mass, volume and... [blue] calculate the mass of... [purple] relate concentration in mol/dm^3 ... [green] describe how to carry out... [blue] calculate the concentrations... [purple] explain how the concentration... [green] identify the balanced equation... [blue] calculate the atom economy... [purple] explain why a particular... [green] explain that the same amount... [blue] calculate the volume of a... [purple] calculate volumes of gases...
	Worked example	Combined Science students need to know and understand all content.
	End of chapter questions	Combined Science students do not need to answer the following questions: 7, 11, 16, 17, 19, 20

Chapter 6	Global challenges	
Student Book only	Chapter Introduction	Combined Science students need to know and understand all content.
6.1	Extraction of metals	Combined Science students need to know and understand all content.
6.2	Using electrolysis to extract metals	Combined Science students need to know and understand all content.
6.3	Alternative methods of metal extraction	Combined Science students need to know and understand all content.
6.4	The Haber process	Combined Science students do not need to know any of the content on this page.
6.5	Production and use of NPK fertilisers	Combined Science students do not need to know any of the content on this page.
6.6	Life cycle assessment and recycling	Combined Science students need to know and understand all content.
6.7	Ways of reducing the use of resources	Combined Science students need to know and understand all content.
6.8	Alloys as useful materials	Combined Science students do not need to know any of the content on this page.
6.9	Corrosion and its prevention	Combined Science students do not need to know any of the content on this page.
6.10	Ceramics, polymers and composites	Combined Science students do not need to know any of the content on this page.
6.11	Maths skills: Translate information between graphical and numerical form	Combined Science students need to know and understand all content.
6.12	Functional groups and homologous series	Combined Science students do not need to know any of the content on this page.
6.13	Structure and formulae of alkenes	Combined Science students do not need to know any of the content on this page.
6.14	Reactions of alkenes	Combined Science students do not need to know any of the content on this page.
6.15	Alcohols	Combined Science students do not need to know any of the content on this page.
6.16	Carboxylic acids	Combined Science students do not need to know any of the content on this page.
6.17	Addition polymerisation	Combined Science students do not need to know any of the content on this page.
6.18	Condensation polymerisation	Combined Science students do not need to know any of the content on this page.
6.19	Amino acids	Combined Science students do not need to know any of the content on this page.
6.20	DNA and other naturally occurring polymers	Combined Science students do not need to know any of the content on this page.
6.21	Crude oil, hydrocarbons and alkanes	Combined Science students need to know and understand all content.

6.22	Fractional distillation and petrochemicals	Combined Science students need to know and understand all content.
6.23	Properties of hydrocarbons	Combined Science students need to know and understand all content.
6.24	Cracking and alkenes	Combined Science students need to know and understand all content.
6.25	Cells and batteries	Combined Science students do not need to know any of the content on this page.
6.26	Fuel cells	Combined Science students do not need to know any of the content on this page.
6.27	Key concept: Intermolecular forces	Combined Science students need to know and understand all content.
6.28	Maths skills: Visualise and represent 3D models	Combined Science students do not need to know any of the content on this page.
6.29	Proportions of gases in the atmosphere	Combined Science students need to know and understand all content.
6.30	The Earth's early atmosphere	Combined Science students need to know and understand all content.
6.31	How oxygen increased	Combined Science students need to know and understand all content.
6.32	Key concept: Greenhouse gases	Combined Science students need to know and understand all content.
6.33	Human activities	Combined Science students need to know and understand all content.
6.34	Global climate change	Combined Science students need to know and understand all content.
6.35	Carbon footprint and its reduction	Combined Science students need to know and understand all content.
6.36	Limitations on carbon footprint reduction	Combined Science students need to know and understand all content.
6.37	Atmospheric pollutants from fuels	Combined Science students need to know and understand all content.
6.38	Properties and effects of atmospheric pollutants	Combined Science students need to know and understand all content.
6.39	Potable water	Combined Science students need to know and understand all content.
6.40	Waste water treatment	Combined Science students need to know and understand all content.
6.41	Practical: Analysis and purification of water samples from different sources, including pH, dissolved solids and distillation	Combined Science students need to know and understand all content.
6.42	Maths skills: Use ratios, fractions and percentages	Combined Science students need to know and understand all content.

Check your progress	Combined Science students do not need to know the content in the following boxes: [green] describe the composition of... [blue] interpret the composition... [purple] evaluate the uses of other alloys... [green] compare quantitatively the... [blue] compare properties of... [purple] explain how the properties of... [green] apply the principles of dynamic... [blue] explain the trade-off... [purple] explain how the commercially... [green] explain how a voltage can... [blue] evaluate the uses of cells... [purple] interpret data for the relative...
Worked example	Combined Science students do not need to know the following examples: 4, 5
End of chapter questions	Combined Science students do not need to answer the following questions: 1, 3, 6, 9, 12, 15, 16, 18, 19, 20

© HarperCollins*Publishers* Limited 2016

ACKNOWLEDGEMENTS

The Publishers gratefully acknowledge the permissions to reproduce copyright material in this book. Every effort has been made to contact the holders of copyright material, but if any have been inadvertently overlooked, the publisher will be pleased to make necessary arrangements at the first opportunity.

Worksheets

Worksheet 1.2.1 blackboard1965/Shutterstock, Georgios Kollidas/Shutterstock, Archive Pics/Alamy Stock Photo, Archive Pics/Alamy Stock Photo, ITAR-TASS Photo Agency/Alamy Stock Photo; Worksheet 1.2.3 ITAR-TASS Photo Agency/Alamy Stock Photo, Georgios Kollidas/Shutterstock, Archive Pics/Alamy Stock Photo, Archive Pics/Alamy Stock Photo; Worksheet 2.4.2 Natan86/Shutterstock; Worksheet 2.20.2 Denys Prykhodov/Shutterstock; Worksheet 2.26.2 nobeastsofierce/Shutterstock; Worksheet 6.34.1 Jan Martin Will/Shutterstock, Silken Photography/Shutterstock, idiz/Shutterstock, Khoroshunova Olga/Shutterstock, Khoroshunova Olga/Shutterstock, tcsaba/Shutterstock; Worksheet 6.38.1 OPgrapher/Shutterstock, RIRF Stock/Shutterstock, Laurence Gough/Shutterstock, Hung Chung Chih/Shutterstock, Alexander Raths/Shutterstock.

PowerPoints

51034_chem_ch3_pres2.2 GIPhotoStock/SCIENCE PHOTO LIBRARY, GIPhotoStock/SCIENCE PHOTO LIBRARY; 51034_chem_ch3_pres8.1 MARTYN F. CHILLMAID/SCIENCE PHOTO LIBRARY, MARTYN F. CHILLMAID/SCIENCE PHOTO LIBRARY; 51034_chem_ch3_pres14.1 GIPhotoStock/SCIENCE PHOTO LIBRARY; 51034_chem_ch3_pres18.2 MARTYN F. CHILLMAID/SCIENCE PHOTO LIBRARY; 51034_chem_ch3_pres19.1 Dorling Kindersley/UIG/SCIENCE PHOTO LIBRARY; 51034_chem_ch3_pres23.1 CHARLES D. WINTERS/SCIENCE PHOTO LIBRARY; 51034_chem_ch4_pres3.1 ER Degginger/SCIENCE PHOTO LIBRARY; 51034_chem_ch4_pres5.1 ANDREW LAMBERT PHOTOGRAPHY/SCIENCE PHOTO LIBRARY, GIPhotoStock/SCIENCE PHOTO LIBRARY; 51034_chem_ch4_pres6.1 GIPhotoStock/SCIENCE PHOTO LIBRARY; 51034_chem_ch4_pres6.2 MARTYN F. CHILLMAID/SCIENCE PHOTO LIBRARY; 51034_chem_ch4_pres7.1 GIPhotoStock/SCIENCE PHOTO LIBRARY; 51034_chem_ch5_pres8.1 MARTYN F. CHILLMAID/SCIENCE PHOTO LIBRARY; 51034_chem_ch5_pres8.2 MARTYN F. CHILLMAID/SCIENCE PHOTO LIBRARY; 51034_chem_ch5_pres10.1 MARTYN F. CHILLMAID/SCIENCE PHOTO LIBRARY; All other images © Shutterstock.